White Gold: The Commercialisation of Rice
Farming in the Lower Mekong Basin

Rob Cramb
Editor

White Gold: The Commercialisation of Rice Farming in the Lower Mekong Basin

palgrave
macmillan

Editor
Rob Cramb
St Lucia, QLD, Australia

ISBN 978-981-15-0997-1   ISBN 978-981-15-0998-8   (eBook)
https://doi.org/10.1007/978-981-15-0998-8

Cover Image: © Peter Stuckings / Getty Images

This Palgrave Macmillan imprint is published by the registered company Springer Nature Singapore Pte Ltd.
The registered company address is: 152 Beach Road, #21-01/04 Gateway East, Singapore 189721, Singapore

To Harry and Rose

FOREWORD

The development story told of Southeast Asia usually focuses on processes of urbanisation, industrialisation, and rapid sectoral change, which have propelled economic growth and thus delivered rising incomes, improving standards of living, and declining poverty. Where, however, does farming and agriculture, and in particular, the region's signature crop, rice, fit into this story? It is not a simple one, because many of the trends anticipated by scholars and policy-makers have not materialised, while others have worked out far more rapidly than anyone expected. Indeed, some of the trends, or the absence of them, appear on first sight to be puzzlingly at odds.

Landholdings have not—generally—amalgamated into larger units of production, which might drive labour productivity increases. Mechanisation of some aspects of rice agriculture has proceeded rapidly, even in countries that remain poor and seemingly in rural labour surplus. Questions and concerns regarding food security stand alongside the disintensification of some aspects of production, even land abandonment. Most rice farms are sub-livelihood in size, but living standards in the countryside continue to improve and poverty to decline. Parents make huge sacrifices to educate their children so that they can escape the drudgery of rice farming, but nonetheless stay rooted in—and to—their natal lands. Production is increasingly commercialised, but farmers in some areas seem to adopt semi-subsistence mindsets in their approach to rice farming.

This volume, then, comes at a particularly important moment in Southeast Asia's agrarian history. How do we interpret these contradictory trends and how they might work out in the years to come? *White Gold* considers these questions and issues in the context of the Lower Mekong

Basin. This region of one river and four countries encompasses more than half a million square kilometres and a population of 66 million, produces 50 million tons of paddy rice each year, and contributes one-quarter of the world's rice exports. It is also home to some of the earliest rice-growing cultures and the great rice-based civilisation of Angkor, and was a pivotal area in the colonial rice export economy. Where better to consider the past, present, and future of "white gold"?

Bristol, UK   Jonathan Rigg
May 2019

PREFACE

Vietnamese farmers have for centuries regarded rice as "white gold" (*vàng trắng*), reflecting its vital importance to household food security and livelihoods. Farmers throughout the Lower Mekong Basin have a similar view of rice as the traditional basis of their wealth and well-being. A household able to produce abundant supplies of rice was not only secure economically but achieved social and political status within the village community. The frequent depredations of floods and droughts on the one hand and extractive state regimes on the other only heightened the value placed on the household's rice supplies.

In the past four decades, rice has also become a commercial crop of great importance to Lower Mekong farmers, augmenting but not replacing its role in securing their subsistence. Particularly in Northeast Thailand and the Mekong Delta in Vietnam, rice farming has become a major export industry, spurring a process of rural development that has helped lift many households out of poverty. Farmers in Cambodia and Laos have also increased their output to such a level that both countries have become self-sufficient in rice and are entering into export markets, particularly through cross-border trade with Vietnam and Thailand. Significantly, the Cambodian government adopted the term "white gold" in 2010 to epitomise the country's push into high-quality rice exports.

This book is the outcome of a collaborative research effort to understand the current status of this process of commercialisation in the rice sector of the Lower Mekong Basin, with a view to identifying prospects and policy issues for the coming decade. This involved studying not just change in rice-based farming systems but in the value chains through which farmers gain access to resources and inputs and market their outputs, and the institutional arrangements governing those farming systems and value chains. The focus was on the rainfed and irrigated lowlands of the Basin rather than the sloping uplands as it is in the former environments that the commercialisation of rice farming has unfolded so dramatically, whereas rice cultivation in the uplands has been increasingly constrained, both technically and politically.

This publication has been made possible with support from the Australian Government through the Australian Centre for International Agricultural Research (ACIAR).

The main body of this research was conducted as part of an ACIAR-funded project—"Developing agricultural policies for rice-based farming systems in Lao PDR and Cambodia" (ASEM/2009/023). This project was co-led by Rob Cramb of the University of Queensland (UQ), Silinthone Sacklokham of the National University of Laos (NUOL), Theng Vuthy of the Cambodia Development Resources Institute (CDRI), Benchaphun Ekasingh of Chiang Mai University (CMU) in Thailand, and Dao The Anh of the Centre for Agrarian Systems Research and Development (CASRAD) in Vietnam.

The findings from this project were supplemented by socio-economic studies undertaken as part of a second project—"Developing improved farming and marketing systems in rainfed regions of southern Lao PDR" (CSE/2009/004)—involving Rob Cramb and Jonathan Newby (then with UQ), Silinthone Sacklokham (NUOL), and Vongpaphane Manivong (then with the National Agricultural and Forestry Research Institute [NAFRI] of Laos). The results of a third ACIAR project involving Rob Cramb and Jonathan Newby—"Review of rice-based farming systems in Mainland Southeast Asia" (C2012/229)—were also drawn upon in writing this book.

In addition, ACIAR provided John Allwright Fellowships for Chea Sareth (of the Cambodian Agricultural Research and Development Institute, CARDI) and Vongpaphane Manivong (NAFRI) to undertake PhD studies at the University of Queensland on topics closely related to the themes of this book. Nguyen Van Kien and Nguyen Hoang Han of An Giang University contributed Chap. 17 based largely on their research. Dao The Anh would like to acknowledge that Chap. 18 is based on research supported by the Asian Development Bank under Regional Research and Development Technical Assistance (R-RDTA) Project TA-7648.

ACIAR also provided a grant for the book to be available through Open Access.

We are grateful to Jonathan Rigg for kindly agreeing to write the Foreword to the book, to CartoGIS of the Australian National University for permission to reproduce the maps in Figs. 1.1, 2.1, 5.1, 11.1, and 17.1, and to the Mekong River Commission for permission to reproduce the maps in Figs. 1.4 and 1.5.

Both local currencies and United States Dollars (USD) are used in the book. Exchange rates have fluctuated over the 2010s, but the mean rates for the period 2010–2018 are a good guide to orders of magnitude: 1 USD = 32.5 Thai Baht (THB) = 4063 Cambodian Riel (KHR) = 8143 Lao Kip (LAK) = 21,227 Vietnamese Dong (VND).

QLD, Australia   Rob Cramb
June 2019

CONTENTS

Part I   Introduction   1

1   The Evolution of Rice Farming in the Lower Mekong
    Basin   3
    Rob Cramb

Part II   A Fragrant Aroma   37

2   Commercialisation of Rice Farming in Northeast
    Thailand   39
    Pornsiri Suebpongsang, Benchaphun Ekasingh, and Rob
    Cramb

3   Evolution of Rice Farming in Ubon Ratchathani Province   69
    Prathanthip Kramol and Benchaphun Ekasingh

4   Farmer Organizations in Ubon Ratchathani Province   85
    Prathanthip Kramol, Pornsiri Suebpongsang, and Benchaphun
    Ekasingh

Part III   A Sticky Situation   101

 5 **From Subsistence to Commercial Rice Production in Laos**   103
    Vongpaphane Manivong and Rob Cramb

 6 **Adapting the Green Revolution for Laos**   121
    Liana Williams and Rob Cramb

 7 **Rainfed and Irrigated Rice Farming on the Savannakhet
    Plain**   151
    Silinthone Sacklokham, Lytoua Chialue, and Fue Yang

 8 **The Supply of Inputs to Rice Farmers in Savannakhet**   169
    Chitpasong Kousonsavath and Silinthone Sacklokham

 9 **Rice Marketing and Cross-Border Trade in Savannakhet**   187
    Phengkhouane Manivong and Silinthone Sacklokham

10 **Economic Constraints to the Intensification of Rainfed
    Lowland Rice in Laos**   201
    Jonathan Newby, Vongpaphane Manivong, and Rob Cramb

Part IV   In Pursuit of White Gold   225

11 **The Commercialisation of Rice Farming in Cambodia**   227
    Rob Cramb, Chea Sareth, and Theng Vuthy

12 **The Production, Marketing, and Export of Rice in Takeo**   247
    Chhim Chhun, Theng Vuthy, and Nou Keosothea

13 **The Role of Irrigation in the Commercialisation of Rice
    Farming in Southern Cambodia**   261
    Chea Sareth, Rob Cramb, and Shu Fukai

14   The Supply of Fertiliser for Rice Farming in Takeo   291
     Theng Vuthy

15   The Use of Credit by Rice Farmers in Takeo   309
     Kem Sothorn

16   Contract Farming of High-Quality Rice in Kampong
     Speu   327
     Nou Keosothea and Heng Molyaneth

Part V   The Overflowing Rice Bowl   345

17   Trends in Rice-Based Farming Systems in the Mekong
     Delta   347
     Nguyen Van Kien, Nguyen Hoang Han, and Rob Cramb

18   The Domestic Rice Value Chain in the Mekong Delta   375
     Dao The Anh, Thai Van Tinh, and Nguyen Ngoc Vang

19   The Cross-Border Trade in Rice from Cambodia to
     Vietnam   397
     Dao The Anh and Thai Van Tinh

20   Cross-Border Trade in Sticky Rice from Central Laos to
     North Central Vietnam   413
     Dao The Anh and Pham Cong Nghiep

Part VI   Conclusion   423

21   Issues of Rice Policy in the Lower Mekong Basin   425
     Rob Cramb

LIST OF CONTRIBUTORS

Chhim Chhun Cambodia Development Resource Institute, Phnom Penh, Cambodia

Lytoua Chialue Faculty of Agriculture, Department of Rural Economics and Food Technology, National University of Laos, Vientiane, Laos

Rob Cramb School of Agriculture and Food Sciences, University of Queensland, St Lucia, QLD, Australia

Dao The Anh Vietnam Academy of Agricultural Sciences, Hanoi, Vietnam

Benchaphun Ekasingh Faculty of Agriculture, Department of Agricultural Economy and Development, Chiang Mai University, Chiang Mai, Thailand

Chea Sareth Cambodian Agricultural Research and Development Institute, Phnom Penh, Cambodia

Shu Fukai School of Agriculture and Food Sciences, University of Queensland, St Lucia, QLD, Australia

Heng Molyaneth Faculty of Development Studies, Royal University of Phnom Penh, Phnom Penh, Cambodia

Kem Sothorn Parliamentary Institute of Cambodia, Phnom Penh, Cambodia

Chitpasong Kousonsavath Faculty of Agriculture, Department of Rural Economics and Food Technology, National University of Laos, Vientiane, Laos

Prathanthip Kramol Faculty of Agriculture, Department of Agricultural Economy and Development, Chiang Mai University, Chiang Mai, Thailand

Vongpaphane Manivong Ministry of Agriculture and Forestry, Vientiane, Laos

Phengkhouane Manivong Faculty of Agriculture, National University of Laos, Vientiane, Laos

Jonathan Newby International Center for Tropical Agriculture (CIAT), Vientiane, Laos

Nguyen Hoang Han Charles Sturt University, Wagga Wagga, NSW, Australia

Nguyen Ngoc Vang An Giang University, Long Xuyen, Vietnam

Nguyen Van Kien An Giang University, Long Xuyen, Vietnam
Fenner School of Environment and Society, Australian National University, Canberra, ACT, Australia

Nou Keosothea National Committee, Economic and Social Commission for Asia and the Pacific, Ministry of Foreign Affairs and International Cooperation, Phnom Penh, Cambodia

Pham Cong Nghiep Center for Agrarian Systems Research and Development, Vietnam Academy of Agricultural Sciences, Hanoi, Vietnam

Silinthone Sacklokham SEAMEO Regional Centre for Community Education Development, Vientiane, Laos

Pornsiri Suebpongsang Faculty of Agriculture, Department of Agricultural Economy and Development, Chiang Mai University, Chiang Mai, Thailand

Thai Van Tinh Center for Agricultural Policy, Institute of Policy and Strategy for Agriculture and Rural Development, Ministry of Agriculture and Rural Development, Hanoi, Vietnam

Theng Vuthy Office of Food Security and Environment, USAID, Phnom Penh, Cambodia

Liana Williams Commonwealth Scientific and Industrial Research Organisation (CSIRO), Brisbane, QLD, Australia

Fue Yang Faculty of Agriculture, National University of Laos, Vientiane, Laos

List of Figures

Fig. 1.1   Mekong River Basin. (Source: CartoGIS, Australian National University)   6

Fig. 1.2   Mean monthly rainfall and temperature for Laos, 1991–2016. (Source: Climate Research Unit, University of East Anglia)   7

Fig. 1.3   Average monthly mainstream flow at Pakse, Laos, 1960–2004 (cubic metres per second). (Source: Cosslett and Cosslett 2018)   8

Fig. 1.4   Land use in the Lower Mekong Basin. (Source: Mekong River Commission)   10

Fig. 1.5   Area planted with rice in Lower Mekong Basin in wet season (July). (Source: Mekong River Commission)   20

Fig. 1.6   Area planted with rice in Lower Mekong Basin in dry season (January). (Source: Mekong River Commission)   21

Fig. 1.7   Schematic outline of evolving rice value chains in Lower Mekong Basin   24

Fig. 1.8   Rice exports from Thailand, Vietnam, Cambodia, and Laos, 1975–2016. (Source: FAOSTAT)   27

Fig. 1.9   Influence of government policies on rice production   28

Fig. 2.1   Northeast Thailand. (Source: CartoGIS, Australian National University)   41

Fig. 2.2   Annual rice production in Northeast Thailand and remainder of country, 1980–2015 (million t)   45

Fig. 2.3   Marketing chain for rice produced in Northeast Thailand. (Note: Numbers refer to million t of paddy or milled rice ACI 2005)   47

Fig. 2.4   Average export price of Thai white and fragrant rice, 2000–2018 (USD/t, FOB). (Source: IRRI—World Rice Statistics Online (http://ricestat.irri.org:8080/wrsv3/entrypoint.htm))   58

Fig. 5.1   Laos with provinces and provincial capitals, 2012. (Source: CartoGIS Services, College of Asia and the Pacific, The Australian National University)   105

Fig. 5.2   Lao farmer showing field trial on his paddy field. (Source: Rob Cramb)   112

Fig. 5.3   Paddy area, output, and yield in Laos, 1985–2017. (Source: Agricultural Statistics Yearbooks (various years), Department of Planning and Finance, Ministry of Agriculture and Forestry (MAF), Vientiane)   112

Fig. 5.4   Paddy output in Laos by production system, 1985–2017 ('000 t). (Source: Agricultural Statistics Yearbooks (various years), Department of Planning and Finance, Ministry of Agriculture and Forestry (MAF), Vientiane)   113

Fig. 5.5   Value of agricultural exports from Laos, 2013–2017. (Source: International Trade Centre)   115

Fig. 6.1   Rice research centres and locations of Lao-IRRI research activities. (Source: Modified from Shrestha et al. 2006: 38)   126

Fig. 6.2   Savannakhet Province showing Outhoumphone and Champhone districts. (Source: Modified from Manivong et al. 2008: 1)   138

Fig. 7.1   Savannakhet Province showing Champhone and other districts and lowland rice-growing areas. (Source: Thavone Inthavong)   153

Fig. 7.2   Sun-drying paddy before storing for household consumption. (Source: Rob Cramb)   161

Fig. 7.3   Incidence of selling rice throughout the year (% of those selling, $n = 141$)   165

Fig. 7.4   Duration of rice deficit (% of households with deficit, $n = 43$)   165

Fig. 8.1   Fertilizer supply chain for Savannakhet Province   171

Fig. 8.2   Farmers' sources of fertilizer (% of survey respondents)   175

Fig. 8.3   The seed supply chain in Savannakhet. ªThasano Centre and PAFO/DAFO buy back seeds from the seed production groups at 10% above the market price for paddy rice   179

Fig. 8.4   Farmer in paddy field planted with improved variety in Savannakhet Province. (Source: Rob Cramb)   182

Fig. 9.1   International border points in Savannakhet Province: Savannakhet-Mukdahan (left); Dansavanh-Lao Bao (right)   188

Fig. 9.2   Mapping of rice marketing and trade in Savannakhet Province   190

Fig. 9.3   Procedure to export rice from Savannakhet Province. (Source: Trade Division, PICO Savannakhet)   197

Fig. 10.1   Household rice status in Champasak for 2010, by district and
            village   206
Fig. 10.2   Yield-area combinations by household rice status   208
Fig. 10.3   Cumulative adoption of improved varieties by paddy area   211
Fig. 11.1   Cambodia, showing provinces and terrain. (Source: CartoGIS,
            Australian National University)   228
Fig. 11.2   Area, yield, and output of rice in Cambodia, 1990–2017.
            (Source: FAOSTAT)   234
Fig. 11.3   Population pyramid for Takeo and Cambodia. (Source:
            Cambodian Census 2008)   240
Fig. 12.1   Map of rice value chain in Takeo   251
Fig. 13.1   Locations of the three study districts in Takeo and Kampong
            Speu Provinces. (Source: Cambodian Agricultural Research and
            Development Institute)   262
Fig. 13.2   Farm pond with portable pump in Takeo. (Source: Rob Cramb)   266
Fig. 13.3   Farmer in Takeo with tube well and pump. (Source: Rob
            Cramb)   267
Fig. 13.4   Farmer in Snao preparing paddy field for radish cultivation in
            the dry season. (Source: Rob Cramb)   281
Fig. 14.1   Percentage of households using chemical fertilisers and
            pesticides in Takeo Province by district, 2010. (Source:
            Commune Database, 2010)   293
Fig. 14.2   Fertiliser distribution channels in Takeo   294
Fig. 14.3   Yearly average nominal retail prices of major fertilisers in Takeo,
            2002–2010 (KHR/bag). (Source: Agricultural Marketing
            Office 2002–2010 (USD 1 = KHR 4000))   298
Fig. 15.1   Growth of MFI clients and loans in Cambodia, 2005–2014.
            (Source: CMA 2015)   312
Fig. 15.2   Number of MFIs by province, 2011. (Source: Constructed
            from CMA data)   312
Fig. 15.3   Number of MFI borrowers by province, 2011. (Source:
            Constructed from CMA data)   313
Fig. 15.4   Total amount of MFI outstanding loans in Takeo in 2011 by
            district (KHR million). (Note: USD 1 = KHR 4000; Source:
            constructed from CMA data)   314
Fig. 15.5   Number of MFI borrowers in Takeo in 2011 by district. (Note:
            USD 1 = KHR 4000; Source: constructed from CMA data)   315
Fig. 15.6   Pattern of access to credit by type of rice farmer. (Source: Field
            interviews)   316
Fig. 17.1   Mekong Delta showing provinces and agro-ecological zones.
            (Source: Base map by CartoGIS Services, College of Asia and
            the Pacific, Australian National University)   348

Fig. 17.2   Mean monthly rainfall and temperature at Can Tho. (Source: Climatic Research Unit, University of East Anglia)   349

Fig. 17.3   Average farm area in Mekong Delta by land use and agro-ecological zone, 2005. (Source: Derived from data in Nguyen, D. C. et al. (2007))   353

Fig. 17.4   Planted area of rice in the Mekong Delta by season, 1995–2016. (Source: General Statistics Office 2016)   356

Fig. 17.5   Paddy yields by season and total paddy production in the Mekong Delta, 1995–2016. (Source: General Statistics Office 2016)   357

Fig. 17.6   Cattle being fattened for sale in a farmyard shed in My An Commune, Cho Moi District, An Giang Province. (Source: Nguyen Van Kien, September 2017)   362

Fig. 17.7   Rice-fish-poultry system in My Phu Dong Commune, Thoai Son District, An Giang Province. (Source: Nguyen Van Kien, September 2017)   364

Fig. 17.8   Area, yield, and production of freshwater shrimp and fish in An Giang Province. (Source: Statistical Office of An Giang Province (2016))   364

Fig. 17.9   Freshwater shrimp farming in Vinh Thanh Trung Commune, Chau Phu District, An Giang Province. (Source: Nguyen Van Kien)   365

Fig. 18.1   Area, yield, and production of rice in Vietnam, 1980–2017. (Source: FAOSTAT)   377

Fig. 18.2   Rice value chain in Mekong Delta   378

Fig. 18.3   Trader transporting paddy in Can Tho Province. (Photo: Dao The Anh)   384

Fig. 18.4   Large rice mill in Can Tho Province. (Photo: Dao The Anh)   386

Fig. 19.1   Cross-border rice value chain between Cambodia and Vietnam   400

Fig. 20.1   The Lao Bao International Border Gate between Quang Tri and Savannakhet. (Source: Bùi Thuỵ Đào Nguyên, https://commons.wikimedia.org/w/index.php?curid=17356460)   414

Fig. 20.2   Value chain for sticky rice and paddy imported from Laos through Lao Bao Border Gate   418

Fig. 21.1   Nominal farm-gate price of paddy in Lower Mekong Basin countries, 2000–2017 (USD/t). (Source: FAOSTAT)   435

LIST OF TABLES

Table 1.1   Land, population, and rice production in the Lower Mekong Basin, 2014   9

Table 1.2   Demographic and economic data for Lower Mekong countries, 2018   11

Table 2.1   Number and area of agricultural holdings in Thailand by region, 2013   43

Table 2.2   Agricultural land use in Thailand by region, 2013 (% of area)   43

Table 2.3   Average returns to white rice and fragrant rice production in the 2018 wet season   44

Table 2.4   Area, output, and yield of rice in Northeast Thailand compared with Thailand as a whole, by season (2018)   46

Table 3.1   Fertiliser use and paddy yield under different fertiliser regimes   74

Table 3.2   Use of machinery in the study villages   75

Table 3.3   Tractors used for land preparation in the study villages   76

Table 3.4   Cost of purchasing and renting machinery for land preparation   77

Table 3.5   Harvesting/threshing cost by method of harvesting   78

Table 3.6   Costs and returns for wet-season rice production in the study villages   80

Table 5.1   Main farming systems in Laos   108

Table 5.2   Three major rice-based farming systems   109

Table 5.3   Seasonal rice cropping calendar for different farming systems   109

Table 6.1   Release of Lao improved varieties from 1993 to 2005   133

Table 6.2   Area of paddy land planted by seed type (%)   137

Table 6.3   Village characteristics   139

Table 7.1   Characteristics of survey villages   154

Table 7.2   Distribution of survey households by farm size   157

Table 7.3	Cropping calendar for wet-season and dry-season rice production	158
Table 7.4	Mean area and yield of wet-season rice in survey villages, 2011	158
Table 7.5	Rice varieties used by respondents in wet and dry seasons, 2011–2012	159
Table 7.6	Representative enterprise budget for one hectare of wet-season rice	160
Table 7.7	Representative enterprise budget for one hectare of dry-season rice	163
Table 7.8	Quantity of rice sold by farm size and season	164
Table 8.1	Number of interviewees by type of actor	170
Table 8.2	Estimated annual imports of fertilizer per supplier	174
Table 8.3	Incidence of farmers paying cash or using credit for fertilizer purchases, by type of supplier	176
Table 8.4	Comparison of fertilizer prices between cash payment and credit	176
Table 8.5	Marketing margins for fertilizer suppliers	177
Table 8.6	Sources of seed reported by survey farmers	181
Table 8.7	Production costs and selling prices of rice seeds by source	183
Table 9.1	Number of interviewees by category	189
Table 9.2	Farm-gate prices of paddy rice by month, 2009–2011 (LAK/kg)	191
Table 9.3	Matrix of trading networks and linkages in rice market chain	193
Table 9.4	Main products exchanged via international border points of Savannakhet Province, 2010	195
Table 9.5	Value of cross-border trade in Savannakhet Province, 2010 and 2011	195
Table 9.6	Cross-border trade in rice, Savannakhet Province, 2009–2011	197
Table 10.1	Status of rice-growing in surveyed villages, 2010 (n = 360)	205
Table 10.2	Mode of land preparation by paddy area and district	210
Table 10.3	Average nutrient application rate by village (kg/ha)	213
Table 10.4	Assumptions for budget scenarios	215
Table 10.5	Economic analysis of fertiliser-input scenarios for a hectare of WS rice	216
Table 10.6	Sensitivity analysis of fertiliser costs and wage rates	218
Table 10.7	Sensitivity analysis for low and high paddy prices	219
Table 11.1	Rice production data for Cambodia and Takeo Province, 2017–2018	235
Table 12.1	Gross margin analysis for rice farming (1 ha)	249
Table 12.2	Margins in value chain for wet-season paddy (USD/t)	252
Table 12.3	DS rice marketing value chain in Takeo province (USD/t)	253

Table 12.4   Rice prices in Cambodia, Thailand, and Vietnam, August–
              October 2011 (USD/t)   255
Table 13.1   Major characteristics of the case-study villages   264
Table 13.2   Characteristics of WS rice cultivation in the case-study villages   270
Table 13.3   Average material and labour inputs for WS rice cultivation in
              the case-study villages   272
Table 13.4   Average unit costs and returns for WS rice production in the
              case-study villages   273
Table 13.5   Characteristics of EWS rice cultivation in the case-study
              villages   276
Table 13.6   Material and labour inputs for EWS rice cultivation in the
              case-study villages   277
Table 13.7   Average unit costs and returns for EWS rice production in the
              case-study villages   278
Table 13.8   Characteristics of DS non-rice crop cultivation in the case-
              study villages   280
Table 13.9   Material and labour inputs per ha for DS crop cultivation in
              the case-study villages   282
Table 13.10   Average unit costs and returns for DS non-rice crop
              production in the case-study villages   282
Table 13.11   Annual inputs, outputs, and net cash flow of representative
              cropping systems in the case-study villages   285
Table 14.1   Analysis of marketing margins for imported di-ammonium
              phosphate   299
Table 14.2   Nutrient analysis of selected fertiliser samples, mid-2010 300
Table 14.3   Estimated value of production losses due to use of fake
              fertilisers in Takeo in 2011   304
Table 15.1   Evolution of formal credit sector in Cambodia since 1995   311
Table 15.2   Enterprise budget for semi-commercial rice farming   318
Table 15.3   Enterprise budget for commercial dry-season rice farming   320
Table 16.1   Estimated costs and returns for a one-hectare rice farm, by
              type of farmer   335
Table 17.1   Characteristics of households in the Mekong Delta by agro-
              ecological zone, 2005   352
Table 17.2   Mean household income by source and agro-ecological zone,
              2005 (USD/year)   354
Table 17.3   Trends in farming systems in major landform units, 1976–
              2016   359
Table 17.4   MARD land-use plan for paddy land in Mekong Delta for
              2014–2020 period   361

Table 17.5   Economic returns to rice-freshwater shrimp farming in An Giang, 2006 — 366

Table 18.1   Number and type of value chain actors interviewed — 376

Table 18.2   Status of input use by farmers in study area (n = 300) — 379

Table 18.3   Availability of machinery services in study communes in 2011 (n = 20) — 380

Table 18.4   Extension offices and cooperatives in communes — 382

Table 18.5   Characteristics of rice producers in study area (n = 300) — 382

Table 18.6   Characteristics of rice traders in study area (n = 60) — 384

Table 18.7   Types of rice processor in study area (n = 70) — 385

Table 18.8   Characteristics of rice processors in study area (n = 70) — 386

Table 18.9   Characteristics of rice wholesalers and retailers in study area (n = 180) — 387

Table 18.10   Cost of paddy rice production (VND $\times$ 10^3 per ha) — 389

Table 18.11   Costs and margins in domestic and export rice value chains (VND per kg rice) — 390

Table 19.1   Area, production, and yield of paddy in An Giang Province in 2011 — 399

Table 19.2   Paddy prices in the border region between An Giang and Takeo — 403

Table 19.3   Characteristics of Vietnamese traders engaged in cross-border paddy trade — 404

Table 19.4   Characteristics of milling/polishing factories — 406

Table 19.5   Characteristics of urban rice wholesalers — 407

Table 19.6   Characteristics of rice retailers in urban areas — 408

Table 19.7   Mode of selling rice by traditional retailers — 408

Table 19.8   Retail prices of Cambodian specialty rice in Vietnamese market — 409

Table 20.1   Estimated production and use of sticky rice in Quang Tri Province in 2011 — 416

PART I

Introduction

CHAPTER 1

The Evolution of Rice Farming in the Lower Mekong Basin

Rob Cramb

In their definitive review of the Asian rice economy in the 1970s, Barker and Herdt wrote: "Most Asian rice farms are small ... and employ intensive labour practices in place of mechanisation ... [R]ainfall is the dominant climatic variable, and the rice crop is normally limited to the rainy season ... Rice dominates not only production and consumption patterns, but is also inextricably woven into the social and economic fabric of life. More farmers are engaged in rice production than in any other single activity, with rice absorbing more than half of the farm labour force in many countries ... [Most] Asian rice economies lacked the capacity for technical change that would permit rapid growth in rice production to create the food surpluses needed for economic development" (Barker and Herdt 1985: 1–2).

That description certainly applied to the millions of rice farmers in the Lower Mekong Basin, where small-scale, labour-intensive, low-productivity, semi-subsistence farming systems predominated. While in parts of Asia, such as Central Luzon in the Philippines, Java in Indonesia, and the Central Plain in Thailand, rice farmers were widely adopting

R. Cramb (✉)
School of Agriculture and Food Sciences, University of Queensland,
St Lucia, QLD, Australia
e-mail: r.cramb@uq.edu.au

© The Author(s) 2020
R. Cramb (ed.), *White Gold: The Commercialisation of Rice Farming in the Lower Mekong Basin*,
https://doi.org/10.1007/978-981-15-0998-8_1

3

modern, high-yielding varieties, in the Lower Mekong traditional, low-yielding varieties still predominated (Barker and Herdt 1985: 63). The low productivity and subsistence orientation of Lower Mekong farmers not only reflected the persistence of traditional farming norms and practices but, in the case of Indochina (Vietnam, Laos, and Cambodia), the havoc wreaked on the rural population and landscape by decades of war, and the disincentives and hardship subsequently introduced through the imposition of collective forms of agriculture. Rural poverty and the threat of famine were rife.

In the 40 years since, rice farming in the Lower Mekong has undergone a dramatic transformation. This transformation can be characterised as "commercialisation" in the broadest sense, meaning the opening up of semi-subsistence rice farming to domestic and international input and output markets and the corresponding adaptation of farmers to the associated opportunities and risks. The process of commercialisation has thus included:

- the increased utilisation of externally produced inputs, including high-yielding seed, fertilisers, pesticides, irrigation, energy, machinery, and machinery services, as well as the credit needed to finance many of these purchases;
- the increased production of a marketable surplus, hence the choice of rice varieties, cropping systems, and processing technologies to meet the requirements of domestic and export markets;
- the progressive removal of state-imposed controls on rice farming in the socialist states of Indochina, including collectivisation of production, forced deliveries, land-use controls, price controls, and yield and production targets;
- the greater role of commercial decisions in farm management—for some farmers entailing specialisation in intensive, commercial rice production and, for others, prompting diversification away from rice production to field crops, tree crops, horticulture, and aquaculture, as well as non-farm pursuits.

This book is about understanding the processes involved in this transformation and the commercial opportunities and challenges of rice-based farming systems in the Lower Mekong in the 2010s, with a view to outlining prospects for the 2020s. It is the result of a collaboration between agricultural economists working in the four principal Lower Mekong countries—Thailand, Laos, Cambodia, and Vietnam. The motivation for this research was to (a) compare the current situation and trajectories of rice farmers within and between different regions of the Lower Mekong, (b) explore the value chains linking rice farmers with input and output markets within and across national borders, and (c) understand the changing role of government policies in facilitating the on-going evolution of commercial rice farming. The role of this chapter is to set the scene for the specific studies that follow. Subsequent sections of the book deal in turn with studies of rice farming, value chains, and policies in Thailand's Northeast Region, the Central and Southern Regions of Laos, the Central Plain of Cambodia, and the Mekong Delta in Vietnam. The setting for these studies and the methods used are described in each section. A final section draws together the findings and implications of the research for rice policies in the region as a whole.

The Lower Mekong Basin

The Mekong River runs for 4500 km from the Tibetan Plateau to the South China Sea and Gulf of Thailand, draining an area of 810,000 km² that takes in parts of Yunnan Province in China and Myanmar, Laos, Thailand, Cambodia, and Vietnam in Southeast Asia (Fig. 1.1). This drainage basin is generally divided into the Upper Mekong (or Lancang Basin) in China, accounting for 20% of the catchment, and the Lower Mekong in Southeast Asia—the region with which this book is concerned—accounting for 80% of the catchment (Cosslett and Cosslett 2018; MRC 2019).

The Lower Mekong Basin comprises four physiographic zones (Cosslett and Cosslett 2018; MRC 2019):

- The Northern Highlands include upland regions in eastern Myanmar, northern Thailand, and northern Laos. Major left-bank tributaries include the Nam Ou, Nam Soung, Nam Khan, and Nam Ngum in Laos and right-bank tributaries include the Nam Mae Kok and Nam Mae Ing in Northern Thailand.
- The Khorat Plateau is a large area of low-lying terrain with sandy soils mainly in north-eastern Thailand but including the lowlands of central and southern Laos. Left-bank tributaries include the Nam Ca Dinh, Se Bang Fai, and Se Bang Hiang in Laos and right-bank tributaries are the Songkhram and Mun Rivers in Thailand. The left-bank

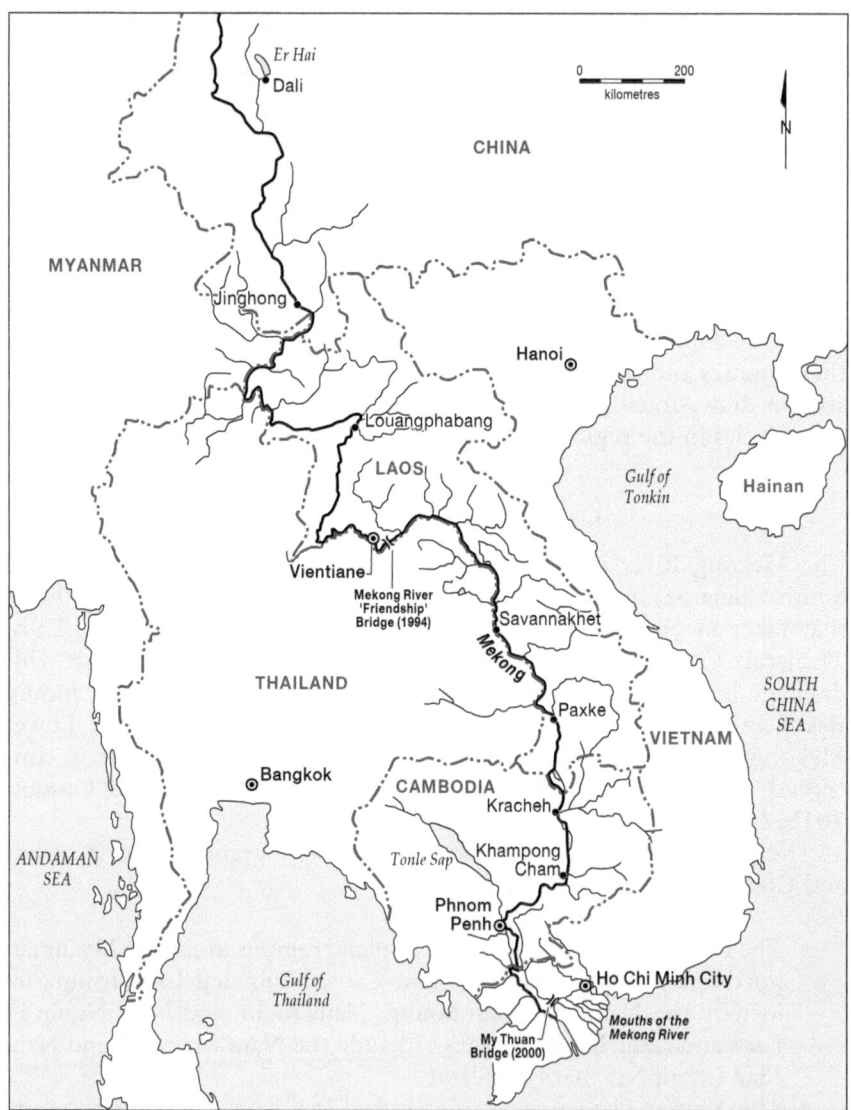

Fig. 1.1 Mekong River Basin. (Source: CartoGIS, Australian National University)

tributaries drain high-rainfall areas and contribute to major wet-season flows, while the right-bank tributaries drain low-relief areas of lower rainfall.

- The Tonle Sap Basin is a large alluvial plain that begins in southern Laos and takes in most of Cambodia. At the eastern edge of the Basin, the main river breaks up into a complex network of channels. The Tonle Sap River and Lake make up the central and western parts of the Basin. During the dry season the Tonle Sap Lake drains into the Mekong via the Tonle Sap River, while during the wet season the high flows in the Mekong cause the Tonle Sap River to reverse direction so that the Lake increases sixfold in area and 40–50 times in volume.
- The Mekong Delta begins near Phnom Penh where the Bassac River, the largest distributary, splits from the main river. The Mekong and Bassac Rivers then split into a number of smaller watercourses as the delta expands into a wedge-shaped plain that covers an area of almost 50,000 km², nearly 80% of which is within southern Vietnam.

The Lower Mekong Basin has a tropical monsoonal climate, with high temperatures throughout the year and distinct wet and dry seasons. The climate of Laos is illustrative of the seasonal pattern (Fig. 1.2). There is a

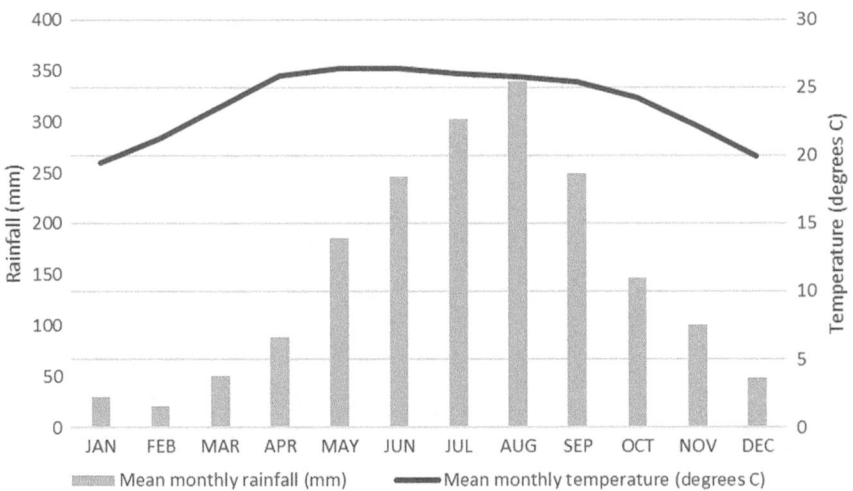

Fig. 1.2 Mean monthly rainfall and temperature for Laos, 1991–2016. (Source: Climate Research Unit, University of East Anglia)

hot wet period from roughly June to mid-October, under the influence of the southwest monsoon; a cooler dry period from mid-October to mid-February, under the influence of the northeast monsoon; and a hot dry period from mid-February to May, encompassing the transition from the northeast to the southwest monsoon. The wet season starts and ends somewhat earlier in the northern part of the Lower Mekong than in the south, with corresponding adjustments in planting times. There is also a declining rainfall gradient from east to west, such that Northeast Thailand and Cambodia experience lower rainfall than Laos and Vietnam. Rainfed rice is frequently affected by drought early in the wet season due to variability in the transition between monsoons, and again late in the wet season if the regular monsoon rains end early (Schiller et al. 2006).

Reflecting this monsoonal climate, the flow down the Mekong follows a regular seasonal pattern that has been part of the rhythm of life along the river for millennia, with high flows and flooding during the wet season from June to November, peaking in August–September, and low flows during the dry season from December to May (Fig. 1.3). The flood season accounts for 80–90% of the total annual flow (MRC 2019). Most of the seasonal flooding occurs along the left tributaries in Laos and Cambodia, which drain mountainous areas of higher rainfall, as well as in the Tonle

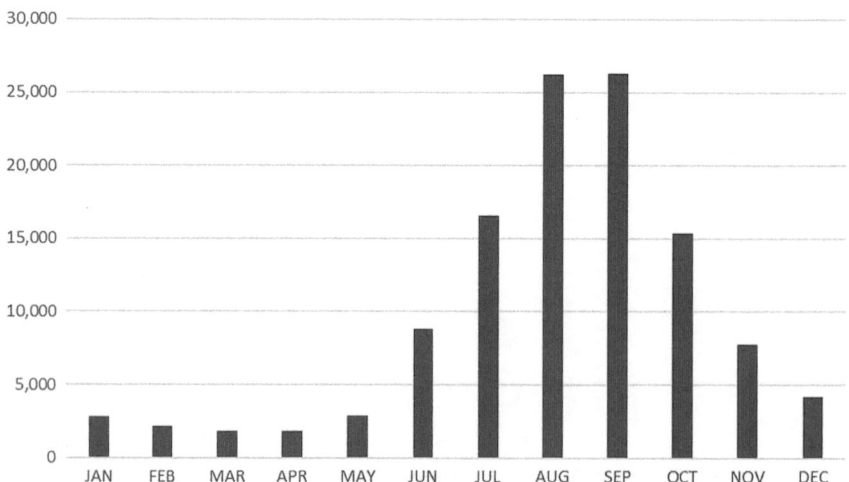

Fig. 1.3 Average monthly mainstream flow at Pakse, Laos, 1960–2004 (cubic metres per second). (Source: Cosslett and Cosslett 2018)

Sap and the Delta. Cosslett and Cosslett (2018) highlight that, since 2000, extreme floods and droughts have become more common, along with sea-level rise, saline intrusion, and changes in runoff, attributable to natural climate variability, climate change, and the construction of hydropower dams in the Upper Mekong. In a comprehensive review of the hydrological impact of hydropower dams throughout the Mekong, Hecht et al. (2019) confirm that the effect of the dramatic increase in mainstream, run-of-the-river dams since 2010 is to reduce and delay maximum flows in the wet season (hence the extent of flooding) and increase flows in the dry season, while reducing the overall delivery of sediment to the Mekong floodplain.

Most of the area of the Lower Mekong Basin falls in Laos (32%) and Thailand (29%), with Cambodia embracing 25% and Vietnam only 15% (Table 1.1). However, the population within the Basin is concentrated in Thailand (37%) and Vietnam (35%), with population densities of 132 and 279 persons per square kilometre respectively, compared with only 28 persons per square kilometre in Laos. Land use in the Basin is dominated by paddy fields (Fig. 1.4). Northeast Thailand accounts for just over half of the agricultural area of the Basin and just under half of the paddy land (Table 1.1). However, Vietnam has the highest proportion of agricultural

Table 1.1 Land, population, and rice production in the Lower Mekong Basin, 2014

Variable	Thailand	Laos	Cambodia	Vietnam	Total
Area in LMB (km^2 × 10^3)	184.0	202.0	161.0	95.0	642.0
Area in LMB (%)	28.7	31.5	25.0	14.8	100.0
Population in LMB (2014) (× 10^6)	24.2	6.1	12.5	23.0	65.8
Population in LMB (2014) (%)	36.7	9.3	19.0	35.0	100.0
Population density (persons/km^2)	132	28	78	279	103
Agricultural area in LMB (ha × 10^3)	10,300	1900	3100	4610	19,910
Paddy area in LMB (ha × 10^3)	4647	631	1647	2606	9531
Paddy area as % of agric. area	45.1	33.2	53.1	56.5	47.9
Irrigated paddy area (ha × 10^3)	1425	172	505	1921	4023
Irrigated area as % of paddy area	30.7	27.3	30.7	73.7	42.2
Paddy prodn. (2014) (t × 10^6)	14.7	3.9	8.7	25.2	52.5
% growth of prodn. (2000–2014)	2.5	4.5	6.1	3.0	3.4
Average yield (2014) (t/ha)	2.6	4.3	3.1	5.9	3.8
Prodn. as % of country total	45	98	94	56	57

Source: Cosslett and Cosslett (2018, Tables 5.2, 5.3, 5.4 and 5.5)

10   R. CRAMB

Land use

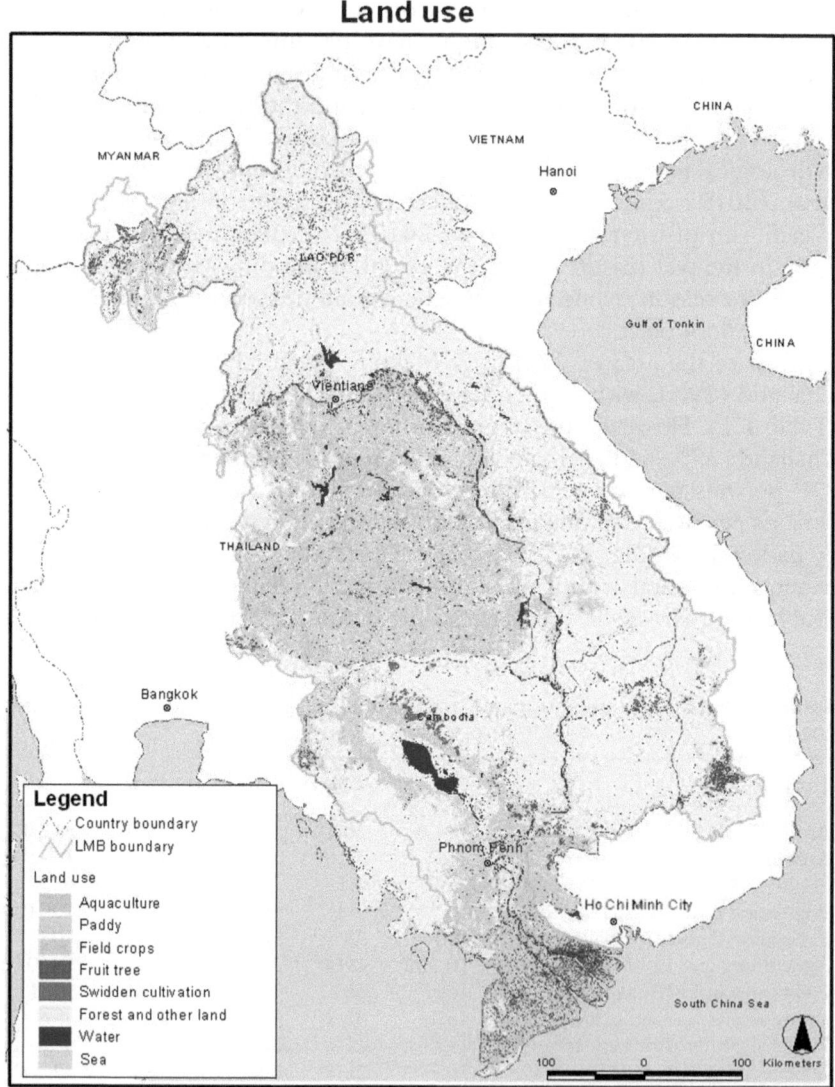

Fig. 1.4 Land use in the Lower Mekong Basin. (Source: Mekong River Commission)

land in paddy production (57%), the highest proportion of paddy land irrigated (74%), and, with average yields of 6 t/ha, accounts for nearly half of total paddy production from the Basin. Production has grown steadily at 2.5 to 3% per annum in the major paddy regions of the Basin in Vietnam and Thailand, but has been accelerating in Laos (4.5%) and Cambodia (6.1%). Paddy production within the Basin accounts for around half of total production in Vietnam (56%) and Thailand (45%) but over 90% of total production in Cambodia and Laos. While the Basin contributes less than 10% of global rice production (being dwarfed by China and India), it accounts for just over a quarter of rice exports.

Demographic and economic change in these four countries has had a profound influence on the commercialisation of rice farming within the Lower Mekong Basin (Table 1.2). Fertility has dropped to below replacement rate in Thailand and Vietnam, and population growth has slowed to less than 1%, approaching zero in Thailand. The growth of population has also slowed to around 1.5% in Laos and Cambodia. Urbanisation of the population has increased to almost 50% in Thailand and over 33% in Laos and Vietnam. These changes have created a growing labour scarcity in rice farming.

All four countries have experienced rapid economic growth, beginning with Thailand, then Vietnam, and now Laos and Cambodia. While

Table 1.2 Demographic and economic data for Lower Mekong countries, 2018

Variable	Thailand	Laos	Cambodia	Vietnam
Population (millions)	69.037	6.858	16.005	95.541
Population density (persons per sq. km)	135.1	29.8	91.1	294.2
Population growth (%)	0.18	1.48	1.46	0.97
Fertility (births/woman)	1.46	2.62	2.52	1.95
Urban population (%)	49.2	34.4	23.0	35.2
Rice consumption (kg/person) (2011)	112	162	159	145
Gross national income (GNI) (USD billion)	410.5	15.6	19.8	206.7
GNI per capita	5700	2270	1230	2160
GDP growth (%)	4.1	6.5	7.1	6.8
Agriculture value added as % of GDP	8.7	16.2	23.4	15.3
Employment in agriculture (% of total)	30.7	68.0	30.4	39.8
Poverty headcount (%)	8.6	23.4	17.7	9.8

Source: World Bank Data, FAOSTAT, ILOSTAT

Thailand's growth has slowed to 4%, the other three countries have among the fastest growth rates in the world at around 7%. All countries are thus going through the agricultural transition associated with modern economic growth, with the agricultural sector increasing in absolute terms while its share of GDP has declined to 9% in Thailand, 15% in Vietnam, 18% in Cambodia, and 23% in Laos. Agricultural employment has fallen to 30–40% of total employment, except in Laos, where it remains high at 68%. Agricultural and economic growth has resulted in a decline in poverty, especially in Thailand and Vietnam, where the overall incidence is under 10%.

Increased incomes and urbanisation have brought about a decline in average rice consumption per capita in Thailand and Vietnam as households diversify their diets, and this tendency appears to be beginning in Laos and Cambodia. Nevertheless the growth in urban populations has increased the aggregate domestic demand for a marketed rice surplus, as well as for higher-quality rice. Similar changes in the rice-deficit countries of Asia have led to a corresponding growth in demand for rice exports from the Lower Mekong.

Origins of Rice Farming in the Lower Mekong

Archaeological evidence indicates that rice (*Oryza sativa*) was fully domesticated and had become a staple in the lower and middle Yangtze by 4500 BCE (Fuller et al. 2010; Higham 2014). This subsequently led to the growth and spread of rice-growing populations into southern China around 3000–2000 BCE, and from there into Mainland Southeast Asia. It is probable that these early Southeast Asian rice farmers spoke languages of the Austroasiatic (Mon-Khmer) family, including the precursors of modern Khmer and Vietnamese. There is evidence for both a coastal expansion route, from southeastern China (modern Guangxi) to the Red (Hong) River and down the coast of Vietnam, and a riverine route, from southwestern China (modern Yunnan) down the Mekong to sites in the Khorat Plateau, the Tonle Sap Basin, and the Delta. These migrants brought with them a cultural package that included rice and millet; domesticated dogs, pigs, and possibly chickens; the preparation of yarn for weaving; a distinctive form of decorated pottery; and particular burial practices (Higham 2014). The rice they brought with them was of the *japonica* sub-species developed in the Yangtze, which they probably cultivated under upland conditions, that is, without bunded paddy fields (Bellwood 2011; Castillo 2011; Castillo et al. 2016).

Daic or Tai populations moved into the Lower Mekong Basin by various routes beginning in the first millennium CE, initially in response to the expansion of Chinese imperial control in southeastern China (Baker 2002; Stuart-Fox 2006). Originating in what is now Guangxi, some groups migrated westward into the northern arc of the Annamite Range, moving gradually across low ridges and into tributaries of the Red (Hong) and Black (Da) Rivers and of the Mekong. Others migrated further west into modern Yunnan, thence down the Mekong, Chao Phraya, and Salween valleys. Though escaping conflict was a factor, one of the prime motivators for these migrations was the search for good rice land (Baker 2002). Tai farmers had developed irrigation techniques suited to broad inland valleys, enabling streams to be diverted into a sequence of bunded and sometimes terraced paddy fields, before rejoining the main river. This assured the water supply in the wet season and, where streams flowed year-round, permitted dry-season cropping. Such sites had already been occupied by Austroasiatic farmers such as the Khmu, who were gradually absorbed by the incoming Tai or displaced into the surrounding hills and mountains in the Northern Highlands and the Annamite Range, or the interior of the Khorat Plateau, though the Khmer remained dominant in the Tonle Sap Basin and the Delta (Evans 2002).

The *indica* sub-species of rice, which had evolved in the Ganges Basin through hybridisation with *japonica* rice from the Yangtze, was dispersed through Iron Age trade networks into Southeast Asia from around 500 BCE and eventually came to dominate lowland rice farming in the Lower Mekong, though *japonica* varieties persisted in upland sites, to which they were adapted (Castillo 2011). Mutations and farmer selection for preferred traits gave rise to thousands of *indica* landraces with varying heights, growing periods, resistances, and grain qualities, including the glutinous rices that became the preferred staple of Tai peoples in the Northern Highlands and Khorat Plateau and the fragrant rices that have formed the basis of high-value rice exports in recent decades.[1] By 500 CE, lowland rice farmers in the Mekong, whether Tai or Mon-Khmer speakers, were planting *indica* rices in bunded paddy fields, cultivated with animal-drawn, iron-tipped ploughs—a technology that had been developed in the rice-growing heartland in the Yangtze but was also now prevalent in the Ganges Basin.

The diverse landraces cultivated by Mekong farmers were incorporated in a range of cropping systems to suit different agro-ecosystems, and these have persisted into the modern era (Barker and Herdt 1985, chap. 3; Javier 1997; Dao 2010; Haefele and Gummert 2015; Cramb 2017):

- Upland rice systems are practised on level to sloping land with no standing water, utilising medium to tall varieties of varying duration. These swidden or shifting cultivation systems are typically found in the Northern Highlands and along the Annamite Range, as well as in the northern uplands of the Tonle Sap Basin (Fig. 1.4).
- Rainfed lowland rice is the most widespread system in the Lower Mekong, involving bunded paddy fields with 5–50 cm of standing water in the wet season (subject to flooding or drought), utilising medium to tall varieties of varying duration. This is the dominant system in the open plains of the Khorat Plateau and the Tonle Sap Basin.
- Irrigated lowland rice includes (a) traditional gravity-fed irrigation using weirs to divert streams into adjacent paddy fields or dams (Northern Highlands, Khorat Plateau, Tonle Sap Basin); (b) lifting water from streams or canals to supply paddy fields using traditional devices such as waterwheels and scoops or (more recently) mechanical pumps, which are also increasingly used to tap groundwater (Khorat Plateau, Tonle Sap Basin, the Delta); and (c) tidal irrigation and drainage (middle reaches of the Delta). Irrigation can be used to supplement rainfall in the wet season and/or to enable dry-season production. Shorter-duration, photoperiod-insensitive varieties are preferred for the dry season.
- Deepwater/floating rice systems have been practised traditionally in areas that are deeply flooded in the wet season, such as around the Tonle Sap or the Plain of Reeds in the upper Delta. These systems utilise medium to tall varieties that elongate to 2–3 m in the case of deepwater rice or 5–6 m in the case of floating rice.
- Flood-recession dry-season rice systems are practised in areas that are continuously flooded in the wet season (such as around the Tonle Sap River and Lake and in the Mekong and Bassac branches of the upper Delta) and so not suitable for conventional rainfed or irrigated rice. Rather, the receding floodwaters are trapped by embankments and in ponds and dams that are used to irrigate a dry-season crop using canals and/or pumps.

HISTORICAL PERIODS OF SURPLUS PRODUCTION

These cropping systems provided subsistence for generations of small farming communities scattered throughout the Lower Mekong and were sufficiently productive to support the early Khmer, Cham, and Tai states that grew up in the first millennium CE and contended for power over the peoples and resources of the region in subsequent centuries (Evans 2002; Chandler 2008; Higham 2014). These states were all dependent on controlling labour and acquiring surplus rice to support state functionaries and invest in public works. Higham (2014: 390) refers to the fundamental importance of rice productivity, especially through permanent rice fields, ploughing, and irrigation, enabling the extraction of a rice surplus through taxation. On the other hand, Scott (2010) argues that swidden agriculture in the uplands provided a way for many to escape the exactions of centralised paddy states.

Funan, an early trading state in the Mekong Delta (c. 50–550 CE), and its successor, Chenla (c. 500–850 CE), likely depended on farmers producing flood-recession dry-season rice in sites such as Angkor Borei in what is now Takeo Province in Cambodia (Higham 2014: 278–285). Fox and Ledgerwood (1999) estimate that a farm workforce of 80,000 practising this system of cultivation could have supported an additional 40,000 people within a 10 km radius of Angkor Borei, utilising the system of canals that linked the rice-growing areas with the harbour at Oc Eo, adjacent to the Gulf of Thailand.

Angkor, the powerful Khmer state that expanded to control the Lower Mekong and beyond from 800 to 1350 CE, also depended in part on surplus production from deepwater/floating and flood-recession rice around Tonle Sap (Fox and Ledgerwood 1999). However, state-directed construction of large reservoirs (*baray*) and an extensive system of canals feeding into bunded paddy fields made irrigated rice possible, generating a large surplus (Helmers 1997; Higham 2014: 349–407). An official Chinese visitor to Angkor in 1296–1297 noted "the cultivation of three to four rice crops a year" (Higham 2014: 390), perhaps referring to the combined crops from irrigated and flood-recession environments. A century earlier it was recorded that, under Jayavarman VII, Angkor's 102 hospitals were supplied with 11,370 t of rice provided by 81,640 people residing in 838 villages, meaning the farmers produced nearly double their subsistence requirements in order to meet their tax obligations. The development of irrigated rice through construction of reservoirs and canals was extended to other centres under Khmer dominance, including Champassak in what is now Southern Laos (Schiller et al. 2006).

The early Tai states (*muang*) established in the Lower Mekong from around 700 CE were also dependent on harnessing sites capable of producing surplus rice, such as the inland valley of Luang Prabang and the Vientiane floodplain. These sites were sufficiently productive to support the Tai state of Lan Xang that stretched across the Northern Highlands and the northern and eastern parts of the Khorat Plateau in the sixteenth century. However, the valleys controlled by Lan Xang had less productive capacity than the vast central plain of the Chao Phraya to the southwest, which supported the rise of Sukhotkai and then Ayudhya, ultimately at the expense of Lan Xang. Moreover, "there is little evidence that the [Lan Xang] state ever sponsored irrigation as a way of augmenting its economic surplus. The construction of dams and irrigation networks was left to local communities. The relatively small surpluses restricted the taxes and corvée (labour) that could be levied on the peasantry and thus the scale of public works that could be carried out, whether it be building roads or major temple complexes and cities" (Evans 2002: 12–13).

In the second half of the nineteenth century, the imposition of colonial rule and the surge in global demand for rice and other tropical commodities created a new set of circumstances favouring the production of rice surpluses. In the Lower Mekong Basin the growth in rice exports was based on surplus production in two regions of French Indochina— Cochinchina (embracing the Delta) and, to a much smaller extent, Battambang Province in western Cambodia. Exports of rice through the port of Cholon (now part of Ho Chi Minh City) averaged 157,000 t over the period 1863–1871, rising to 793,000 t in 1902–1911 and 1,314,000 t in 1930–1934 (Robertson 1936; Owen 1971), an average annual growth rate of 3% over 65 years.

Over 90% of these exports came from the Delta. The growth was stimulated by global demand, which led French and Chinese businesses to construct rice mills and Chinese traders to fan out into the Delta to purchase paddy from farmers. These farmers responded by producing increasing surpluses for sale—not through increased yields, which remained low at around 1.1 t/ha in the 1930s (Robertson 1936), but by expanding the area cultivated. The colonial regime invested in opening up the southern part of the Delta (the Trans Bassac) through construction of canals, encouraging in-migration of workers from poorer parts of Cochinchina as tenant farmers and labourers (Biggs 2012; Biggs et al. 2009). Thus the area planted in Cochinchina increased from 200,000 ha in 1868–1870 to 1.7 million ha in 1911–1914, with the Trans Bassac increasing its share of planted area from 8% in 1872 to 37% in 1908 (Owen 1971). The population of Cochinchina grew from 1.2 million in 1867 to 3 million in 1910 (Owen 1971), a growth rate of 2.2%. This rapid growth reflected the influx of Vietnamese rice farmers as well as Chinese workers in the trading, milling, and exporting sectors of the industry.

In Cambodia, the French regime gave land concessions to French settlers for the establishment of large rice plantations in Battambang Province (Helmers 1997). These concessions occupied over 16,000 ha and made use of hired labour to produce around 30,000 t of paddy per year. The government supported the plantations with irrigation infrastructure and a railway link to Phnom Penh, from where the paddy was shipped to Cholon for processing and export. The smallholder sector also contributed to the growth in exports, not through any increase in yields but through areal expansion. In the boom conditions of the 1920s, Khmer smallholders earned good incomes from rice sales but with the collapse in prices in the early 1930s, they responded by reducing the area cultivated by 60%. Over the first half of the twentieth century, the French regime obtained exports from Cambodia of from 50,000 to 200,000 t of paddy per year, mostly from smallholders.

The more isolated regions of the Lower Mekong in Laos and Northeast Thailand, which produced mainly glutinous rice for subsistence and the local market, contributed little or nothing to the pre-war export boom. Rice exports from Northeast Thailand accounted for only 7% of the country's rice exports in 1925 and 18% in 1935 (Ekasingh et al. 2007). For much of the colonial period, Laos was a net importer of rice, with only the Champassak area consistently producing a surplus (Schiller et al. 2006).

From the 1940s to the 1970s, war was the overriding factor affecting rice farming in the Lower Mekong. In Vietnam, under Japanese rule, the great famine of 1944–1945 resulted in between one and two million deaths due to failed harvests in the north and the forced acquisition and export to Japan of over a million tonnes of rice per year from the Delta (Gunn 2011). From 1945 to 1975, the First and Second Indochina Wars devastated the rural sector, despite attempts to boost rice production through land reforms and (in the south) the US-funded introduction of high-yielding varieties, fertilisers, and mechanisation. By the end of the war in 1975, there was a nation-wide production deficit of 2–3 million t of paddy (Le Coq et al. 2001). In Laos, too, despite high levels of US assistance and the introduction and distribution of some improved variet-ies, rice production received little support and the escalating war disrupted and destroyed rural livelihoods.

In Cambodia, in the first decade after obtaining independence from France in 1953, and with support from United States Agency for International Development (USAID), paddy production increased to around 2.3 million t and rice exports to 250,000–400,000 t (Helmers 1997). From 1964, rice exports were nationalised and the government mounted campaigns to forcibly collect rice at the low official price, prompting armed rebellions by farmers in 1967 and 1968. From 1970 to 1975, Cambodia was caught up in the war, devastating rice production, which fell by 84%. Under the Khmer Rouge regime (1975 to 1979), despite a fanatical focus on developing intensive irrigated rice production through the mobilisation of labour in collective farms, the programme failed and the country was devastated, leaving the surviving population under threat of widespread famine by 1979 (Helmers 1997).[2]

During this period, Northeast Thailand was a remote and impoverished region but its strategic importance during the Indochina conflict led to substantial US-funded investment in roads, communications, irrigation, agricultural extension, and other forms of rural development. In particu-lar, the Friendship Highway for the first time provided the region with a road link to Bangkok. These investments laid the foundation for the com-mercialisation of agriculture and diversification of livelihoods in the 1980s and 1990s (Ekasingh et al. 2007).

RECENT CHANGES IN THE TECHNOLOGY OF RICE FARMING

The cropping systems that supported small communities, large empires, and colonial economies for two millennia, with little change in technol-ogy, have undergone significant changes since the mid-1970s, notably in (a) their relative importance, (b) the productive potential of the varieties cultivated, and (c) the extent of mechanisation (Cramb and Newby 2015).

Upland rice systems have declined in extent, partly through govern-ment policies directed at eliminating shifting cultivation and partly due to declining productivity and the economic attraction of alternative crops (Cramb et al. 2009). Deepwater and floating rice systems have also declined in importance. However, rainfed lowland systems have continued to dominate throughout the Khorat Plateau and the Tonle Sap Basin (Fukai and Ouk 2012). There has been increasing use of on-farm irrigation in some of these rainfed lowlands through digging small ponds and sinking tubewells, enabling supplementary irrigation of wet-season rice and dry-season cultivation of non-rice crops on a part of the paddy field. Particularly in Northeast Thailand, there has also been a shift in the use of the more drought-prone upper-level paddies to field crops such as cassava and sugarcane (Barnaud et al. 2006; Grandstaff et al. 2008).

Full-scale irrigated systems have expanded with public investment in irrigation infrastructure, especially in Thailand and Vietnam (Hoanh et al. 2009; Floch and Molle 2013; Schiller et al. 2006). While pump-irrigation schemes in the Khorat Plateau (both in Thailand and Laos) have not delivered the intended expansion in dry-season rice production, the development of flood control and irrigation infrastructure in the Vietnamese Delta has enabled the expansion of double and triple cropping of rice and, more recently, diversification into non-rice crops. Figures 1.5 and 1.6 show the current extent of rice cultivation in the wet and dry (irrigated) seasons.

Rice farming has also been transformed by the dissemination of modern varieties, giving higher and/or more stable yields, particularly in association with increased fertiliser use (Fukai and Basnayake 2001; Haefele and Gummert 2015). While the International Rice Research Institute (IRRI) had been working in Thailand from 1966, formal collaboration with the countries of Indochina did not begin until 1978 in Vietnam, 1986 in Cambodia, and 1987 in Laos. The succeeding decades of collaborative rice research in these countries and the growth of national rice breeding programmes have had a major impact on the development of locally adapted modern varieties.

The first high-yielding semi-dwarf variety, IR8, was made available in the Delta soon after its release in 1966.[3] It was widely displaced by the more resistant IR36 in the 1980s and then by IR64 in the 1990s. With its wide adaptation, early maturity, and improved eating quality, IR64 was the ideal variety for commercial production. While IR64 is still widely planted, many more varieties with specific adaptations (e.g., flood tolerance, salinity tolerance) have been developed by local plant breeders and are being taken up by farmers (Bui and Nguyen 2017).

In Thailand, IR8 was not adopted because of it low eating quality, but the semi-dwarf gene in IR8 was incorporated in a series of locally bred varieties (labelled RD for Rice Department) that were widely adopted in the irrigated areas of the Central Plain. For the Northeast, the major breakthroughs were the selection of a line of Thai fragrant rice (*hom mali*,

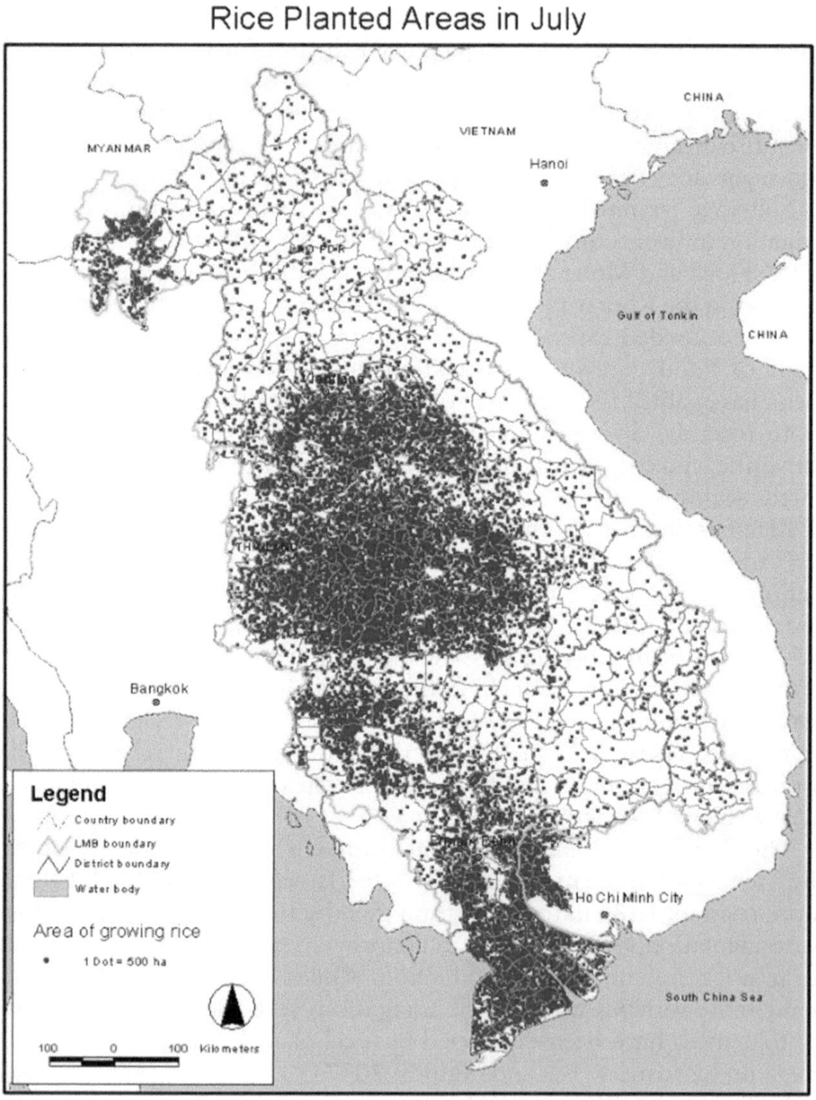

Fig. 1.5 Area planted with rice in Lower Mekong Basin in wet season (July). (Source: Mekong River Commission)

Rice Planted Areas in January

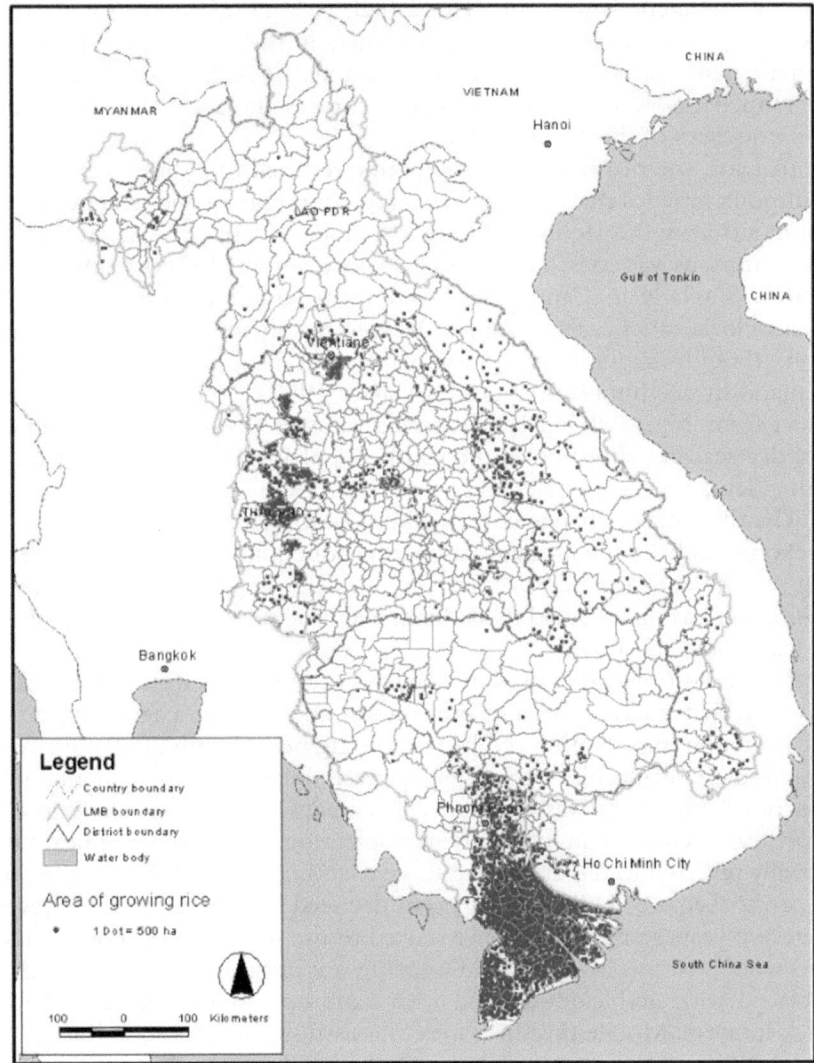

Fig. 1.6 Area planted with rice in Lower Mekong Basin in dry season (January). (Source: Mekong River Commission)

KDML105) that since the 1990s has become the major commercial crop, and its mutagenesis to form RD6, a high-yielding glutinous variety that meets the subsistence needs of Lao farmers in the Northeast. These two varieties have formed the basis of what Grandstaff et al. (2008) have called a "rainfed revolution".[4] Rambo (2017) traces the social and economic consequences of this revolution.

In Laos, the rice breeding programme resulted in a suite of improved glutinous varieties that were widely adopted in rainfed and irrigated environments from the 1990s, resulting in a modest increase in yields, though the glutinous varieties have limited export potential (Inthapanya et al. 2006). Similarly in Cambodia, breeding programmes released selected lines of local varieties from the 1990s, used mainly for domestic consumption, though fragrant non-glutinous Cambodian varieties are also in demand in neighbouring countries (Javier 1997). Nevertheless, it is the short-term, high-yielding IRRI-derived varieties that dominate commercial dry-season cultivation in the south, supplying the cross-border trade with Vietnam (Wang et al. 2012).

The third major change in the technology of rice farming in the Lower Mekong has been the mechanisation of production, driven by the increasing scarcity and rising cost of farm labour. This began in Thailand in the 1960s but has since spread to Vietnam and is beginning to have an impact in Cambodia and Laos.

The earliest machines used in Thailand in the 1960s were locally made two-wheeled tractors for land preparation and low-lift axial-flow pumps for irrigation, mostly powered by tractor engines (Cramb 2019). Farmers acquired these machines themselves, given their low cost and multiple functions, but there was also some localised renting, particularly in the Northeast. Rather than mechanise transplanting, Thai farmers almost universally reverted to direct seeding to save labour, using hand broadcasting of pre-soaked seed in irrigated areas or dry seed in rainfed areas. However, in recent years some farmers have started to use seed drills or hire contractors who use transplanters. Hand weeding was progressively augmented or replaced with herbicides applied with hand-operated or powered backpack sprayers. Mobile threshers were successfully introduced in the 1970s and 1980s, mostly on a contract service basis. However, these were superseded from the 1990s by combine harvesters, also operated by contractors. The use of combines has entailed the delivery of harvested grain directly to mills, which have installed mechanical driers to deal with the high moisture content.

In Vietnam there was a parallel development of small-scale mechanisation, beginning in the mid-1960s with the local invention and rapid adoption of the engine-driven shrimp-tail pump, used for irrigation and drainage as well as to power boats (Biggs 2012). The success of the high-yielding variety IR8 gave added incentive to acquiring the pumps, especially in the context of the deteriorating hydraulic infrastructure in the Delta. However, small-scale mechanisation stalled after 1976 with the return of population to the countryside alleviating labour shortages, the collectivisation of machinery and other assets, and a renewed emphasis on large-scale, centrally controlled mechanisation (Le Coq et al. 2001; Biggs 2012). Moreover, commercial rice production was not remunerative, given the imposition of fixed supply contracts at low official prices. With market liberalisation from 1986, large farmers could purchase equipment such as pumps, hand tractors, and axial-flow threshers and provide contract services to poorer farmers. Mechanisation spread in the 1990s and 2000s so that by 2013 land preparation for rice was 95% mechanised, 50% of the rice crop was mechanically threshed, and 50% was harvested by small combines (Tran 2016).

Farmers in the rainfed and irrigated lowlands of Laos and Cambodia are beginning to adopt two-wheeled tractors, low-lift pumps, and combine harvesters, typically of Thai or Vietnamese manufacture. In Cambodia, much of the dry-season crop in Battambang in the west is mechanically harvested for immediate export to Thailand, and in Takeo in the south for export to Vietnam.

The Evolution of Rice Value Chains

The changes in rice production systems have been associated with major changes in rice value chains (ACI 2005; Purcell et al. 2008; Reardon et al. 2014; Haefele and Gummert 2015; Swinnen and Kuijpers 2019), as summarised in Fig. 1.7. In the 1970s the value chain was relatively simple. Most of the inputs for rice production were supplied by the farm household itself or by neighbouring farmers, including seed, manure, draught animal power, and labour. To the extent that industrial inputs such as fertiliser or pesticides were purchased, these were typically provided on credit by a village trader who deducted the cost of the loan from the purchase of the crop. In some cases, especially in Thailand, government agencies provided these inputs, including credit. Paddy for household consumption was stored in the home compound and dehusked manually

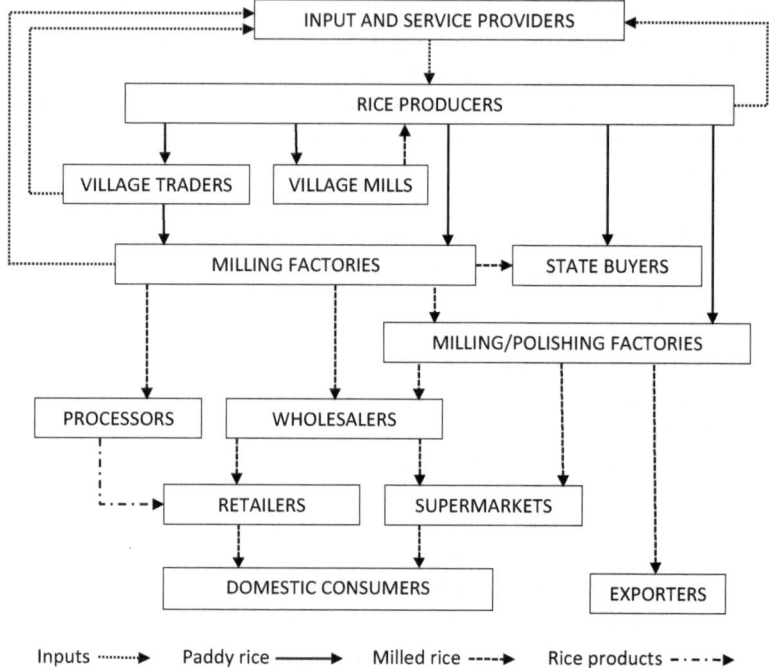

Fig. 1.7 Schematic outline of evolving rice value chains in Lower Mekong Basin

as required or taken to a small-scale village rice mill. Paddy for sale was almost all purchased by a village collector who transported the crop to small or large commercial rice mills. Some of the crop may have been acquired by state purchasing agencies, whether to supply the bureaucracy or military, to accumulate disaster reserves, or to intervene in the market in an attempt to stabilise or support farm-gate prices. Once milled, rice was sold in bulk to wholesalers and then to retailers in urban markets, where it was sold loose to consumers and food outlets. In Thailand, large modern mills sold high-quality rice to export companies but in the 1970s and 1980s in Laos, Cambodia, and Vietnam the overriding concern was to produce enough rice for domestic consumption.

While these features persist in many parts of the Lower Mekong in the 2010s, there has been a "quiet revolution" in rice value chains that is still incipient in more remote regions but proceeding rapidly in the major surplus-producing zones (Reardon et al. 2014). Some of the key changes are sketched here and explored in more detail in subsequent chapters (Fig. 1.7).

- Input and service providers have expanded to include seed, fertilisers, agrochemicals, irrigation equipment (pipes, hoses, tubewells), machinery (pumps, sprayers, tractors), and machinery services (tractor-hire, harvesting, digging ponds, sinking tubewells). These are increasingly provided by specialised suppliers (e.g., local contractors, dealers) and paid for in cash, through dealer finance, or using bank loans or microfinance rather than through tied credit from a local trader.
- The role of the village trader has declined, especially in Northeast Thailand, with increasing incidence of direct sales from farmers to medium-large mills. However, in the Delta, where harvested paddy is mainly transported by a network of waterways, local collectors still predominate, as they do in Laos and Cambodia. Contract farming of rice, whereby the mill provides seed, inputs, harvesting, and processing, has been introduced in some areas but with limited success.
- Rice mills are increasingly privately owned and financed rather than cooperatively or state-owned. Small mills are in decline, apart from their traditional function of custom milling paddy for local consumption, while farmers and traders increasingly sell to medium-large mills, implying transportation over longer distances on improved infrastructure. Larger mills, particularly in Vietnam and Thailand, have invested in expanding and upgrading milling equipment, enabling them to handle greater throughput, polish rice, and produce higher grades for both domestic and export markets. In Cambodia and Laos, though modern mills have been constructed in recent years, milling capacity remains a constraint.
- There is increasing coordination between large mills and urban wholesalers and (in Thailand and Vietnam) supermarkets. This provides a basis for sorting, packaging, labelling, and branding to meet the requirements of middle-class consumers, particularly in Thailand, for greater product differentiation and identification. There has also been growth in the processing of rice for both traditional food products such as rice flour and noodles and convenience foods such as rice crackers, though the supply of rice as a staple food remains the dominant chain.

- State procurement of paddy and rice for contingencies and price stabilisation continues to be a feature of the value chain in all countries.
  However, in Thailand this was taken to unprecedented lengths in the
  2010s in an effort to support farm-gate prices and influence the
  world market, with disastrous economic and political consequences.
  Government-held stocks reached record levels by 2013 (13 million
  t) and have had to be progressively sold off at discount prices
  (Welcher 2017).
- The most remarkable development in the value chain has been not
  only the achievement of rice self-sufficiency in Vietnam, Cambodia,
  and Laos, but the overall growth in exports (Fig. 1.7), such that
  exports from the Lower Mekong Basin account for over 25% of
  global exports by volume. Exports from Thailand as a whole have
  increased from under 1 million t in 1975 to around 10 million t in
  2016, over half of it now derived from expanded production in the
  Northeast. In Vietnam, exports recommenced within three years of
  the 1986 *doi moi* economic reforms, rising to an average of 6 million
  t in the 2010s (five times the volume of exports in the 1930s), over
  90% of which is produced in the Delta. Cambodia and Laos have
  begun exporting on a much smaller scale in the past decade.
- An interesting aspect of the export value chain is the growth in cross-
  border trade in both paddy and rice between the four Mekong countries. This trade is two-way but is dominated by the flow of paddy
  from Cambodia into Thailand and Vietnam, where it is processed for
  both domestic and export markets.

SHIFTS IN RICE POLICY

While much of the process of commercialisation over the past 40 years has
been driven by private actors throughout the value chain, shifts in government policy have been crucial (Byerlee et al. 2009; Chang 2009). The
overriding concerns of governments in the Lower Mekong countries have
been to achieve national food security (viewed as self-sufficiency in rice)
and reduce rural poverty. In the 1970s these two goals coincided, given
that subsistence was widely under threat (especially in the war zones of
Indochina) and that impoverished rice farmers made up most of the population. However, given several decades of economic development, the two
goals have increasingly diverged such that a continued emphasis on rice
intensification can be at odds with the goal of poverty reduction. Farm

households in much of the Lower Mekong are now interested in more profitable non-rice crops and non-farm sources of livelihood. Moreover, the national self-sufficiency goal has been achieved in all jurisdictions, along with the infrastructure to ensure that rice-deficit areas can access supplies from surplus-producing areas. Hence, in the 2010s, government policies have gradually come to allow and even encourage greater farm diversification and to treat rice production as primarily a commercial activity, with the focus on upgrading value chains and promoting exports rather than merely attaining yield and production targets.

The range of policies pursued over this period can be conveniently broken down into: (a) those affecting *access* to resources (land, water, draught animals, machinery) and inputs (seed, fertiliser, services, information, credit); (b) those directly regulating farm *activities* (the organisation of production and the choice of crops, varieties, and cropping systems); and (c) those affecting the *appropriation* of the ensuing product (whether retained for subsistence, requisitioned by the state, sold at market prices, or taxed) (Ellis 1992; Chang 2009; Fig. 1.9).

In Northeast Thailand the emphasis was primarily on reducing rural poverty as Thailand as a whole was a rice-surplus country throughout the period under consideration (Fig. 1.8). The focus was on public investment

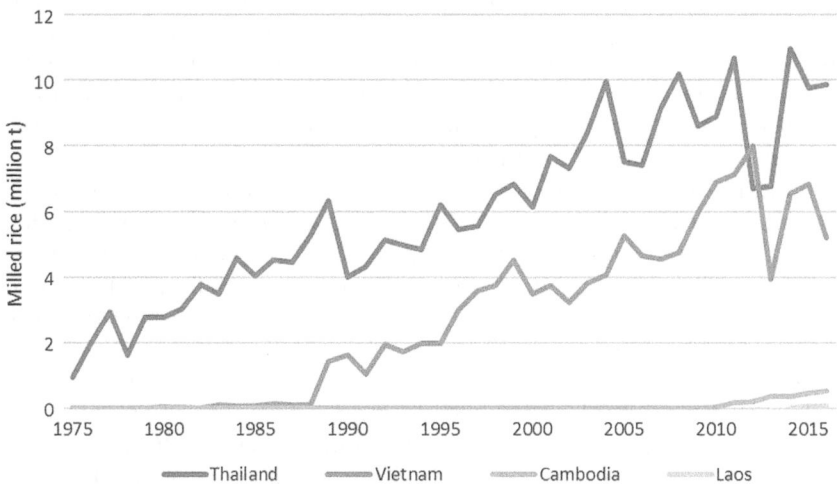

Fig. 1.8 Rice exports from Thailand, Vietnam, Cambodia, and Laos, 1975–2016. (Source: FAOSTAT)

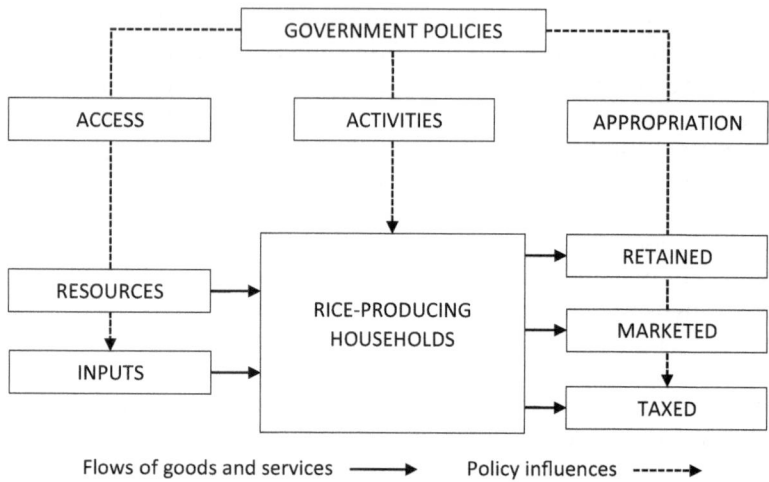

Fig. 1.9 Influence of government policies on rice production

to provide access to the inputs (improved seed, fertiliser, extension, credit) and infrastructure (transport and irrigation) needed for independent smallholders to intensify rice production, to both safeguard their subsistence and generate a marketable surplus (Ekasingh et al. 2007). There was no attempt by the state to organise the production activities of farm households directly, though the establishment of marketing cooperatives was encouraged. With regard to the appropriation of farm income, 1976 marked a shift from taxing to supporting rice farmers. In that year the export tax on rice was abolished and the first price support programmes were instituted. The price support policy developed into a rice buffer stock scheme from 1981, intended to raise farm-gate paddy prices, and a paddy mortgage programme from 1983, financed by the Bank for Agriculture and Agricultural Cooperatives (BAAC) and using public warehouses or on-farm storage to carry over paddy stocks. With increasing subsidisation, more farmers participated (Ekasingh et al. 2007). However, as noted above, aggressive government intervention to support prices through this mechanism from 2011 to 2014 led to the accumulation of record stocks and the eventual collapse of the programme (Welcher 2017), though mortgaging newly harvested paddy for seasonal price stabilisation has recently been reinstated. A significant shift in government policy in the past two decades has been the promotion of sustainable and self-sufficient agriculture within cooperative groups, including incentives for farmers to switch to organic rice production (Amekawa 2010), though this new emphasis has not had a great impact on the overall extent of commercial rice farming.

The economies of Vietnam, Laos, and Cambodia were centrally planned from 1975. Faced with an urgent need to raise rice production to avert famine, all three governments attempted to intensify rice production by exercising state control over access to resources and inputs, collectivising production activities, and appropriating much of the output at low official prices.[5] The disincentives and inefficiencies created by this regime meant that rice production stagnated. In 1986 Vietnam introduced the *doi moi* reforms, which were soon emulated in Laos and Cambodia, whose 1986–1990 five-year plans were closely coordinated with Vietnam's.[6] These allowed farm households to access inputs from the private sector (including imports), manage their own production activities, and sell surplus production at market prices. Though area, yield, and production targets are still a feature of government policy in Laos, and land-use controls to keep land in paddy production have persisted in Vietnam, the role of government has largely reverted to the Thai model of providing public goods through research, extension, and rural infrastructure (roads, canals, irrigation, electrification), subsidising key inputs (seed, fertiliser, water, electricity), and attempting to support or at least stabilise the farm-gate price of paddy while controlling the retail price of rice. However, the Vietnam state still plays a major role in rice marketing and exports through the Vietnam Food Association and state-owned enterprises, and has used floor prices, export quotas, and export bans in an attempt to control domestic stocks and prices (Nguyen and Talbot 2014; Tran and Dinh 2015; Nguyen et al. 2017; VNA 2018). Laos and Cambodia have also used export bans at times of high world prices with the intention of safeguarding food security.

Notes

1. Genetic studies indicate that "glutinous *indica* landraces in Laos were generated through repeated natural crossing with glutinous-*japonica* landraces and severe selection by local farmers" (Muto et al. 2016: 580).
2. Despite the population being subjected to starvation rations, the Khmer Rouge regime appropriated rice for export, e.g., 150,000 t in 1976 (Chandler 2008).

3. Taiwanese advisors in the Delta with links to Taiwanese colleagues in IRRI were responsible for establishing experimental plots of IR8, and quantities of IR8 seed were distributed by the US military when replacement seed was urgently needed after flooding wiped out seedlings in a valley north of Saigon in 1967 (Biggs 2012).

4. In 1995, RD6 accounted for 40% of the total wet-season rice area in Northeast Thailand and 83% of the glutinous rice area (Ekasingh et al. 2007).

5. Collectivization in southern Vietnam after 1975 was incomplete—only 25% of farm households belonged to a cooperative in 1980, compared with 97% in northern Vietnam (Tsukada 2011). In Laos, too, the campaign to form producer cooperatives after 1975 had limited success, with most regarded as "pseudo cooperatives … really only labour exchange groups" (Evans 1988: 76). In Cambodia, under the People's Republic of Kampuchea (PRK), by 1986 "97 per cent of the rural population were in the collective sector which was composed of more than 100,000 solidarity groups each of which consisted of seven to fifteen families" (Sokty and Luyna n.d.). However, here too the effective extent of collective control over production activities is questionable.

6. In Laos the 1986 policy shift to a market-based economy was termed the New Economic Mechanism (NEM).

References

ACI, 2005. *Northeast Thailand Rice Value Chain Study.* Project Brief. Khon Kaen: Agrifood Consulting International.

Amekawa, Yuichiro, 2010. Rethinking sustainable agriculture in Thailand: a governance perspective. *Journal of Sustainable Agriculture* 34: 389–416.

Baker, C., 2002. From Yue to Tai. *Journal of the Siam Society* 90: 1–26.

Barker, R., and Herdt, R. W., 1985. *The Rice Economy of Asia.* Washington, DC: Resources for the Future.

Barnaud, C., Trebuil, G., and Dufumier, M., 2006. Rural poverty and diversification of farming systems in upper Northeast Thailand. *Moussons* 9–10: 157–187.

Bellwood, P., 2011. The checkered prehistory of rice movement southwards as a domesticated cereal—from the Yangzi to the Equator. *Rice* 4: 93–103.

Biggs, D., 2012. Small machines in the garden: everyday technology and revolution in the Mekong Delta. *Modern Asian Studies* 46: 47–70.

Biggs, D., Miller, F., Chu, T. H., and Molle, F., 2009. The delta machine: water management in the Vietnamese Mekong Delta in historical and contemporary perspectives. In François Molle, Tira Foran, Mira Kakonen, eds. *Contested*

Waterscapes in the Mekong Region: Hydropower, Livelihoods and Governance, pp. 203–225. London: Earthscan.

Bui, C. B., and Nguyen, T. L., 2017. New rice varieties adapted to climate change in the Mekong River Delta of Vietnam. *Vietnam Journal of Science, Technology and Engineering* 60: 30–33.

Byerlee, D., de Janvry, A., and Sadoulet, E., 2009. Agriculture for development: toward a new paradigm. *Annual Review of Resource Economics* 1: 15–31.

Castillo, C., 2011. Rice in Thailand: the archaeobotanical contribution. *Rice* 4: 114–120.

Castillo, C., Tanaka, K., Sato, Y., Ishiwawa, R., Bellina, B., Higham, C., Chang, N., Mohanty, R., Kajale, M., and Fuller D. Q., 2016. Archaeogenetic study of prehistoric rice remains from Thailand and India: evidence of early japonica in South and Southeast Asia. *Archaeological and Anthropological Sciences* 8: 523–543.

Chandler, D., 2008. *A History of Cambodia*, 4th ed. Chiang Mai: Silkworm Books.

Chang, H. J., 2009. Rethinking public policy in agriculture: lessons from history, distant and recent. *Journal of Peasant Studies* 36: 477–515.

Cosslett, T. L., and Cosslett, P. D., 2018. *Sustainable Development of Rice and Water Resources in Mainland Southeast Asia and Mekong River Basin*. Singapore: Springer Nature.

Cramb, R. A., 2017. Shifting cultivation and human interaction with the forest. In P. Hirsch, ed. *Routledge Handbook of the Environment in Southeast Asia*, pp. 180–203. London and New York: Routledge.

Cramb, R. A., 2019. Evolution of agricultural mechanization in Thailand. In Xinshen Diao, Hiroyuki Takeshima, and Xiaobo Zhang, eds., *A New Paradigm of Agricultural Mechanization Development: How Much Can Africa Learn from Asia?*, pp. 148–185. Washington, DC: International Food Policy Research Institute.

Cramb, R. A., and Newby, J. C., 2015. Trajectories of rice-farming households in Mainland Southeast Asia. In R. A. Cramb, ed. *Trajectories of Rice-Based Farming Systems in Mainland Southeast Asia*, pp. 35–72. Canberra: Australian Centre for International Agricultural Research.

Cramb, R. A., Colfer, C. J. P., Dressler, W., Laungaramsri, P., Le, Q. T., Mulyoutami, E., Peluso, N. L., and Wadley, R. L., 2009. Swidden transformations and rural livelihoods in Southeast Asia. *Human Ecology* 37: 323–346.

Dao The Anh, 2010. The rice-based cropping systems in Vietnam. In Bui Ba Bong, Nguyen Van Bo, and Bui Chi Buu, eds. *Vietnam: Fifty Years of Rice Research and Development*, pp. 317–330. Hanoi: Agriculture Publishing House.

Ekasingh, B., Sungkapitux, C., Kitchaicharoen, J., and Suebpongsang, P., 2007. *Competitive Commercial Agriculture in the Northeast of Thailand*. Chiang Mai:

Department of Agricultural Economics and the Multiple Cropping Centre, Faculty of Agriculture, Chiang Mai University.

Ellis, F., 1992. *Agricultural Policies in Developing Countries*. Cambridge: Cambridge University Press.

Evans, G., 1988. *Agrarian Change in Communist Laos*. ISEAS Occasional Paper No. 85. Singapore: Institute of Southeast Asian Studies.

Evans, G., 2002. *Laos: The Land in Between*. Chiang Mai: Silkworm Books.

Floch, P., and Molle, F., 2013. Irrigated agriculture and rural change in Northeast Thailand: reflections on present developments. In R. Daniel, L. Lebel, and K. Manorom, eds., *Governing the Mekong: Engaging in the Politics of Knowledge*, pp. 185–212. Petaling Jaya, Malaysia: Strategic Information and Research Development Centre (SIRD).

Fox, J., and Ledgerwood, J., 1999. Dryseason flood-recession rice in the Mekong Delta: two thousand years of sustainable agriculture? *Asian Perspectives* 38: 37–50.

Fukai, S., and Basnayake, J., eds., 2001. *Increased Lowland Rice Production in the Mekong Region*. ACIAR Proceedings No. 101. Canberra: Australian Centre for International Agricultural Research.

Fukai, S., and Ouk, M., 2012. Increased productivity of rainfed lowland rice cropping systems of the Mekong region. *Crop and Pasture Science* 63: 944–973.

Fuller, D. Q., Weisskopf, A. R., and Castillo, C. C., 2010. Pathways of rice diversification across Asia. *Archaeology International* 19: 84–96.

Grandstaff, T. B., Grandstaff, S., Limpinuntana, V., and Suphanchaimat, N., 2008. Rainfed revolution in Northeast Thailand. *Southeast Asian Studies* 46: 289–376.

Gunn, G., 2011. The great Vietnamese Famine of 1944–45 revisited. Online Encyclopedia of Mass Violence, published 12 May 2011. Available at https://www.sciencespo.fr/mass-violence-war-massacre-resistance/en/document/great-vietnamese-famine-1944-45-revisited (accessed 3 April 2019), ISSN 1961-9898.

Haefele, S., and Gummert, M., 2015. Trajectories in lowland rice-based cropping systems in Mainland Southeast Asia. In R. A. Cramb, ed. *Trajectories of Rice-Based Farming Systems in Mainland Southeast Asia*, pp. 73–110. Canberra: Australian Centre for International Agricultural Research.

Hecht, J. S., Lacombe, G., Arias, M. E., Thanh, D. D., and Piman, T., 2019. Hydropower dams on the Mekomg River basin: a review of their hydrological impacts. *Journal of Hydrology* 568: 285–300.

Helmers, K., 1997. Rice in the Cambodian economy: past and present. In H. J. Nesbitt, ed., *Rice Production in Cambodia*, pp. 1–14. Manila: International Rice Research Institute.

Higham, C., 2014. *Early Mainland Southeast Asia: From First Humans to Angkor*. Bangkok: River Books.

Hoanh, Chu Thai, Facon, Thierry, Thuon, Try, Bastakoti, R. C., Molle, F., and Phengphaengsy, G., 2009. Irrigation in the Lower Mekong Basin countries: the beginning of a new era? In François Molle, Tira Foran, and Mira Kakonen, eds. *Contested Waterscapes in the Mekong Region: Hydropower, Livelihoods and Governance*, pp. 143–171. London: Earthscan.

Inthapanya, P., Boualaphanh, C., Hatsadong, and Schiller, J. M., 2006. The history of lowland rice variety improvement in Laos. In J. M. Schiller, M. B. Chanphengxay, B. Linquist, and S. Appa Rao, eds., *Rice in Laos*, pp. 325–348. Los Baños, Philippines: International Rice Research Institute.

Javier, E. L., 1997. Rice ecosystems and varieties. In H. J. Nesbitt, ed., *Rice Production in Cambodia*, pp. 39–81. Manila: International Rice Research Institute.

Le Coq, J. F., Dufumier, M., and Trebuil, G., 2001. *History of Rice Production in the Mekong Delta*. Paper presented at Third EUROSEAS Conference, London, 6–8 September 2001.

MRC, 2019. Mekong Basin. Mekong River Commission for Sustainable Development website. Available at http://www.mrcmekong.org/mekong-basin/ (accessed 21 February 2019).

Muto, Chiaki, Ishikawa, Ryuji, Olsen, Kenneth M., Kawano, Kazuaki, Bounphanousay, Chay, Matoh, Toru, and Sato, Yo-Ichiro, 2016. Genetic diversity of the *wx* flanking region in rice landraces in northern Laos. *Breeding Science* 66(4): 580–590.

Nguyen, H. T. M., Do, H., Kay, A., Kompas, T., Nguyen, C. N., and Tran, C. T., 2017. *The Political Economy of Policy Exceptionalism during Economic Transition: The Case of Rice Policy in Vietnam*. Crawford School Working Papers 1713. Canberra: Crawford School of Public Policy, Australian National University.

Nguyen Manh Hai, and Talbot, T., 2014. The political economy of food price policy in Vietnam. In P. Pinstrup-Andersen, ed. *Food Price Policy in an Era of Market Instability: A Political Economy Analysis*. Oxford: Oxford University Press.

Owen, N. G., 1971. The rice industry of Mainland Southeast Asia 1850–1914. *Journal of the Siam Society* 59: 78–143.

Purcell, T., Gniel, S., and Van Gent, R., 2008. *Making Value Chains Work Better for the Poor*. Phnom Penh: Agricultural Development International, for UK Department for International Development.

Rambo, A. T., 2017. The agrarian transformation in Northeastern Thailand: a review of recent research. *Southeast Asian Studies* 6: 211–245.

Reardon, T., Chen, K. Z., Minten, B., Adriano, L., Dao, T. A., Wang, J., and Gupta, S. D., 2014. The quiet revolution in Asia's rice value chains. *Annals of the New York Academy of Sciences* 1331: 106–118.

Robertson, C. J., 1936. The rice export from Burma, Siam and French Indo-China. *Pacific Affairs* 9: 243–253.

Schiller, J. M., Hatsadong, and Doungsila, K., 2006. A history of rice in Laos. In J. M. Schiller, M. B. Chanphengxay, B. Linquist, and S. Appa Rao, eds., *Rice in Laos*, pp. 9–28. Los Baños, Philippines: International Rice Research Institute.

Scott, J. C., 2010. *The Art of Not Being Governed: An Anarchist History of Mainland Southeast Asia*. Singapore: NUS Press.

Sokty Chhair, and Luyna Ung, n.d. *Economic History of Industrialization in Cambodia*. Learning to Compete Working Paper No. 7. Learning to Compete Program of the Africa Growth Initiative at Brookings Institution, the African Development Bank, and the United Nations University World Institute for Development Economics Research.

Stuart-Fox, M., 2006. Population diversity and rice in Laos. In J. M. Schiller, M. B. Chanphengxay, B. Linquist, and S. Appa Rao, eds., *Rice in Laos*, pp. 1–8. Los Baños, Philippines: International Rice Research Institute.

Swinnen, J., and Kuijpers, R., 2019. Value chain innovations for technology transfer in developing and emerging economies: conceptual issues, typology, and policy implications. *Food Policy* 83: 298–309.

Tran Cong Thang and Dinh Thi Bao Linh, 2015. Rice policy review in Vietnam. Unpublished paper.

Tran Duc Tuan, 2016. *Status of Agricultural Mechanization and Testing in Vietnam*. Policy Brief 9. Beijing: Centre for Sustainable Agricultural Mechanization, United Nations Economic and Social Commission for Asia and the Pacific. Available at http://www.un-csam.org/Publication/PB201601.pdf.

Tsukada, K., 2011. Vietnam: Food security in a rice-exporting country. In S. Shigetomi, K. Kubo, and K. Tsukada, eds., *The World Food Crisis and the Strategies of Asian Rice Exporters*, pp. 53–72. Chiba: IDE-JETRO.

VNA, 2018. Vietnam Food Association risks losing rice monopoly. Vietnam News Agency, *Vietnam Plus*. Available at https://en.vietnamplus.vn/vietnam-food-association-risks-losing-rice-monopoly/129262.vnp (accessed 12 April 2019).

Wang, H., Pandey, S., Velarde, O., and Hardy, B., 2012. *Patterns of Varietal Adoption and Economics of Rice Production in Asia*. Los Banos: International Rice Research Institute.

Welcher, P., 2017. *Thailand: Rice Market and Policy Changes over the Past Decade*. GAIN Report No. TH7011. Bangkok: USDA Foreign Agricultural Service.

PART II

A Fragrant Aroma

Commercialisation of Rice Farming in Northeast Thailand

Pornsiri Suebpongsang, Benchaphun Ekasingh, and Rob Cramb

Rice has been central to the culture, economy, and politics of Thailand for more than a millennium. Thailand has long been ranked as the sixth largest producer of rice after China, India, Indonesia, Bangladesh, and Vietnam (FAO 2019), all of which have much larger populations. The dominance of rice in Thailand's agricultural economy reflects both the suitability of the natural environment for rice production and the historical origins of Thai agriculture in the long-term migrations of rice-growing populations from southern China (Falvey 2000; Chap. 1). In the 1950s and 1960s, over two thirds of the population lived in rice-producing households and a significant percentage of the remainder was involved in rice trading, transporting, and milling (Behrman 1968). Despite the

P. Suebpongsang (✉) • B. Ekasingh
Faculty of Agriculture, Department of Agricultural Economy and Development, Chiang Mai University, Chiang Mai, Thailand
e-mail: pornsiri.s@cmu.ac.th

R. Cramb
School of Agriculture and Food Sciences, University of Queensland, St Lucia, QLD, Australia
e-mail: r.cramb@uq.edu.au

© The Author(s) 2020   39
R. Cramb (ed.), *White Gold: The Commercialisation of Rice Farming in the Lower Mekong Basin,*
https://doi.org/10.1007/978-981-15-0998-8_2

growth of other crop and livestock industries in subsequent decades and a decline in rice consumption per capita as incomes have grown, rice remains the dominant agricultural industry, accounting for 51% of cultivated area and contributing 15% of agricultural GDP (Pongsrihadulchai 2018).

Not only is rice still a vital food crop domestically but it is also a major export, second only to rubber in value. Thailand became a major exporter of rice following the opening up of the country to global trade in the 1850s (Owen 1971).[1] The volume of rice exports increased fivefold between 1870 and 1905, virtually all coming from the central plain of the Chao Phraya Basin with its fertile soils, developed canal system, and close proximity to the port at Bangkok (Dohrs 1988). Rice exports surged again in the post-war period. In the 1950s and 1960s, rice accounted for 43% of total export revenue (Behrman 1968) and rice export taxes contributed more than 11% of government revenues or about USD 40 million per year (Falvey 2000). For most of the post-war period Thailand has been the world's largest exporter of rice, until being overtaken by India in 2017. Thai rice is renowned for its quality, including conventional white rice and Thai fragrant or jasmine rice (*kao hom mali* or *kao dok mali*), derived from the local variety, KDML105.

Whereas the Central Region remains the largest producer of rice for the domestic and export markets, this section focuses on the Northeast Region, or Isan, which lies within the Lower Mekong Basin (Fig. 2.1). White rice is produced mainly in the Central Region whereas jasmine rice predominates in the Northeast Region to which it is more suited. It is the high profitability of the KDML105 variety and the productivity of a related glutinous variety (RD6) that has helped spread a "fragrant aroma" over the countryside of the Northeast since the 1980s, permitting widespread commercialisation that has helped lift many rural households out of poverty and spurred wider economic development in what has long been regarded as a backward region (Barnaud et al. 2006; Grandstaff et al. 2008; Rigg et al. 2012; Rambo 2017).

In this chapter we analyse the broad trends in the commercialisation of rice farming in the Northeast in the context of the country as a whole, considering production, marketing, and policy dimensions. The following two chapters present aspects of rice farming in Ubon Ratchathani Province, located in the southeastern corner of the Northeast Region on the floodplain of the Mun-Chi river system just before it enters the Mekong (Fig. 2.1). This is one of the leading rice-producing provinces in Thailand and encapsulates the diverging trends among Thai rice farmers in the 2010s.

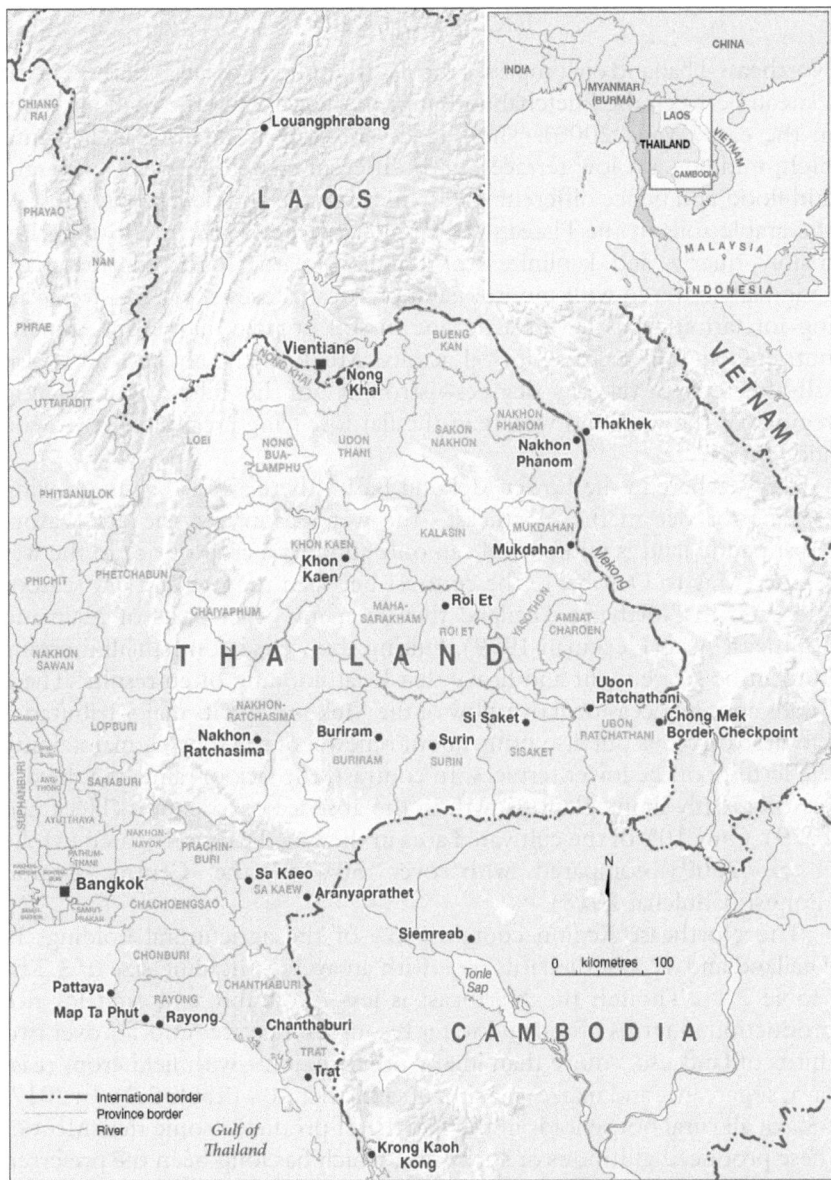

Fig. 2.1 Northeast Thailand. (Source: CartoGIS, Australian National University)

RICE PRODUCTION

Northeast Thailand encompasses the flat to undulating lands of the Khorat Plateau between the Phetchabun Range in the west and the Mekong River in the east (Wada 2005). These lands are conventionally classified into high, middle, and low terraces, with different susceptibilities to drought and flood and hence different suitabilities for rice and field crops. Most of the arable soils in the Plateau are sandy, acidic, and infertile, comprising mainly quartz and kaolinite from highly weathered parent materials. Originally covered with monsoonal dipterocarp forests, progressive clearing for farming has led to a decline in soil organic matter and mineral nutrients and an increase in soil acidity. Much of the arable land is also salt-affected to varying degrees—strongly in the hilly to undulating regions in the west and weakly in the flat low-lying areas along the Mun and Chi Rivers.

As elsewhere in the Lower Mekong Basin, there are two main growing seasons for rice in the Northeast—the wet season and the dry season. Most paddy land is rainfed and can only support a crop of rice in the wet season (May to October). The contrast between the wet and dry seasons is greater in Northeast Thailand than in many other parts of Mainland Southeast Asia (Heckman 1979). During the wet season rainfall is erratic but can be so frequent and heavy that local flooding often results. These floods and the occasional overflow of the Mekong and its major tributaries can destroy crops but also bring about renewal of the topsoil, maintaining the fertility of the lower terraces. In contrast, the lack of rainfall in the dry season greatly limits plant growth in the absence of irrigation (Heckman 1979). Only 10% of the cultivated area in the Northeast is irrigated (Molle et al. 2009) compared with over 50% in the Central Region (Pongsrihadulchai 2018).

The Northeast Region contains 46% of the agricultural holdings in Thailand and 47% of the farm area, with an average holding size of 3.2 ha (Table 2.1). Though the Northeast is less favourably endowed for rice production in terms of soil and water resources, rice accounts for over two thirds of land use—more than in any other region—with field crops (cassava, sugarcane, and maize) making up a further 20% (Table 2.2). In 2017, 63% of all rural households in the Northeast produced some rice. Most of these produced glutinous or sticky rice, which has long been the preferred staple of the predominantly ethnic Lao population.[2] Almost half the glutinous rice produced is consumed by the household, compared to no more

Table 2.1 Number and area of agricultural holdings in Thailand by region, 2013

Region	No. of holdings		Area of holdings		Area per holding
	No.	%	ha	%	ha
Central	847,163	14.3	3,136,686	16.8	3.7
North	1,298,468	22.0	4,401,677	23.6	3.4
Northeast	2,744,457	46.4	8,737,201	46.8	3.2
South	1,021,479	17.3	2,384,222	12.8	2.3
Total	5,911,567	100.0	18,659,786	100.0	3.2

Source: Agricultural Census 2013

Table 2.2 Agricultural land use in Thailand by region, 2013 (% of area)

Land use	Central	North	Northeast	South	Total (2003)	Total (2013)
Rice	42.4	50.1	67.5	5.7	52.9	51.3
Field crops	27.2	34.8	20.5	0.3	18.5	22.4
Rubber	8.5	3.2	8.2	66.7	8.9	14.5
Permanent crops	13.1	7.7	0.9	24.2	10.5	7.5
Forest (planted)	1.5	1.5	0.6	0.2	1.0	0.9
Horticulture	2.0	1.1	0.3	0.6	1.4	0.8
Fish culture	2.9	0.2	0.2	0.4	1.1	0.7
Pasture	0.4	0.3	0.2	0.1	1.0	0.2
Other	2.0	1.1	1.6	1.9	5.1	1.6
Total	100.0	100.0	100.0	100.0	100.0	100.0

Source: Agricultural Census 2013

than 20% of non-glutinous rice (ACI 2005). Most of the remainder is marketed within the region, especially to farmers who specialise in planting the non-glutinous jasmine rice. Around 10% of glutinous rice production is exported from the Northeast to neighbouring countries, especially Laos, where glutinous rice is also the preferred staple (Chap. 5).

The improved glutinous variety RD6, which was released in 1978, gives a higher and more stable yield than traditional glutinous varieties and was widely adopted from the late 1980s. In 1995 RD6 accounted for 83% of the area planted with glutinous rice in the Northeast and about 40% of the total wet-season area of all types of rice (Agrifood Consulting International 2005). The higher yield from RD6 meant that farmers needed less land to meet their subsistence targets and so could devote more land to commercial rice, notably the fragrant variety KDML105 from which jasmine rice is produced, or other crops such as cassava and sugarcane. This was the basis of the growing prosperity in the Northeast (Grandstaff et al. 2008; Rambo 2017).

As noted above, most of the non-aromatic white rice for the domestic and export markets is produced in the irrigated areas of the Central Region, whereas most of the fragrant KDML105 rice is produced under rainfed conditions in the Northeast. Though lower-yielding than the main white rice varieties, KDML105 attracts a price premium, which can be two or three times the price of white rice. The contrast between the economics of non-aromatic white rice and the fragrant KDML105 can be seen in Table 2.3, showing the average costs and returns for the 2018 wet season. The yield of KDML105 was 70% lower but this was offset by a 60% higher price, resulting in a higher gross revenue and net return. In the case of white rice, the computed net return was negative. This probably reflects the inclusion of a market wage for family labour in the total cost, which typically overstates the true opportunity cost of labour.

The "rainfed revolution" in the Northeast has involved not only the widespread adoption of improved varieties but the rapid mechanisation of production, lagging the Central Plain by perhaps a decade (Grandstaff et al. 2008; Viboon and Chamsing 2009; Rambo 2017). In the 1980s, multipurpose two-wheeled tractors began to replace buffaloes as the source of draught power for land preparation. Now medium-sized four-wheeled tractors are becoming more common. By the 1990s combine harvesters were being widely used, superseding the use of mechanical threshers. At first mechanical harvesting services were provided by contractors from the Central Region but soon a local contracting business emerged. The peak labour requirement for transplanting was overcome by

Table 2.3 Average returns to white rice and fragrant rice production in the 2018 wet season

Variable	White rice	KDML105
Yield (t/ha)	2.56	1.81
Farm-gate price (THB/t)	9482	15,267
Gross return (THB/ha)	24,274	27,633
Total costs (THB/ha)	24,945	22,480
Net return (loss) (THB/ha)	(671)	5153

Source: Office of Agricultural Economics (2018a)

the use of broadcasting in rainfed areas, whereas in irrigated areas transplanting machines have begun to be used. The release of labour from these tasks gave households more scope to engage in off-farm and non-farm work, generating additional income to enable the purchase of inputs and hiring of machinery services.

In addition, beginning in the 1990s, many farmers have dug small ponds adjacent to their paddy fields to store water for supplementary irrigation using portable diesel pumps (Grandstaff et al. 2008; Rambo 2017). By the end of that decade there were 65,000 farm ponds and in 2004 the government set up a revolving fund with the aim to increase the number to 450,000. The ponds enabled farmers to irrigate the wet-season crop during drought periods, thus helping to stabilise wet-season yields. This in turn enabled them to reduce the area planted to their subsistence crop of glutinous rice and allocate more land to KDML105 or to commercial field crops. In addition, the ponds permitted cultivation of short-term, high-value horticultural or field crops in the dry season on a small part of the paddy field, typically generating a higher return to labour and capital than a crop of dry-season rice.

The annual output of paddy in the Northeast has more than doubled over the past four decades, from 5.8 million t in 1980 to a peak of 15.1 million t in 2011, dropping back to 12–13 million t in subsequent years (Fig. 2.2). The growth in output since the 1980s was due to an increase in wet-season area from 4.5 million ha in 1980 to 5.9 million ha in 2018

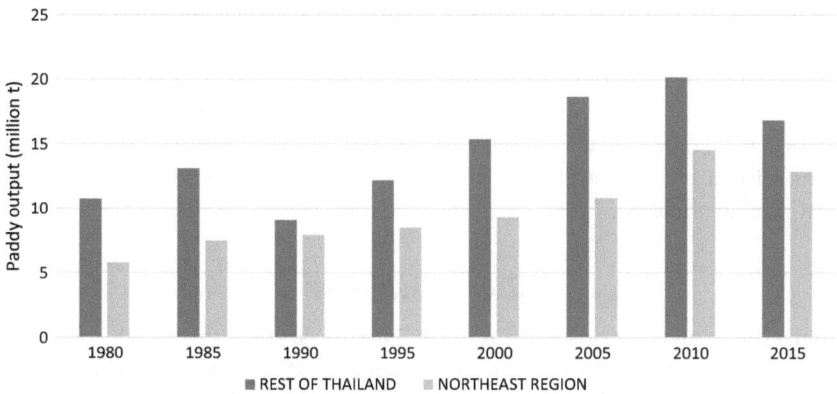

Fig. 2.2 Annual rice production in Northeast Thailand and remainder of country, 1980–2015 (million t)

Table 2.4   Area, output, and yield of rice in Northeast Thailand compared with Thailand as a whole, by season (2018)

Variable	Thailand	Northeast Region	Northeast %
Wet-season area (million ha)	9.47	5.85	61.8
Wet-season output (million t)	24.22	11.02	45.5
Wet-season yield (t/ha)	2.56	1.88	73.4
Dry-season area (million ha)	1.93	0.28	14.5
Dry-season output (million t)	7.96	1.02	12.8
Dry-season yield (t/ha)	4.12	3.60	87.4
Annual area (million ha)	11.43	6.13	53.6
Annual output (million t)	32.18	12.04	37.4
Average yield (t/ha)	2.82	1.96	69.5

Source: Office of Agricultural Economics (2018b)

as farmers cleared more village forest lands for paddy fields; an increase in dry-season (irrigated) area from 11,000 ha to 282,000 ha over the same period, reflecting public investment in mainly pump-irrigation schemes; and an increase in yields, from 1.3 to 1.9 t/ha in the wet season and from 2.2 to 3.6 t/ha in the dry season, reflecting the wider use of improved varieties and greater use of fertilisers (Table 2.4).

The Northeast contributed 35% of total paddy output in 1980, rising to 47% in 1990 before dropping to between 37% and 43% in subsequent years (Fig. 2.2). In 2018, the Northeast accounted for 62% of wet-season planted area and 46% of wet-season output, reflecting that the yield was only 73% of the national average (Table 2.4). The contribution of the Northeast to national dry-season production was less, given the lower proportion of irrigated paddy land, amounting to 15% of total dry-season area and 13% of output. Dry-season yields were higher at 87% of the national average. On an annual basis, in 2018 the Northeast accounted for 54% of planted area and 37% of output, with annual yields at 70% of the national average.

Rice Marketing

With the growth of production in the Northeast since the 1980s, the marketing of rice has expanded and developed. A study analysing the flow of paddy and rice in the 2000s gives an indication of how the marketing chain had evolved to that point (ACI 2005; Fig. 2.3). The 9.5 million t of

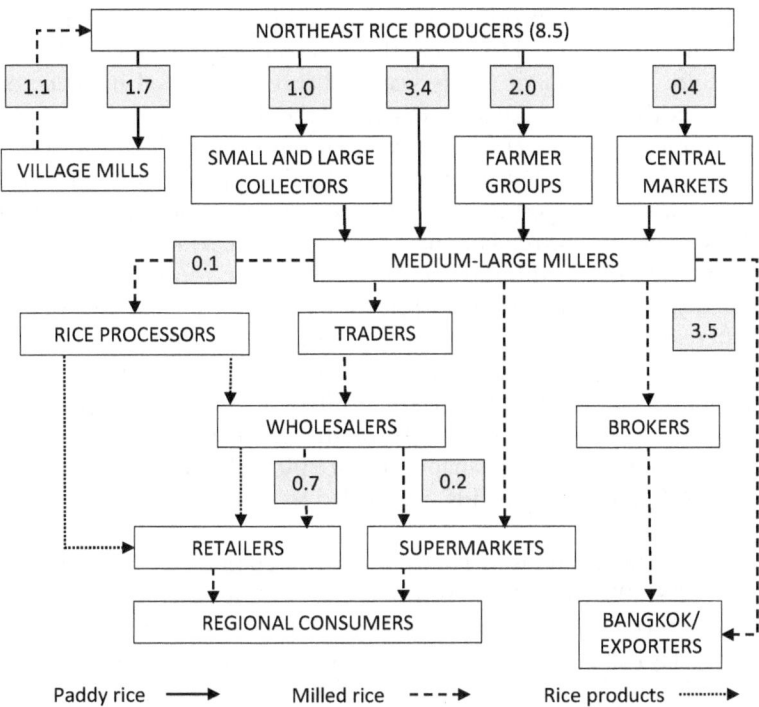

Fig. 2.3 Marketing chain for rice produced in Northeast Thailand. (Note: Numbers refer to million t of paddy or milled rice ACI 2005)

paddy produced in 2003 was traced along two channels—a subsistence channel and a market channel. Of the 8.5 million t of paddy available after deducting seed, animal feed, and losses, about one fifth was retained by farmers for household consumption (mostly glutinous rice) and four fifths was marketed commercially (mostly KDML105 and other non-glutinous rices). Farmers milled paddy for their own consumption at small village mills. Given an estimated recovery rate of 65%, about 1.1 million t of milled rice was consumed by producing households.

About 6.8 million t of paddy was thus available for marketing. There were several pathways for farmers to dispose of their marketable surplus. The traditional method was to sell to primary collectors, including small-scale village collectors without warehouses and larger-scale collectors at the district and provincial levels who had their own warehouses. Collectors were located close to producing areas and transported paddy in pick-ups or six-wheeled trucks. However, these accounted for only 15% of the paddy sold. About 5% of paddy was sold through central markets, operated by both public agencies (such as the Bank for Agriculture and Agricultural Cooperatives) and private entrepreneurs.[3] These served as a meeting place to conduct transactions. Facilities provided could include labour, moisture gauges, drying yards, warehouses, and loans, depending on the size of the centre. The market owners were sometimes also assemblers and/or millers, though the owner of a large central market usually refrained from trading to avoid price interference, relying on fees, rent, and interest from loans. A much larger share of the surplus (80%) was sold directly to millers, representing the most striking change that has occurred with commercialisation. This included 50% sold by farmers operating on their own account and 30% sold through farmer groups or cooperatives (Wiboonpongse and Chaovanapoonphol 2001). Using a recovery rate of 65%, the 6.8 million t of paddy channelled to mills produced 4.4 million t of milled rice.

Over half the rice mills in Thailand are located in the Northeast Region, totalling 24,000 in 2003 (Agrifood Consulting International 2005). However, about 98% of these were small-scale or village custom mills (less than 5 t/day) and only 2% were medium- to large-scale private or cooperative mills. Only about 1% of mills in the region had a capacity of more than 50 t/day; two mills could operate at 1000 t/day. Large mills with storage capacity could stockpile paddy during the harvesting period when there was excess supply and process and sell it when the price increased (Rabobank 2003). While village mills still predominate, investment in large, modern mills in the Northeast has increased in the decade or so since this study.

After milling, rice followed different channels for the domestic and export markets (ACI 2005). The milled rice for the domestic channel was sold to traders, wholesalers, and directly to retailers. There were also various markets for the by-products of milling such as rice bran. Of the 4.4 million t of milled rice produced in 2003, about 1 million t (23%) went into the regional market, whether for retail sales (90%) or further processing into rice-based products (10%). Much of the domestic market was provided by the 40% of Northeast farmers who specialised in producing KDML105 rice for the export market and therefore needed to buy glutinous rice for their own consumption (ACI 2005). The remaining 3.4 million t of milled rice went through traders and wholesalers to markets outside the region, including both Bangkok and surrounding areas and the export market. Most millers, except for a few large firms, sold rice to wholesalers and exporters through a broker. The broker was commissioned to search for quantities of rice of specified qualities to meet the requirements of wholesalers and exporters, charging a brokerage fee of 2–3% of the sale value.

The retail distribution system for food, including rice, is undergoing rapid change in Thailand. Alongside the traditional food markets, modern distribution systems are emerging, including supermarkets, superstores, and convenience store chains. Whereas in the traditional retail market rice is sold loose and customers can purchase any quantity, in supermarkets rice is sold in labelled packages of fixed sizes such as 5 kg bags. In the 2000s, 75% of the retail food trade was still through traditional markets (ACI 2005), with modern distribution channels mostly limited to major urban centres. Nevertheless, the number of supermarkets and other modern retail outlets in the Northeast is increasing.

Most exporters are located in Bangkok and surrounding areas (ACI 2005). Many exporters have their own rice mills, while those without mills deal directly with large mills or depend on brokers who can guarantee an adequate supply of a specified type and quality of rice. Most exports are handled by private companies, who must be registered under the provisions of the Rice Trading Act. Only around 5% of exports are handled by state agencies. Each exporter has connections to particular importers demanding a given type and quality of rice. For example, exporters of Thai jasmine rice have standing arrangements with importers in Hong Kong, Singapore, and the Middle East. Thailand exports mostly to Africa and Asia, with the African market the fastest growing. In 2005 Africa accounted for 47% of total rice exports (3.4 million t), followed by Asia (28% or 2.0 million t), and the Middle East (12% or 0.9 million t). Europe accounted for only 5% (0.3 million t) and the US 6% (0.4 million t). Exports of Thai jasmine rice, mainly grown in the Northeast, amounted to 2.9 million t, or 26.1% of the total volume, in 2018. Glutinous rice exports were only 180,159 t or 2% of the total.

RICE POLICY

The commercialisation of rice farming in the Northeast has been strongly influenced by state policies affecting farmers' access to resources (land and water) and inputs (seed, fertilisers, extension, and credit), their management of farm activities, and their appropriation of farm output and revenue (Fig. 1.8).

Land

Private rights in land have been recognised since the nineteenth century and a land titles system was introduced in 1901, administered by the Department of Lands (DOL) (Burns 2004). Land titling was initially concentrated in the Central Plain but was gradually extended to the other parts of the country, including the Northeast, where the DOL had provincial and district offices. Title deeds were issued based on a cadastral survey. However, studies for the fifth National Social and Economic Development Plan (NSEDP, 1981–1985) found that only about 12% of the 23.7 million ha of occupied agricultural land was held by title deeds, a further 49% was held by lesser (unsurveyed) documents, 18% was occupied by claimants without documentation, and 21% was illegally occupied forest land. The Plan set out a strategy to grant secure tenure to agricultural landholders, noting that this would improve access to institutional credit and thus provide the basis for long-term investment by farmers. A land titling project was initiated with support from the World Bank and by 2001 a further 8.5 million titles had been issued over about 4.9 million ha (Burns 2004).

Alongside this initiative, an agricultural land reform programme was started in 1975 to redistribute land to landless farmers and provide titles to "squatters" on public lands (as much of the agricultural expansion had occurred through clearing of forest on what was regarded as public land). The Agricultural Land Reform Office (ALRO) in the Ministry of Agriculture and Cooperatives (MOAC) was responsible for land allocation and titling, and for providing domestic water supplies, village roads, farm ponds, and small-scale reservoirs and irrigation facilities (ALRO 2006). The ALRO also established Agricultural Land Reform Cooperatives, provided agricultural credit and production inputs, supported on-farm and off-farm occupations, and promoted community-based conservation of natural resources (Ekasingh et al. 2008). From 1975 to 2005, 2.1 million ha were allocated to 762,170 households in the Northeast—52% of the total area nationally. This represented 24% of the area of farm holdings in the Northeast, as recorded in the 2013 Census, and 28% of the number of holdings.

Land has also been made available through research to improve its capability. Research by the Department of Land Development (DLD) within MOAC led to the amelioration of saline soil over a large area known as Thung Kula Rong Hai, mainly in Roi Et Province (Fig. 1.2). This enabled the rehabilitation of 88,000 ha to become a major area of jasmine rice production, benefitting 14,280 farm households (Ekasingh et al. 2008).

The effect of this long-term land policy has been to provide secure tenure for large numbers of rice-growing smallholders in the Northeast. While population growth and the closure of the agricultural frontier in the 1980s meant that the size of these smallholdings declined—from an average of 4.5 ha in 1980 to 4.3 ha in 1990, 3.6 ha in 2000, and 3.2 ha in 2013—the area per household member and per worker has remained stable, reflecting the trend to smaller families and the outmigration of younger household members (Grandstaff et al. 2008). This, combined with improvements in infrastructure, has enabled millions of smallholder farmers to capitalise on the opportunities for commercial rice farming since the 1990s. While in some areas farmers are consolidating their paddy fields into larger units to permit land levelling and greater field efficiency in the use of machinery (Rambo 2017), there has not been significant differentiation in the ownership of land. Thus in 2013 51% of landholdings were within the bracket 1.6 to 6.2 ha, accounting for 51% of the total farm area. As Rigg et al. (2012) have highlighted, land ownership and paddy farming remain central to household livelihood strategies, even as non-farm sources of income become dominant.

Water

Access to irrigation has been a less important element in the commercialisation of rice farming, despite official rhetoric about "greening Isan". There has been a long-term policy focus on the development of water resources for agriculture in the Northeast as a way to intensify rice farming and reduce poverty (Molle et al. 2009; Floch and Molle 2013). Much of this has involved investment in small- and large-scale pump-irrigation schemes, managed by the Royal Irrigation Department (RID) of MOAC, to permit dry-season cropping of rice. However, investment in the required canal system has often lagged and, even where the distributional infrastructure has been satisfactorily completed, the utilisation for dry-season cropping has been much lower than predicted due to the low and variable profitability of dry-season rice (KDML105 is a wet-season variety) and the increasing scarcity of farm labour. The total irrigable area in Northeast

Thailand is only 1.2 million ha, with limited utilisation in the dry season (Floch and Molle 2009). Thus most rice production still takes place under rainfed conditions, though increasingly with the benefit of supplementary irrigation from small ponds, as noted above.

Seed

The crucial input that has enabled the commercialisation of rice farming in the Northeast has been the selection, breeding, and dissemination of improved varieties by public agencies. The term "improved", rather than "high-yielding", is appropriate as the varieties adopted by farmers were better adapted to local conditions and/or produced higher-quality rice rather than simply increasing yields. The International Rice Research Institute (IRRI) began working in Thailand from 1966, earlier than in the other countries of the Lower Mekong Basin, but the first high-yielding, semi-dwarf variety, IR8, was not adopted because of it low eating quality. However the Rice Department (RD) of MOAC incorporated the semi-dwarf gene in IR8 in a series of locally bred varieties (labelled RD) that were widely adopted in the irrigated areas of the Central Plain. RD1 gave a 50% higher yield but was not adopted due its poor eating quality and high input requirements. Subsequently developed photoperiod-insensitive, high-yielding varieties, especially RD7, RD15, and RD23, became dominant in the irrigated areas of the Central Region, underpinning the rapid rise in output. However, most of this additional production was not consumed domestically but exported as lower-quality white rice or used to make parboiled rice, also for export.

In the 1980s, with the emergence of Vietnam as a major exporter of low-quality rice, there was downward pressure on the international price for this type of rice, whereas the price of higher-quality rice showed an upward trend (Setboonsang 1996). This prompted rice researchers in MOAC to revive earlier efforts to select and breed for rice quality. The major breakthroughs were, as noted above, the selection and promotion of a line of local fragrant rice labelled *Kao Dok Mali 105* (KDML105) and its mutagenesis to form RD6, a higher-yielding glutinous variety. Both are medium-term, photoperiod-sensitive varieties that are well adapted to the soil and climatic conditions of the Northeast and respond well to additional inputs. RD6 was rapidly adopted and, by 1995, accounted for 40% of the total wet-season rice area in the Northeast and 83% of the glutinous rice area (Ekasingh et al. 2008). By the same year, KDML105 accounted for 72% of the non-glutinous rice area, increasing to 80% in 2002–2005 (Grandstaff et al. 2008: 332). The ability to secure subsistence with RD6 enabled farmers to allocate more land to production of KDML105 for the domestic and export markets, generating a much higher return.

The provision of improved seed to farmers was facilitated by the Department of Agricultural Extension (DOAE), which provides a linkage between researchers and extension staff. The DOAE's crop promotion project aimed for increased yields by disseminating good-quality seed of improved varieties (Ekasingh et al. 2008). One approach used was the "seed exchange method", implemented from 1982 to 1998 for rice, by which a kilogram of farmers' seed was exchanged with a kilogram of good-quality seed of an improved variety. In poorer areas, the "free rice seed" approach was used. The DOAE also initiated community seed production centres in 65 rice-producing provinces to produce good-quality seed for local dissemination.

Fertilisers

As part of the intensification of rice farming and other crop production, the use of synthetic fertilisers has increased more than tenfold, from around 20 kg/ha in 1980 to 250 kg/ha in 2008, though the rate of increase has slowed. More than 95% of the synthetic fertilisers used in agricultural production are imported (OAE 2011). Urea (46-0-0), used mainly for rice and vegetable crops, is the most important imported fertiliser, accounting for 35% of imports by value, while other widely used fertilisers are ammonium phosphate (16–20), accounting for 9%, and the NPK compound fertiliser 15-15-15 (6%). The import, mixing, and distribution of fertilisers is in private hands and is not subsidised, though the government monitors and regulates retail prices to avoid price spikes which would adversely affect farmers (Chitibut et al. 2014). This regulatory imposition has led to the emergence of a parallel informal market distributing products of variable quality.

Despite the increase in use of synthetic fertilisers, alternative approaches to nutrient management and crop protection have been promoted by the government, particularly since the 2000s. Farmers are encouraged to apply compost, green manure, animal manure, and other organic fertilisers instead of or together with synthetic fertilisers. Organic fertiliser has been adopted for many types of agricultural production, including rice farming. A survey by the Office of Agricultural Economics published in

2010 found that about 48% of farms applied organic fertiliser together with synthetic fertiliser, while 41% used only synthetic fertiliser and 6% used only organic fertiliser. A recent industry analysis concludes: "The organic fertilizers and biofertilizers market is expected to grow as a result of the increased demand for organic food products. Government encouragement by providing subsidies and incentives for the use of organic fertilizers will drive the growth of organic fertilizers" (Mordor Intelligence 2019). The report adds that the government is encouraging private sector production of organic fertilisers and bio-fertilisers in part to reduce the dependence on imported fertilisers.

Extension

For agricultural extension, the Department of Agricultural Extension (DOAE) established in 1967 is mainly responsible for extension of crop production whereas Department of Livestock Development (DLD) and Department of Fisheries are for livestock production and aquaculture, respectively. The DOAE is directly responsible for integrating the concepts and strategies of crop promotion by cooperating with research institutes, universities, agricultural credit, marketing organisations, and other related agencies. Its tasks are to provide extension services and technology transfer to farmers to help increase farm productivity, both qualitatively and quantitatively, and to meet market demands and standards. The DOAE has established a regional office in all regions and also provincial and district offices in all provinces and all districts for the whole country. In each district, there are extension officers to work closely with farmers. District and sub-district agricultural extension officers have a duty to convey knowledge and technologies which have already tested for local adaptability from research institutions to farmers and get feedback regarding the problems and constraints, either technical or biological, being faced by farmers and farmers' attitudes, and proposed to researchers by extension officers.

In 1999, the DOAE established a new extension system in response to the new constitution law in 1997 and the 9th National Social and Economic Development Plan for giving priority to human development. The new extension system has a principle that farmers will determine the development pathway by themselves and extension officers will be facilitators and coordinators as well as learning partner of farmers. The Agricultural Technology Transfer and Service Center (ATSC) in each sub-district have been established throughout the whole country to be mechanisms for working with farmers as well as other related institutions such as local government, farmer associations, NGOs, and private sector in the agricultural development process. The ATSC is formulated to develop one-stop service centres for farmers and communities in the areas of agricultural development, agricultural production, market development, and natural resources management. The ATSC implementation was carried out on the basis of community-based development by providing opportunity to farmers, enabling them to participate, and promoting their potential to plan and solve existing problems by themselves. Thus, the establishment of ATSC paved the way to decentralisation and empowerment for community development (Panee Boonyaguakul and Surangsri Wapet, 2005, cited in Ekasingh et al. 2008).

Credit

As farmers became more commercialised in their activities, the demand for credit increased. There have been many sources of formal and informal credit available to farmers but the major sources have been the Bank of Agriculture and Agricultural Cooperatives (BAAC), whose operations have been strongly supported by government policies since the 1970s. By 1982, the BAAC provided credit totalling THB 12 billion to about half of the farm households in Thailand. While BAAC met the most obvious credit needs of medium-scale farmers, it had some difficulty in reaching the poorest farmers (Falvey 2000). The use of group-guaranteed loans has been an effective alternative means to lend to poor farmers who lack collateral. Credit has been used by farmers to purchase improved seeds, equipment, fertilisers, and pesticides. Out of 5.8 million farm households in 2008, 3.5 million (59.9%) had agricultural debt. Of these, 63.5% borrowed from the BAAC, 9.2% from cooperatives, 9.9% from village funds, and 7.4% from informal sources. The average agricultural debt was THB 104,640 per household (National Statistics Office 2008).

Farm Management

Farmers in the Thailand are not subject to state direction regarding their choice of crops or production techniques. However, government policy since the 1990s has sought to persuade and subsidise farmers to shift towards "sustainable agriculture", which has been variously interpreted as integrated farming systems, diversified farming, good agricultural practices (GAP), or organic farming. This push has been given strong ideological support from the late King Bhumibol Adulyadej's "new theory of agriculture"—part of his philosophy of a "sufficiency economy" (Kasem and Thapa 2012). This promotes the need for farm management to provide food self-sufficiency or food security for farmers.[4]

The emphasis on sustainable agriculture encourages crop diversification, a shift away from agricultural intensification, reduced use of inorganic fertilisers and pesticides, and promotion of organic agriculture and healthier food. A number of programmes have been implemented in support of sustainable agriculture, including subsidised credit and training programmes for farmers who are willing to participate. In 2016, the military government's National Rice Policy and Management Committee (NRPMC) initiated a scheme that paid qualified farmers subsidies to stop planting rice in areas deemed to be unsuitable and to develop integrated farming systems, with on-farm irrigation, fish, and livestock. According to official maps, most land deemed unsuitable for growing rice is located in the Northeast, though there are doubts about the accuracy of the zoning (Sunsuk 2016).[5]

Concerns about the intensive use of chemical inputs in Thailand have underpinned policies promoting good agricultural practices (GAP) and food safety. GAP-certified farmers are required to use organic fertilisers together with inorganic fertilisers to ensure high-quality produce. They must also use bio-pesticides to control pest outbreaks, though they can apply inorganic pesticides if approved by the business they are contracted to supply (Kasem and Thapa 2012). The area of certified GAP land was 366,000 ha in 2008, considerably more than the 22,000 ha certified as organic. Nevertheless, Thailand is the world's largest exporter of organic rice and some villages in the Northeast are embracing this version of sustainable agriculture (Chaps. 3 and 4).

Marketing and Pricing

The overwhelming policy intervention in Thailand's rice industry has been the state's involvement in the purchase and storage of paddy. In 1976 the export tax on rice was abolished and the first price support programmes were instituted. Price support developed into a rice buffer stock scheme from 1981, intended to stabilise consumer prices and raise farm-gate paddy prices, and a paddy mortgage or pledging programme from

1983, financed by the Bank for Agriculture and Agricultural Cooperatives. The latter used public warehouses or on-farm storage to carry over paddy stocks so that farmers did not have to sell immediately after harvest when prices were low. With increasing subsidisation, more farmers participated (Ekasingh et al. 2008). However, aggressive government intervention to support prices through this mechanism, first in the mid-2000s but most ambitiously from 2011 to 2014 under the government of Yingluck Shinawatra, led to the accumulation of record stocks of up to 18 million t and the eventual collapse of the programme (Welcher 2017) and of the government. At its height, the scheme offered farmers prices 50–60% above the market price, with no upper limit on purchases, thus squeezing out commercial traders, millers, and exporters and severely disrupting the export market. Stocks are still being progressively disposed of in government-to-government deals and auctioned to private buyers at discount prices, but the domestic and export markets have been able to return to some normality (Chuasuwan 2018).

Figure 2.4 traces the FOB export price of Thai A1 Super white broken rice (the reference price for the world market) and of Thai Fragrant Rice over the past two decades, showing the impact of both government policies and global shocks. The figure shows the doubling in the price of white rice as part of the global food price crisis in 2008 and the subsequent even greater price spike brought about by the Thai Government's 2011–2014 stockpiling. The unavoidable release of carryover stocks combined with the continued expansion of exports from India has brought the world price down again to just above pre-2008 levels. The figure also shows the growing premium in the world market for high-quality rice, encouraging countries such as Vietnam and Cambodia to increase their share of this segment of the market (as discussed in subsequent chapters).

In 2016 the military government announced it would no longer continue with the rice pledging and income insurance programmes. Instead, short-term measures were introduced, including the "Farmer Loans to Delay the Sales of Paddy" (known as "On-Farm Paddy Pledging"), aimed at stabilising the farm-gate prices of fragrant and glutinous paddy rice. The government also approved a budget of THB 45 billion (USD 1.3 billion) under the Rice Farmer Assistance Measure to finance direct payments to farmers adversely affected by drought. Farmers in this programme also are eligible to have their BAAC debt suspended for two years at a reduced interest rate (3% instead of the normal 7%). Moreover, farmers who buy

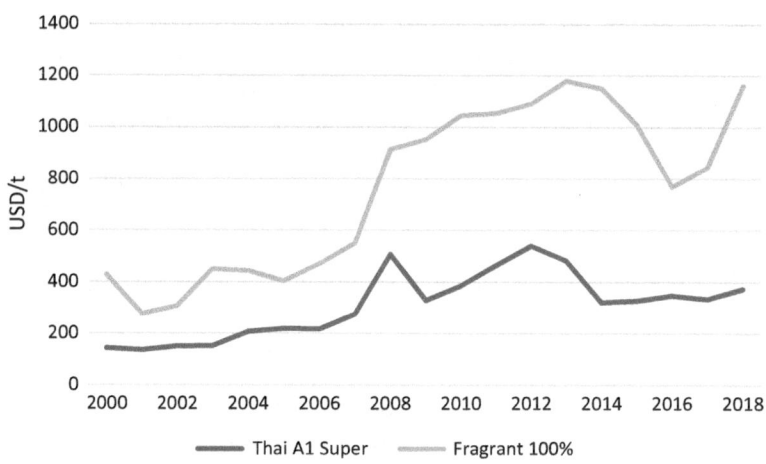

Fig. 2.4 Average export price of Thai white and fragrant rice, 2000–2018 (USD/t, FOB). (Source: IRRI—World Rice Statistics Online (http://ricestat.irri.org:8080/wrsv3/entrypoint.htm))

commercial crop insurance will receive partial compensation from the government for losses from natural disasters (Welcher 2017).

Changes in Rice Farming in a Village in Khon Kaen Province

To understand the long-term changes in rice farming in the Northeast Region, it is useful to consider a case study. Barnaud et al. (2006) studied agricultural change in Ban Hin Lad, a village in Khon Kaen Province 40 km northwest of Khon Kaen City (Fig. 2.1). The village was located in the undulating middle terrace of the Khorat Plateau that accounts for over 70% of the Northeast Region; hence, the farming landscape was a typical combination of uplands and lowlands. Rainfed lowland rice was preferentially cultivated in the shallow depressions with more clayey soils, called the "lower paddies", and secondarily in the adjacent "upper paddies" which were sandier and more drought-prone. Above these paddies were the more hilly "uplands" with very sandy soils, which supported dry dipterocarp forests before being cleared for cash crops, notably cassava and sugarcane. The village did not have access to the irrigation schemes that had been developed in the region since the 1960s (Floch and Molle 2009).

A previous survey of the village had been carried out in 1983, providing a basis for comparison with the diagnostic analysis conducted by the authors in 2002. In 1998 the village had 1937 inhabitants in 383 households (averaging 5.1 residents per household) farming 1250 ha (3.3 ha per household). Barnaud et al. (2006) drew on various sources of data, in particular, interviews with 26 different types of farm household. Their data are supplemented below with data from some more recent studies in the same region.

Up to the 1960s, farming in the study village involved the cultivation of glutinous rice in bunded paddy fields for household subsistence. Each farmer combined three or four varieties of different duration to adapt to variations in the toposequence and the weather, with 28 different varieties recorded in the village as a whole. Large numbers of buffaloes were reared to provide draught power and manure for rice cultivation and as a store of wealth. Hunting and gathering in the forest and fishing in the rivers provided additional subsistence. With no rivers or canals to connect the region with the Central Plain, villagers were cut off from trading networks. In the 1960s this began to change with the construction of the Friendship Highway (Route 2) across the region and the promotion of the long-fibre crops, roselle, and kenaf, for export production. From the mid-1960s there was increased deforestation, driven in part by the expansion of these cash crops in the uplands, though forest clearance was initially limited by reliance on manual tools. The area planted to long-fibre crops began to fall away from around 1970.

Nevertheless, rapid forest clearance continued for another two decades. The area of forest cover in the Northeast fell from around 7 million ha in 1965 to around 2 million ha in 1985, with a corresponding increase in the area farmed. There were two reasons for this. First, the Northeast was experiencing rapid population growth, with the population density increasing from 55 persons per km^2 in 1960 to 100 persons per km^2 in 1980. The increased subsistence needs of the local population were met, not by increasing yields, but by expanding the area cultivated onto the more drought-prone transitional zone between the lowlands and the uplands, creating more "upper paddies". In addition, from around 1970, farmers began to take up cassava cultivation in the uplands in response to the booming European market. Cassava cultivation in the Northeast increased from 10,000 ha in 1969 to almost a million ha in the mid-1980s. The accelerated deforestation was because of the heavy mechanisation of land clearing and tillage for cassava production, using rented four-wheeled tractors, and the higher land-to-labour ratio in cassava production compared with the long-fibre crops (which required very laborious post-harvest operations). As the forested land was "up for grabs", households with more resources who got in first acquired more land, creating an inequality in landholdings that was to persist in subsequent decades, affecting the subsequent pattern of mechanisation.

Despite the expansion of cultivated area, population pressure led to a reduction in farm size. The average farmed area in the study village decreased from 5.5 ha in 1983 to 3.2 ha in 1998. At the same time, non-farm wage rates were rising, both in the Northeast and in Thailand as a whole. Real daily wages in Khon Kaen City rose from about USD 1.50 in 1973 (similar to Bangkok wages) to USD 3 in 2001 (compared to USD 3.50 in Bangkok).[6] In the 1970s, many farmers from the Northeast, particularly younger household members, migrated to the Central Plain to work as hired labourers on larger rice and sugarcane farms. In the next two decades, as Thailand's industrial boom accelerated, they migrated to the Bangkok region to work in factories, on construction sites, and in a wide range of unskilled service industries. From the mid-1990s, with some decentralisation of industry to Khon Kaen and other regional centres, there was more work available within the region, enabling some household members to commute from the village. During this period, off- and non-farm employment came to be the major source of income for most households in the study village, particularly for smallholders with 2 ha or less. While this non-farm income enabled poor households to continue farming, the absence of workers contributed to a labour shortage in the village, adding to the impetus for mechanisation.

By the 1990s the sugar industry had partly relocated from the Central Plain to the Northeast, creating the incentive for farmers in the study village and elsewhere to take up this crop. Government control over and support for sugar prices made sugarcane production more remunerative and less risky than cassava production; hence, the area of cassava began to decline, though it remained a major land use. Unlike cassava, sugarcane was suited to the ecological conditions of the upper paddies; hence, farmers began to plant sugarcane on this land-type, contracting large four-wheeled tractors for tillage, thus reducing the area available for both wet-season rice cultivation and dry-season livestock grazing. In the study village, 55% of the upper paddies were converted to sugarcane production from 1983 to 1998. Sugarcane production was more labour-intensive than cassava, particularly at harvest, adding to the overall demand for labour in the village.

The 1990s also saw widespread investment in two-wheeled tractors for rice cultivation, a decade or two after their spread in the Central Plain. In 1983 there were only two two-wheeled tractors for the 234 households in the village. By 1998 there were 255 among 383 households. According to Barnaud et al. (2006), the main reason for this rapid uptake was the need to decrease the peak demand for labour at transplanting by speeding up the ploughing and puddling of the paddies. Farmers borrowed from the BAAC or sold their water buffaloes to pay for the purchase of the two-wheeled tractors. In Khon Kaen Province as a whole, the buffalo population fell from 380,000 in 1987 to 80,000 in 2000. The disposal of the household's buffaloes also released one family worker from the care of these animals, freeing that person to seek more remunerative non-farm employment. In any case, the area of available grazing land was now in short supply due to the expansion of cassava and sugarcane.

Likewise, though not mentioned by Barnaud et al. (2006), the switch to direct seeding occurred in the Northeast a decade after its spread in the Central Plain. Konchan and Kono (1996) surveyed 100 farmers across different transects in an area to the southeast of Khon Kaen in 1994–1995. They found that, until the 1980s, most fields were transplanted. However, from the early 1990s, labour shortages induced farmers to use direct seeding, first in rainfed fields and then spreading to all field types. Dry seeding was the most common form used, but wet seeding was used in irrigated fields and in rainfed fields with a source of supplementary irrigation (e.g., pumping from a pond).

The growing shortage of labour and the increase in daily wages was also impacting on the costs of harvesting rice. This was exacerbated by revolutionary changes in rice varieties (Grandstaff et al. 2008). As described above, from 1977 the Rice Department introduced RD6, a photoperiod-sensitive glutinous rice variety with wide adaptability and excellent grain quality, which became the preferred subsistence crop, displacing most of the traditional varieties. In the study village only two of the 28 varieties of glutinous rice reported in 1983 were still found in 2002. In addition, the profitability of the non-glutinous jasmine rice (KDML105) meant that it was widely cultivated in the upper paddies as a cash crop. The two varieties are both medium-term varieties of moderate yield that mature at much the same time. Hence the harvest period, previously spread over two to three months, was concentrated (in any one farm) over two to three weeks in late November and early December, requiring farmers with more than 0.75 ha per family worker to hire labour for harvesting. This, combined with increasing wage rates as young people sought off- and non-farm work outside the village, added significantly to production costs, creating a strong incentive for the mechanisation of harvesting.

While combine harvesters had apparently not yet spread to the village at the time of the 2002 study by Barnaud et al. (2006), a more recent study by Poungchompu and Chantanop (2016) helps to bring the mechanisation story up to date. They conducted a survey in all 14 provinces of the Northeast Region in the 2014 wet season and the 2015 dry season, interviewing 85 operators and 729 farmers, including those with rainfed and irrigated rice farms. They found that over 70% of farmers used combine harvesters, all on a contract basis. From the early 2000s, contractors came from Central Thailand, transporting larger machines with a capacity to harvest 6–8 ha per day. These outside operators needed a local broker to assemble the rice area to be harvested to minimise the cost of travel between farms and because there was only a 30- to 35-day window for the wet-season rice harvest.

The profitability of providing harvesting services given the rapidly growing demand prompted local farmers with capital to acquire harvesters to service their neighbours, typically using smaller machines harvesting 2–5 ha per day. These local contractors did not require brokers; farmers requested their services directly and if the contractor proved reliable he would be rehired in subsequent seasons and other contractors would not encroach on his "territory". The higher costs facing outside operators and the government regulation of contract rates meant that the number of service providers from Central Thailand declined and they accounted for only 10% of operators in the Northeast in 2014–2015. Poungchompu and Chantanop (2016) found that replacing hand harvesting with the services of a combine harvester reduced farmers' harvesting costs from USD 270/ha to USD 140/ha and that this was the main motivator for the widespread use of the new technology. The statistical likelihood of using a combine harvester increased with the education of the household head and the area of rice cultivated, and decreased with the size of the household.

From the late 1990s, following the Asian financial crisis and the slow-down in the non-farm economy, government support was provided to dig small, multipurpose farm ponds to promote more integrated and diverse farming systems and retain more people on the land (Barnaud et al. 2006).

In 2002 there were 130 such ponds in the study village, ranging from 0.04 to 0.16 ha. Farmers with two-wheeled tractors could use them to power a low-lift axial-flow water pump to provide supplementary irrigation to rice nurseries and paddies, allowing earlier transplanting in some cases and helping to stabilise yields. The ponds were also used to rear fish, irrigate fruit and vegetables on a small scale, and provide drinking water for cattle.

Barnaud et al. (2006) summarised the agrarian change in the study village by classifying farming households into four main types:

1. "Very small farms with off-farm workers" (accounting for 75% of households) farmed 0.3 to 2 ha—not enough to fully employ family workers or to meet cash requirements. They had no investment capacity; had sold assets including upland fields, cattle, and buffaloes; and were indebted. They owned or rented a two-wheeled tractor for rice cultivation. Between 45 and 75% of their income was from off- or non-farm sources, whether from working on larger farms at the peak times for rice, sugarcane, or cassava, or from working outside the village in casual or permanent employment. Their net income in 2002 was USD 950 per worker.
2. "Small farms" (accounting for 20% of households) had 2–3 ha, including some upland, which was enough to make farming more remunerative than working for wages. They hired out or hired in as little labour as possible and received only 25% of income from non-farm sources. They accumulated little capital but had few debts. These households owned two-wheeled tractors and sometimes rented in a large four-wheeled tractor for cultivation of their upland crops. Their net income in 2002 was USD 1150 per worker.
3. "Large farms" (accounting for 3% of households) had 3–4.5 ha, allowing them to fully employ family labour, meet their cash requirements, and accumulate capital, especially through intensive rearing of beef cattle for sale. They made extensive use of hired labour, owned a two-wheeled tractor, and rented in a four-wheeled tractor for upland cultivation. Nearly 100% of their income was from their own farming activities. Their net income in 2002 was USD 1600 per worker.
4. "Entrepreneurs" (accounting for 2% of households) owned up to 5.5 ha and made extensive use of hired labour to farm their land. They had invested in heavy equipment (trucks and four-wheeled

tractors) which they hired to other villagers, enabling them to fully employ family members, meet the household's cash requirements, and accumulate further capital. Around 80% of their income was from these farm service activities. Their net income in 2002 was USD 2800 per worker.

Thus the supply of manual labour, the ownership, and use of farm machinery, and the provision of contract services were closely related to the agrarian structure in the village that had emerged since the opening up of the region to commercial agriculture in the 1960s and the economic and demographic changes that had occurred in the 1970s and 1980s. Within the village, hired labour was provided by the 75% of poorer, small-farm households for the 5% of better-off large-farm households, while the predominant flow of labour was into non-farm employment outside the village and the region. Two-wheeled tractors for rice farming were mostly owned by the farmers who used them, with some hiring by very small farmers, while combine harvesters, four-wheeled tractors for heavy upland cultivation, and trucks for transporting cash crops were in the hands of a few well-off entrepreneurial farmers specialising in contract service provision.

Notes

1. With the accession of King Mongkut in 1851 and the signing of the Bowring Treaty with Britain in 1855, Siam was opened up to global trade and the export of rice was formally allowed. This coincided with a long, steady rise in the price of rice, reflecting growing demand from other Asian countries. Siam's rice exports were overwhelmingly to the British entrepots of Hong Kong and Singapore (Owen 1971).
2. Isan was a relatively sparsely populated no-man's land between Bangkok and Vientiane until the 1827 revolt of Chao Anou, ruler of Vientiane, against his Thai overlords in Bangkok. After the revolt was crushed, tens of thousands of ethnic Lao from the left bank of the Mekong were forcibly settled in Isan where they could be more readily controlled. Over the subsequent century, more ethnic Lao came to be living in Isan than in Laos (Evans 2002: 25–32).
3. These central markets have since ceased to exist throughout Thailand (Pongsrihadulchai 2018).
4. According to the Chaipattana Foundation (2017), established by Royal Charter in 1988, "Sufficiency Economy is a philosophy based on the fundamental principle of Thai culture. It is a method of development based on

moderation, prudence, and social immunity, one that uses knowledge and virtue as guidelines in living. Significantly, there must be intelligence and perseverance which will lead to real happiness in leading one's life".

5. The official zoning of land into "suitable" and "unsuitable" for rice farming can be viewed online at the following site—http://agri-map-online.moac.go.th/.

6. Wages expressed in 2000 US dollars.

REFERENCES

ACI, 2005. *The North East Thailand Rice Value Chain Study.* Bethesda: Agrifood Consulting International.

ALRO, 2006. *National Report on Agrarian Reform and Rural Development in Thailand.* Submitted by Agricultural Land Reform Office to the International Conference on Agrarian Reform and Rural Development (ICARRD), 6–10 March 2006, Brazil.

Barnaud, C., Trébuil, G., Dufumier, M., and Suphanchaimart, N., 2006. Rural poverty and diversification of farming systems in upper Northeast Thailand. *Moussons* 9–10: 157–187.

Behrman, J. R., 1968. *Supply Response in Underdeveloped Agriculture: A Case Study of Four Major Annual Crops in Thailand, 1937–1963.* Amsterdam: North-Holland Publishing Company.

Burns, Anthony, 2004. *Thailand's 20-Year Program to Title Rural Land.* Background paper prepared for the World Development Report 2005.

Chaipattana Foundation, 2017. *Philosophy of Sufficiency Economy.* Available at http://www.chaipat.or.th/eng/concepts-theories/sufficiency-economy-new-theory.html (accessed 12 June 2019).

Chitibut, W., Poapongsakorn, N., and Aroonkong, D., 2014. *Fertilizer Policy in Thailand.* ReSAKSS Asia Policy Note 3. Washington, DC: IFPRI.

Chuasuwan, C., 2018. *Rice Industry Outlook, 2018–2020.* Bangkok: Krungsiri Research.

Dohrs, S. L., 1988. *Commercial Agriculture and Equitable Development in Thailand.* Southeast Asia Business Papers No. 5. Ann Arbor: Center for South and Southeast Asia Studies, University of Michigan.

Ekasingh, B., Sungkapitux, C., Kitchaicharoen, J., and Suebpongsang, P., 2008. *Competitive Commercial Agriculture in the Northeast of Thailand.* Chiang Mai: Department of Agricultural Economics and the Multiple Cropping Centre, Faculty of Agriculture, Chiang Mai University.

Evans, G., 2002. *A Short History of Laos: The Land in Between.* Sydney: Allen and Unwin.

Falvey, Lindsay, 2000. *Thai Agriculture: Golden Cradle of Millennia.* Bangkok: Kasetsart University.

FAO, 2019. Country fact sheet on food and agriculture policy trends (Thailand). Food and Agriculture Organization of the United Nations. Available at http:// www.fao.org/3/I8683EN/i8683en.pdf (accessed 5 May 2019).

Floch, P., and Molle, F., 2009. *Pump Irrigation Development and Rural Change in Northeast Thailand.* Working Paper, Mekong Program on Water, Environment and Resilience (M-POWER). Chiang Mai: University of Natural Resources and Applied Life Sciences, Institut de Recherche pour le Développement, International Water Management Institute.

Floch, P., and Molle, F., 2013. Irrigated agriculture and rural change in Northeast Thailand: reflections on present developments. In R. Daniel, L. Lebel, and K. Manorom, eds., *Governing the Mekong: Engaging in the Politics of Knowledge*, pp. 188–212. Petaling Jaya: Strategic Information and Research Development Centre.

Grandstaff, T. B., Grandstaff, S., Limpinuntana, V., and Suphanchaimat, N., 2008. Rainfed revolution in Northeast Thailand. *Southeast Asian Studies* 46: 289–376.

Heckman, C. W., 1979. *Rice Field Ecology in Northeastern Thailand: The Effect of Wet and Dry Seasons on a Cultivated Aquatic Ecosystem.* The Hague: Dr. W. Junk B.V., Publishers.

Kasem, Sukallaya, and Thapa, Gopal, 2012. Sustainable development policies and achievements in the context of the agriculture sector in Thailand. *Sustainable Development* 20: 98–114.

Konchan, S., and Kono, Y., 1996. Spread of direct seeded lowland rice in Northeast Thailand: farmers' adaptation to economic growth. *Southeast Asian Studies* 33: 523–546.

Molle, F., Floch, P., Promphakping, B., and Blake, D. J. H., 2009. The 'greening of Isaan': politics, ideology and irrigation development in the northeast of Thailand. In F. Molle, T. Foran, and M. Kakonen, eds. *Contested Waterscapes in the Mekong Region: Hydropower, Livelihoods and Governance*, pp. 253–282. London: Earthscan.

Mordor Intelligence, 2019. *Fertilizers Market—Thailand. Industry Growth, Trends, and Forecasts 2017–2022.* Available at https://www.mordorintelligence.com/industry-reports/thailand-fertilizers-marke (viewed 11 June 2019).

National Statistics Office, 2008. Agriculture inter-censal survey in 2008 for whole kingdom. Available at http://service.nso.go.th/nso/nsopublish/service/ agricult/ais-wk/ais-wk.pdf (accessed 2 July 2012).

OAE, 2011. *Agricultural Economics Outlook 2011–2012.* Thailand: Office of Agricultural Economics. Available at http://www.oae.go.th/bapp/download/gdp/outlook%202011-12.pdf (accessed 12 January 2012).

Office of Agricultural Economics, 2018a. *Agricultural Statistics of Thailand.* Office of Agricultural Economics, Ministry of Agriculture and Cooperatives, Thailand. Available at http://www.oae.go.th/assets/portals/1/files/journal/2562/yearbook2561.pdf (accessed 4 May 2019).

Office of Agricultural Economics, 2018b. *Agricultural Information*. Office of Agricultural Economics, Ministry of Agriculture and Cooperatives, Thailand. Available at http://www.oae.go.th/assets/portals/1/files/jounal/2562/commodity2561.pdf (accessed 4 May 2019).

Owen, N. G., 1971. The rice industry of Mainland Southeast Asia 1850–1914. *Journal of the Siam Society* 59: 78–143.

Pongsrihadulchai, A., 2018. *Thailand's Rice Industry and Current Policies towards High Value Rice Products*. Paper submitted for International Seminar on Promoting Rice Farmers' Market Through Value-Adding Activities, 6–7 June 2018, Kasetsart University, Bangkok.

Poungchompu, S., and Chantanop, S., 2016. Economic aspects of rice combine harvesting service for farmers in Northeast Thailand. *Asian Social Science* 12: 210–211.

Rabobank, 2003. *Thai Rice: Time for a Sustainable Development*. Food and Agribusiness Research. Bangkok: Rabobank International.

Rambo, A. T., 2017. The agrarian transformation in Northeastern Thailand: a review of recent research. *Southeast Asian Studies* 6: 211–245.

Rigg, J., Salamanca, A., and Parnwell, M., 2012. Joining the dots of agrarian change in Asia: a 25 year view from Thailand. *World Development* 40: 1469–1481.

Setboonsang, S., 1996. Rice research priorities in Thailand. In R. E. Evenson, R. W. Herdt, and M. Hossain, eds. *Rice Research in Asia: Progress and Priorities*. Los Banos: International Rice Research Institute.

Sunsuk, D., 2016. Isaan farmers and local officials slam failed rice policy. *The Isaan Record*, 25 October 2016.

Viboon, Thepent, and Chamsing, Anucit, 2009. *Agricultural Mechanization Development in Thailand*. Fifth Session of the Technical Committee of APCAEM, 14–16 October 2009. Los Banos: APCAEM.

Wada, H., 2005. Managing sandy soils in Northeast Thailand. In *Proceedings of Conference on Management of Tropical Sandy Soils for Sustainable Agriculture*, 27 November to 2 December 2005, Khon Kaen, Thailand. Available at http://www.fao.org/3/ag125e/AG125E11.htm (accessed 28 May 2019).

Welcher, P., 2017. *Thailand: Rice Market and Policy Changes over the Past Decade*. GAIN Report No. TH7011. Bangkok: USDA Foreign Agricultural Service.

Wiboonpongse, A., and Chaovanapoonphol, Y., 2001. *Agribusiness Research on Marketing System in Thailand*. Chiangmai: Multiple Cropping Center, Chiangmai University.

CHAPTER 3

Evolution of Rice Farming in Ubon Ratchathani Province

Prathanthip Kramol and Benchaphun Ekasingh

INTRODUCTION

Rice production systems in Thailand have changed from traditional to modern practices since the 1980s. The change began in the Central Region where improved irrigation systems were developed, providing a basis for innovations in technology such as the use of high-yielding varieties (HYVs), fertilisers, pesticides, and machinery (Srisompun and Isvilanonda 2012). Previously, growth in rice production was mainly due to increases in the cultivated area (Thepent and Chamsing 2009). However, the widespread adoption of modern technologies has entailed an intensification and commercialisation of rice production, increasing the productivity of both land and labour and the size of the marketable surplus (Chamsing 2007). Apart from the seed-fertiliser technology associated with the green revolution, the major innovation has been the widespread mechanisation of rice farming. Farm machinery was initially imported from East Asia but rapidly came to be manufactured in Thailand and adapted for local conditions. Notable developments were the adoption of

P. Kramol (✉) • B. Ekasingh
Faculty of Agriculture, Department of Agricultural Economy and Development, Chiang Mai University, Chiang Mai, Thailand

© The Author(s) 2020
R. Cramb (ed.), *White Gold: The Commercialisation of Rice Farming in the Lower Mekong Basin*,
https://doi.org/10.1007/978-981-15-0998-8_3

two-wheeled tractors, medium-sized four-wheeled tractors, and combine harvesters (Thepent and Chamsing 2009).

Thai rice farming systems have changed again since the country's economic crisis in 1997. In the wake of the crisis, agricultural input prices increased while product prices were low. Alternative rice farming systems following King Bhumibol's philosophy of the "sufficiency economy" were widely promoted (Chap. 2). In particular, since 2007 the Community Development Department in the Ministry of the Interior initiated a project promoting the model of a "self-sufficiency village". The project aims to strengthen villages' competencies in self-management and develop the economic base on self-sufficiency principles (Prathanchawano 2013). Some commercial farmers are reported to have reverted to being self-sufficient farmers, relying less on external resources, drawing more on local resources, and diversifying into other crops. In addition, to address the affordability of farm inputs, the government has prioritised the provision of more suitable forms of credit. Provision of agricultural credit has always been an important policy for the Thai Government and has been a crucial driver of the commercialisation of agriculture. However, in recent years a number of government projects have been implemented using micro-finance principles. Villages have been designated as grassroots units and targeted in credit projects.

In the Northeast, as described in Chap. 2, rice is an important crop and the major source of income for farmers. The area cultivated with rice was about 6.37 million ha in 2010 out of a total crop area of 10.3 million ha (OAE 2012)—the largest rice area of any of the four regions in Thailand. Rice in the Northeast is also recognised as a high-quality export product because of the fragrant rice variety widely grown in the region called *Khow Hom Mali*. However, farmers' incomes in the Northeast are still low, partly due to low yields, averaging around 2.2 ton/ha. The yield is 0.6 t/ha below the national average and about 1.2 t/ha lower than in Northern and Central Thailand (OAE 2012). Low income from agricultural production has forced some farm household members to migrate seasonally to the big cities.

Hence, in the context of a book about the commercialisation of rice farming in the Mekong, it is of special interest to study how farmers in the Northeast have made use of modern technologies and credit to increase net returns or have adopted the principles of the "sufficiency economy". Three contrasting villages in Ubon Ratchathani Province were selected for this study to investigate the use of rice farming technologies, especially fertiliser and machinery, to estimate the returns to the farm household from rice production, and to identify the problems and potential of different approaches to the development of rice farming in the region. The case studies drew on village statistics, discussions with key informants, and household interviews.

THE CASE STUDY VILLAGES

Ubon Ratchathani Province is located in the southeastern corner of Northeast Thailand, in the lower part of the Mun River catchment (Fig. 2.1). The province borders Laos to the east (in particular the provinces of Saravan and Champasak), Cambodia to the south, Si Sa Ket Province to the west, and Amnat Charoen and Yasothorn Provinces to the north. In terms of pursuing the sufficiency economy programme, there were about 332 farmers' groups in the province but only 180 groups were still active in 2012. According to the Cooperative Promotion Department, about 114 groups had withdrawn registration, 12 had stopped working as a group, four were just forming, and the status of 22 groups could not be ascertained.

The study was conducted in 2014 in Ban Donmoo in Trakan Phuet Phon District, Ban Bua Teang in Sawang Weerarong District, and Ban Nong Bua Hi in Phiboon Mangsahan District. The three villages were selected because most farmers engaged in rice-based farming systems and, unlike most villages in Thailand, they were attempting to follow "sufficiency economy" principles, though in different ways. Ban Donmoo and Ban Bua Teang were officially designated as "model self-sufficiency villages"; in particular, Ban Donmoo won a prize for the best model village in the Province in 2012. However, Ban Nong Bua Hi was quite different from the first two villages. The villages also differed in their distance from Ubon Ratchathani City and in the degree of farm diversification that had occurred. The contrasts between the three villages highlight the varying degrees of change in rice-based farming systems in the Northeast generally.

Ban Donmoo

Ban Donmoo was located far from the main road about 14 km from the district capital and 50 km from Ubon Ratchathani. The village had a population of 776 in 156 households in 2013. The average household size was 5.0, ranging from 1 to 15. Almost half (49%) of the workforce was engaged in the agricultural sector as rice farmers and 18% were wage labourers. Very few were salary-earners or engaged in trade or business. The villagers had a close-knit community and tried to follow the sufficiency economy philosophy. They had formed the Donmoo Group and operated various sub-groups such as the rice mill group, the rice farmer group, the community bank, and a community training group. As mentioned, the village had been given a provincial-level award as a model "sufficiency economy" village. The village's land was rainfed, so farmers usually only cropped in the wet season. Organic rice was the major crop, while a few farmers grew vegetables such as long bean, Chinese cabbage, and cucumber after rice using part of their paddy land and on-farm sources of irrigation such as ponds. The rice yield was around 3.1–3.4 t/ha. In 2013, the average household income was THB 223,343 (USD 6741), of which THB 80,922 (USD 2443), or 36%, was farm income.

Ban Bua Teang

Ban Bua Teang was located near the main road about 16 km from the district town of Sawang Weerawong in one direction and 16 km from the provincial capital, Ubon Ratchathani, in the other. There were 189 households in this village in 2013 with a total population of 808. The average household size was 4.3, ranging from 1 to 10. About a fifth (22%) of the workforce was self-employed in farming, mainly planting rice, while 40% were employed as labourers. Farmers in this village had formed farmers' groups to provide services related to agro-tourism, homestays, and training. The village farming area included irrigated lowlands as well as uplands; hence, farmers could diversify their farms with sugarcane, cassava, rubber, fruit trees, flowers, and vegetables. Nevertheless, rice was the most important crop, with paddy fields accounting for about 70% of the farming area. Thirty per cent of rice farmers grew rice twice a year. Rice yields ranged from 1.6 to 2.2 t/ha. In 2013, the average household income was THB 298,124 (USD 9001), of which THB 174,272 (USD 5262), or 59%, was farm income.

Ban Nong Bua Hi

Ban Nong Bua Hi was located near the main road about 16 km from the district town of Phibun Mangsahan and 62 km in the other direction to Ubon Ratchathani. There were 187 households in 2013 with total population of 643. The average household size was 3.4, ranging from 1 to 9.

About 60% of the workforce was engaged in farming, mainly rice. Labouring, business, trading, and professional activities each accounted for less than 7%. Farmers in this village had been encouraged to form farmers' groups but there was no strong group at the time of the survey. However, there were several agrochemical shops because the village was at the centre of the sub-district. As the village croplands were irrigated, about 80% of farmers practised double cropping of rice. The rice yield averaged about 2.5 t/ha. In 2013, the average household income was THB 233,430 (USD 7048), of which THB 146,092 (USD 4411), or 63%, was farm income.

FERTILISER USE

Overall, farmers in the case study villages used mainly synthetic fertilisers together with some organic fertilisers. However, the use of fertiliser was dependent on farming practices. In Ban Donmoo, 90% of rice farms were organic, so there was no use of synthetic fertilisers or other agrochemicals on these farms. Alternative methods of nutrient management and crop protection were used, such as organic fertiliser, compost, animal manure, green manure, and animal or plant extracts. Most farmers in Ban Bua Teang and Ban Nong Bua Hi applied both synthetic and organic fertilisers; a minority of farmers applied only organic fertilisers while some applied mainly organic fertilisers with a small quantity of synthetic fertilisers.

In Ban Nong Bua Hi there was intensive use of synthetic fertilisers and pesticides. The three agrochemical shops in the village supplied credit to farmers for input purchase. Farmers had used organic fertilisers together with synthetic fertilisers since 2008 when a project promoting organic fertiliser to improve soil conditions was launched in the village through the village fund. Farmers produced organic fertilisers as a group and received a share to use on their farms. The amount of each farmer's share was enough for only 3 rai (0.5 ha) per crop, so the farmers used only synthetic fertilisers on their remaining rice area. Farmers interviewed could not yet see any difference from using organic fertilisers on their farms, apart from the lower cost compared to synthetic fertilisers.

Most farmers in Ban Donmoo and Ban Bua Teang obtained inputs on credit from the local sub-district cooperative, while farmers in Ban Nong Bua Hi obtained credit from both the cooperative and the agricultural chemical stores in the village and district town. The sub-district cooperatives ordered in fertilisers and agricultural chemicals such as pesticides,

Table 3.1   Fertiliser use and paddy yield under different fertiliser regimes

Input/output	Conventional practice	Conventional with organic	Organic practice
NPK Complex rate (kg/ha)	468	125	–
NPK Complex expense (THB/ha)	1500	400	–
Ammonium sulphate rate (kg/ha)	468	106	–
Ammonium sulphate expense (THB/ha)	1500	340	–
Green manure imputed cost (THB/ha)	–	–	250
Organic fertiliser rate (kg/ha)	–	1250	187
Organic fertiliser imputed cost (THB/ha)	–	400	120
Total fertiliser cost (THB/ha)	3000	1140	370
Paddy yield range (t/ha)	1.6–2.5	1.9–2.8	2.5–3.4

Source: Farmer survey and group interviews in each village
Note: NPK Complex comprises NPK in the ratio 15:15:15; ammonium sulphate comprises NPK in the ratio 21:0:0

herbicides, and fungicides from the district- or provincial-level cooperatives, which in turn had direct links to agencies or companies.

Farmers in Ban Donmoo had been cultivating organic rice since 2003
when they experienced failure from commercialised rice production. The
farmers interviewed claimed that they had increased their rice yield from
1.25 t/ha in 2003 to 2.5 t/ha in 2012. For the first crop after going
organic, they applied about 1250 kg/ha of organic fertilisers and obtained
yields of only 1.25–1.6 t/ha. However, in recent years farmers used only
187 kg/ha of organic fertilisers and obtained 2.5 t/ha.

The use of fertilisers and associated rice yields under different farming
practices are summarised in Table 3.1. The survey data suggest that integrating organic fertiliser into the soil fertility regime resulted in higher
yields and lower cost. The organic practice reportedly gave 50% higher
average yield for 12% of the cost. However, other factors would need to be
quantified to confirm this result.

LABOUR AND MACHINERY USE

With increasing scarcity and ageing of farm labour, mechanisation of lowland rice farming has occurred throughout Thailand, first in the Central
Region, where increased cropping intensity due to the expansion of irrigation was a key driver, but now including both irrigated and rainfed systems
in the Northeast. The study villages varied in the extent of irrigation, but

Table 3.2 Use of machinery in the study villages

Machine	Capacity[a]	Year of initial use in village		
		Ban Donmoo	Ban Bua Teang	Ban Nong Bua Hi
Two-wheeled tractor	4–5 rai/day	1993	1993	1993
Small four-wheeled tractor	10–15 rai/day	2010	2010	2010
Medium four-wheeled tractor	20 rai/day	2009	2010	2010
Large four-wheeled tractor	30–40 rai/day	2009	2007	2002
Threshing machine	4–6 tons/hr	1983	1983	1983
Combine harvester	50–70 rai/day	2011	2010	2005

Source: Group and individual interviews
[a]1 day = 8 hours; 6.25 rai = 1 ha

most farmers depended on a single wet-season crop of rainfed lowland rice. Some farmers had invested in small ponds or tubewells and pumps to tap underground water and were able to grow small areas of crops such as vegetables, beans, and chillies following rice. Nevertheless, rice production had become highly mechanised in all villages, though not in all operations (Table 3.2).

Labour shortage in the main rice season was prevalent in the villages as many household members migrated to Bangkok or other urban centres. Typically, migrants came back to the village during peak periods to work on their own farms or as hired labour. However, due to increasing transportation costs and higher urban wages, many farmers preferred to invest in owning or hiring machinery instead of asking family members to incur the financial and opportunity costs of returning for rice production. At the same time, small- and medium-sized machinery suitable for small farms has become widely available in the past two decades, enabling mechanisation of the once-labour-intensive operations of land preparation, planting, and harvesting, including threshing (Table 3.2). In addition, the availability of long-term credit from the Bank for Agriculture and Agricultural Co-operatives (BAAC) has enabled many farmers to purchase farm machinery. The Bank also offers to refinance farmers' loans from finance companies and machinery dealers.

Land Preparation

The use of tractors for land preparation had become the norm in all three villages (Table 3.3). Most farmers owned two-wheeled tractors, which

Table 3.3   Tractors used for land preparation in the study villages

Type of tractor	Advantages	No. of units in village		
		Ban Donmoo	Ban Bua Teang	Ban Nong Bua Hi
Large 4W tractor	Suitable for large areas and heavy work, for example, newly opened land	2	1	1
Medium 4W tractor	Suitable for small areas of paddy land	3	3	20
2W tractor	Suitable for very small areas of paddy land; used as power sprayer	Every farm	Every farm	Every farm

Source: Group and individual interviews

began to be purchased in the early 1990s. However, after improvements in four-wheeled tractor technology, farmers began using them from 2009, particularly in Ban Nong Bua Hi. These medium-sized tractors were light enough to work satisfactorily in paddy fields and could cross the bunds surrounding individual paddy plots without causing damage. The small numbers of large Ford tractors were normally used only for heavy-duty work, particularly in newly opened land.

Farmers ploughed their paddy land twice. For the first ploughing they generally hired a four-wheeled tractor to break up the soil and for the second ploughing they had previously used their own two-wheeled tractor to get an even planting surface. At the time of the study, however, most farmers were hiring medium-sized four-wheeled tractors for all land preparation; only a few still used two-wheeled tractors for this purpose. The capacity of the four-wheeled tractors was two to nine times that of the two-wheeled tractors, depending on the type of work.

The costs of purchasing and renting the different types of tractor are shown in Table 3.4. The cost of a four-wheeled tractor was more than ten times that of a two-wheeled tractor. However, for farmers with no non-farm work and hence less time-pressure, owning a two-wheeled tractor was still the most attractive alternative, with running costs lower than hiring either a two-wheeled or a four-wheeled tractor. Thus two-wheeled tractors were still used for preparing very small paddy fields (and could be modified to be used as mobile power sprayers, as well as having other functions). Farmers' ability to purchase medium-sized four-wheeled tractors usually depended on access to credit from machinery dealers or banks. Farmers who owned a four-wheeled tractor normally had to do contract

Table 3.4 Cost of purchasing and renting machinery for land preparation

Type of tractor	Purchase cost (baht)	Contractor charge (baht/rai)[c]	
		First cultivation	Second cultivation
Large 4W tractor	>1,000,000	200–250	300–400
Medium 4W tractor[a]	550,000–900,000	200–250	300–400
2W tractor	50,000–130,000	250	350
Use of own 2W tractor[b]		163	163

Source: Group and individual interviews
[a]When contracting both first and second cultivation the charge was 600 baht/rai
[b]Imputed labour cost plus fuel cost, excluding depreciation
[c]6.25 rai = 1 ha; USD 1 = THB 33

land preparation for other farmers to earn the money to repay their loans. These farmers mentioned that the payback period was only two years.

Sowing/Planting

Farmers in the three villages did not use any equipment for crop establishment. Some farmers, particularly in Ban Nong Bua Hi, practised broadcasting directly onto the paddy field, while some, especially in Ban Donmoo and Ban Bua Teang, still established nurseries and transplanted seedlings. Rice transplanters are now widely used in Central Thailand and in some parts of the Northern Region. However, few of the farmers in the study villages who practised transplanting had tried using a transplanter. The machine was found not to work well on their gravelly soils and no farmers had adopted them.

Harvesting and Threshing

Mechanical harvesting has been extended to the study villages. However, farmers in Ban Donmoo were producing organic rice and required a high-quality product to get maximum returns, so they harvested manually, though they hired mechanical threshers. Many farmers in Ban Bua Teang also harvested manually because their paddy fields were quite small and located in wetter lowland sites where the combine harvesters could not be used. While farms in Ban Donmoo and Ban Bua Teang were mostly harvested manually, farms in Ban Nong Bua Hi were mostly harvested by combine harvesters.

Table 3.5   Harvesting/threshing cost by method of harvesting

Method	Unit cost (baht/rai)
Daily hired labour	1000
Piece rate	1000
Exchange labour (imputed cost)[a]	1000
Thresher (@ 250 baht/t)	60
Combine harvester	600

Source: Group and individual interviews. 6.25 baht = 1 ha; USD 1 = THB 33
[a]Computed as opportunity cost of labour

Harvesting costs using combine harvesters were much lower than for manual harvesting (Table 3.5). The combine harvesters achieved both harvesting and threshing for only 600 baht/rai. Farmers who hired labour for harvesting incurred about 1000 baht/rai, whether the labour was paid daily wages or a piece rate, and they still had to pay for contract threshing, which cost about 60 baht/rai (assuming a yield of 1.5 t/ha). Hence the demand for combine harvesting had increased in areas where manual harvesting was not necessary.

Nevertheless, there were still no local combine services in the study villages or in the surrounding districts. Combine harvester services were mostly supplied from provinces in Central Thailand such as Suphanburi, Nonthaburi, and Bangkok. There were also contractors from other provinces in the Northeast such as Sisaket and Roi Et. Farmers contacted an agent to obtain harvesting services. The agent checked the paddy fields to determine if they were suitable for their machines. Large, dry paddy fields with no lodging were the preferred conditions. The appropriate time to bring the harvester into the field was also estimated. There was a high demand for harvesting services in the wet season; hence, farmers in Ban Bua Teang could obtain harvesting services only in the dry season.

Summary

Machinery use in rice production varied across the three villages. Farmers in Ban Donmoo and Ban Bua Teang were found to use fewer machines than in Ban Nong Bua Hi. Differences in farming systems played an important role in this variation. Organic farming systems required less machine use, mostly in land preparation. Farmers who grew rice mainly for household consumption tended to keep their cash expenses low and rely on household labour. Two-wheeled tractors were still used by these subsistence-oriented farmers as they were affordable and provided multiple benefits. Commercial farms were found to make greater use of contract machinery services, including medium-sized four-wheeled tractors for land preparation and combine harvesters for harvesting and threshing, because the financial cost was significantly lower than when hiring manual labour.

COMPARATIVE RETURNS TO RICE PRODUCTION

Given the different uses of fertilisers and agrochemicals in the three villages, the rice cropping systems can be termed "organic" in Ban Donmoo, "mixed" in Ban Bua Teang, and "conventional" in Ban Nong Bua Hi. The organic and mixed systems used transplanting for crop establishment, while the conventional system used broadcasting, thus saving on labour. More generally, the conventional system relied more on contracted machinery services (including four-wheeled tractors and combine harvesters) and purchased inputs than the organic system, with the mixed system somewhere in between. These three systems showed different levels of yield, cost, and gross margins (Table 3.6).

The organic rice system had higher average yield (3.2 t/ha) and enjoyed a 33% price premium over the mixed and conventional systems; hence, gross revenue was about 50% higher than the other two at THB 64,000 (USD 1900) per ha. Total paid-out costs were very similar across the three systems at about THB 16,000 (USD 490) per ha. However, there were differences in the importance of individual cost items. The organic system had lower costs for land preparation (because it was mainly done by family labour), higher costs for wage labour (used for transplanting, hand weeding, applying organic fertiliser, and hand harvesting), zero costs for inorganic fertiliser, and higher costs for manure and organic fertiliser.

The conventional system, on the other hand, had higher costs for seed (because of broadcasting) and inorganic fertilisers. The imputed value of family labour used in rice production (valued at the local wage rate of THB 300/day) was around THB 3000 (USD 90) per ha for the organic and mixed system, but much lower for the conventional system, which relied extensively on contractors for land preparation, spraying herbicides, and harvesting. The overall unit cost of production was slightly lower for the organic system at THB 5900 (USD 180) per ton of paddy produced.

Given the organic system's higher gross revenue and similar paid-out costs, the gross margin for organic rice, at THB 48,000 (USD 1500) per

Table 3.6 Costs and returns for wet-season rice production in the study villages

Variable	Village (cropping system)		
	Donmoo (organic)	Bua Teang (mixed)	Nong Bua Hi (conventional)
Paddy production (kg/ha)	3200	2813	2719
Paddy price (baht/kg)	20	15	15
Gross revenue (baht/ha)	64,000	42,188	40,781
Input costs (baht/ha)			
Land preparation	2188	4375	3250
Seed	547	625	3125
Hired labour	4375	2813	846
Inorganic fertiliser	0	2763	4950
Manure/organic fertiliser	3375	200	0
Pesticide	0	0	398
Harvesting and threshing	3750	3750	2500
Other costs	1623	1561	1168
Total paid-out costs	15,857	16,086	16,237
Family labour (@ 300 baht/day)	3008	3109	329
Total costs	18,865	19,195	16,566
Hired labour use (days/ha)	14.6	9.4	2.8
Family labour use (days/ha)	10.0	10.4	1.1
Total labour use (days/ha)	24.6	19.7	3.9
Gross margin 1 (baht/ha)	48,143	26,102	24,545
Gross margin 2 (baht/ha)	45,135	22,992	24,215
Cost per kg of paddy (baht)	5.90	6.82	6.09
Gross margin per kg of paddy (baht)	14.10	8.18	8.91
Return to labour (baht/day)	2135	1468	6511

Source: Household Survey, 2013
Notes: 1 hectare = 6.25 rai; USD 1 = THB 33; Gross margin 1 is excluding the cost of family labour; Gross margin 2 is including the cost of family labour; Return to labour is gross revenue less all non-labour costs divided by the number of days of family and hired labour

ha, was about double that of the mixed and conventional systems. The ranking was similar when the imputed cost of family labour was deducted, with organic rice averaging a gross margin of THB 45,000 (USD 1360) per ha. However, the higher labour input for the organic system meant that the return to labour (family and hired) was THB 2135 (USD 65) per day compared with THB 6511 (USD 197) for the conventional system. The labour requirement was the main reason conventional farmers did not want to follow organic practices.

CONCLUSION

Rice farming in Northeast Thailand has changed significantly in the past few decades, becoming more commercialised and mechanised. This has involved increased use of high-yielding seed, inorganic fertilisers, and machinery, especially for land preparation and harvesting, and lower use of family labour as household workers find more profitable non-farm employment, often outside the district and province. However, the study of three villages in Ubon Ratchathani found different patterns of change depending on both choice and circumstances.

Ban Donmoo was a more remote, close-knit community with somewhat larger households (a mean of 5.0 members) and active farmer groups that followed the "sufficiency economy" approach. With no irrigation, farmers planted a single crop of organic rice in the wet season. They used little or no synthetic fertiliser and made less use of farm machinery. Nevertheless they obtained higher yields and a price premium, while incurring comparable costs to the other two villages, giving them a high return to land and a moderate return to labour. Household income averaged USD 6800, of which only USD 2400 (36%) was farm income, implying a high dependency on non-farm activities.

Ban Bua Teang was near the main road and close to Ubon Ratchathani. Household size was somewhat less (4.3) and farmers' groups had been set up for different activities. Some farmers had irrigated lowlands and 30% grew rice twice a year; they also had upland crops such as sugarcane. Nevertheless, 40% of the population was employed in labouring. With their intermediate approach to rice farming, which was mainly for subsistence, farmers obtained lower gross income than those in Ban Donmoo but incurred similar costs. The return to land was half that in Ban Donmoo and the return to labour about three quarters. Nevertheless, given their greater cropping activity, household income averaged USD 9000 (32% more than Ban Donmoo), of which USD 5300 was farm income (59%), more than twice the farm income of Ban Donmoo.

Ban Nong Bua Hi was also near main road and had several agrochemical shops in the village. Household size was even smaller (3.4) and group formation had not been successful. Most households cultivated rice and, given that most land was irrigated, about 80% of farmers practised double cropping. Rice farming was commercially oriented, relying on synthetic fertilisers and contracted use of machinery such as large and medium four-wheeled tractors and combine harvesters. Farmers in this village needed more capital for inputs and incurred higher debts than in the first two villages. The returns to land were half that of Ban Donmoo but the return to labour was more than three times as high. Household income averaged USD 7100, of which USD 4400 was farm income (63%), almost twice that of Ban Donmoo.

Comparing the three villages shows that, even after several decades of commercialisation in the Northeast, rice farming is following different trajectories and making different contributions to household livelihoods, depending on the goals and circumstances of individual households and communities. Alternative agriculture based on organic production methods can be a viable pathway alongside conventional commercial agriculture. However, in all cases, non-rice and non-farm sources of income are needed to augment income from rice production.

References

Chamsing, Anucit, 2007. *Agricultural Mechanization Status and Energy Consumption for Crop Production in Thailand.* Dissertation No. AE-07-01. Pathum Thani, Thailand: Asian Institute of Technology.

OAE, 2012. *Agricultural Statistics of Thailand 2011.* Bangkok: Office of Agricultural Economics.

Prathanchawano, Pison, 2013. *Community Strength.* Available at http://www.cdd.go.th/director_cddnew/pison2/file.pdf/Strategy.pdf (accessed 31 October 2013).

Srisompun, Orawan, and Isvilanonda, Somporn, 2012. Efficiency change in Thailand rice production: evidence from panel data analysis. *Journal of Development and Agricultural Economics* 4(4): 101–108.

Thepent, Viboon, and Chamsing, Anucit, 2009. *Agricultural Mechanization Development in Thailand.* Fifth Session of the Technical Committee of APCAEM, 14–16 October 2009. Los Banos: APCAEM.

CHAPTER 4

Farmer Organizations in Ubon Ratchathani Province

Prathanthip Kramol, Pornsiri Suebpongsang,
and Benchaphun Ekasingh

INTRODUCTION

Notwithstanding the rapid growth of commercial agriculture in Thailand over the past half century, farmer organizations and community enterprises have been a common and distinctive feature of the rural economy and have been strongly supported by government policies, especially with the promotion of the concept of a "sufficiency economy" introduced by the late King Bhumibol and incorporated in national development plans (Thai Chaipattana Foundation 2013). While informal cooperation such as through labor exchange and rotating savings groups has been a traditional part of village life, the government has consistently promoted more formal organizational arrangements for farmers, alongside mainstream policies for intensification and commercialization of rice and other crops. Farmers are encouraged to form themselves into groups as legal entities to obtain support from outside agencies, especially through government programs.

P. Kramol (✉) • P. Suebpongsang • B. Ekasingh
Faculty of Agriculture, Department of Agricultural Economy and Development, Chiang Mai University, Chiang Mai, Thailand
e-mail: pornsiri.s@cmu.ac.th

© The Author(s) 2020
R. Cramb (ed.), *White Gold: The Commercialisation of Rice Farming in the Lower Mekong Basin,*
https://doi.org/10.1007/978-981-15-0998-8_4

85

These groups are formed to meet members' common needs with regard to the production and marketing of agricultural and non-agricultural products goods and services. This juxtaposition of independent small-holder farming and collective, community-based economic activity is explored in this chapter through case studies of organizations in three villages in Northeast Thailand.

FARMER ORGANIZATIONS AND COMMUNITY ENTERPRISES IN THAILAND

Organizations, groups, and networks are forms of social structure that can provide social, economic, and political benefits to participants and others (Ishihara and Pascual 2013). In Thailand, farmer organizations and community enterprises have been advocated as tools to improve rural livelihoods. They are said to provide the following benefits: (a) Forming groups provide an opportunity for the members to link up with other groups, government, and traders to conduct business or obtain support. (b) Marketing members' products jointly enable them to realize economies of scale and receive better prices. (c) Developing a culture of saving is seen as a means to overcome farmers' chronic lack of finance and inputs and build the autonomy of farmer organizations. (d) Beyond the economic enterprises they undertake, farmer organizations strengthen social bonds and help address social issues such as alcohol abuse, crime, and domestic violence. (e) Organizations make rural development more effective by building close collaboration with local administration, making it easier to voice development concerns. (f) Working in groups requires leadership, financial accounting, and record keeping, ultimately promoting transparency and accountability. (g) Providing equitable access to support services has encouraged women to take up leadership roles and improve their management skills and an open membership policy has encouraged poor and disadvantaged villagers to participate.

Stockbridge et al. (2003) found ten factors associated with the success of farmer organizations in general:

- Participants should be relatively homogeneous in terms of their socio-economic status and cultural values.
- The size of a farmer organization should match the organizational abilities of its members and be appropriate for the type and scale of activities being collectively undertaken.

- The services provided by the organization should reflect the demands of its members and should be matched by the ability of the organization to deliver them.
- The organization should be able to identify and undertake activities that make good business and commercial sense.
- The organization should not be dominated by outsiders (e.g., government, donors, and non-governmental organizations) in pursuit of their own respective agendas and in the long run should not be overly dependent upon outsiders for support and guidance.
- The organization should have financial capacity to support its own activities and not be heavily dependent upon subsidies.
- There should be a minimum level of skills and education among the organization's membership.
- Strong incentives exist for active participation by members in decision making and in the use and/or provision of services.
- The structure of the organization facilitates good governance and effective day-to-day management of the organization and ensures that the leadership is accountable to members.
- The legislative framework within which the organization operates promotes good governance while at the same time avoiding excessive regulations and the harm this can do to autonomous development of the organization.
- Resources should be focused on effectively undertaking a limited number of activities rather than less effectively engaging in a larger number of activities.

These factors have been confirmed in many studies, including by Kassam et al. (2011), who studied the success of the Samroiyod Shrimp Farmers' Cooperative in Thailand, and by Chumsri (2010), who analyzed lessons learned from successful cases of three farmer organizations in Thailand. Asia DHRRA and Agriterra (2002) found that farmer organizations in Thailand had the following strengths: self-management, government-influenced rules and regulations, well-directed strategies, closely government-supervised programs, services responsive to the needs of the people, and adequate resources.

In the Thai context, a "farmer organization" is considered to be a body providing credit, savings, farm supplies, joint marketing, and agricultural extension services to members. Farmer organizations include both "farmer groups" and "farmer cooperatives", the latter having commercial marketing function. Both have the same organizational principles, including voluntary and open membership, democracy, autonomy, independence, cooperation, human resource development, information dissemination, and community spirit. Farmer groups and cooperatives are generally formed within villages and sub-districts (*tambon*) and are linked at district and provincial levels through farmer networks and higher-level organizations.

Rice farmer groups were first established unofficially in 1955. Since 1967, the Government has encouraged the formal establishment and registration of these and other agricultural groups. To form an operational unit, a group needs at least thirty members and has to register with the Registrar of Farmer Groups.[1] In 2012, there were 4277 active farmer groups in Thailand with a total of 642,096 members (Cooperative Promotion Department 2013a).[2] These included groups for rice farmers, field crop farmers, horticultural farmers, fishers, and livestock farmers, with rice farmer groups accounting for 50% of the total. The Northeast Region had the highest number of farmer groups in 2012. In the same year there were 5124 rural cooperatives, of which 74% were agricultural cooperatives. In 2010, the total turnover of all farmer groups was THB 7.1 billion and of agricultural cooperatives, THB 215.3 billion (Cooperative Promotion Department 2013b).

A second strand of cooperation among agricultural households in Thailand is in the form of "community enterprises". These are cooperative activities or micro-enterprises involved in selling products or services and are found in most *tambon*. Community enterprises originated in women's groups which initially produced food and handicraft products for local consumption. After the 1997 financial crisis, small and medium enterprises were promoted to strengthen the grassroots economy and help overcome hardship (Wiboonpongse et al. 2006). Further support came from the launch of the One Tambon One Product (OTOP) program in 2001. Women's groups in a number of villages were targeted to be scaled up into micro-enterprises (Teerakul 2011). The Government has supported these enterprises through the provision of information, technologies, and marketing. Under the OTOP program, about 37,000 villages established community enterprises during 1997–2006 (Kittisataporn 2006).

Community enterprises are based on four core principles: (a) belonging to the community, (b) aiming to meet the needs of the community, (c) being fully operated by the community, and (d) contributing benefits to the community. They aim to embody a learning process and contribute to self-reliance (Teerakul 2011). According to the Community Enterprise Promotion Board (2010), community enterprises can deliver four types of product and six types of service. The products are agricultural goods, handicrafts, processed food, and other products. Agricultural products include crops, livestock, and fish and account for a higher percentage than the other types (Teerakul 2011). Handicrafts include fabrics, artificial flowers, weaving, gifts, souvenirs, jewelry, furniture, leather goods, and pottery. Food products include cottage foods, herb products, and beverages. Other products include machinery and agricultural inputs (Community Enterprise Promotion Board 2010). The services include community grocery stores, community savings groups, tourism, health services, mechanics, and other services (Teerakul 2011).

Case Studies in Ubon Ratchathani

Case studies were conducted in 2014 in the three villages in Ubon Ratchathani Province described in Chap. 3, namely, Ban Donmoo, Ban Bua Teang, and Ban Nong Bua Hi. There were several successful small farmer groups in Ban Donmoo, a few in Ban Bua Teang, and one loosely organized group in Ban Nong Bua Hi (not described further here). The groups had been formed for a variety of purposes, including to obtain access to inputs or credit, obtain support with production technologies, increase market accessibility, increase agricultural incomes, reduce production costs, and increase off-farm incomes. Other purposes of group formation were environmental protection and extension of traditional healing therapies. Key informant interviews and group discussions were used to obtain qualitative data about five different types of group, which were assessed according to the participants' own experience and evaluations.

Ban Donmoo Farmer School Group

The farmer school group in Ban Donmoo consisted of forty-five farmers. The group was established in 2003 to help members solve rice production problems and reduce costs. The group was formed by farmers who attended training provided by the Debt Suspension and Debt Burden Reduction Project of the Bank of Agriculture and Agricultural Cooperatives (BAAC) in 2001. The training course was called "The Truth of Life" and was conducted by the Ratchathani Asoke Group.[3] Farmers learned how to have a sustainable livelihood, in particular by reducing both living costs and farming costs. An alternative farming system, namely, organic farming, was introduced to them. Farmers learned to produce alternative fertilizers and pesticides.

After the BAAC-sponsored training, the group started operations by setting up an organic farming demonstration site of 1.3 rai on a lead farmer's land so that members could learn the appropriate practices for organic rice production. The forty-two farmers who had attended the course started to practice what they had learned. They used the demonstration farm to grow organic rice and this farm became the site of a farmer field school. When the farmers needed particular knowledge, specialists came to share their experience. In recent years, the farmers have been growing organic rice on their own farms but they still came together to produce organic fertilizer and provided credit to members to make organic fertilizer in their own farms.

Initially, the farmers found organic farming quite complicated. They needed to spend more time in their fields, the alternative pesticides were not as effective as purchased agrochemical inputs, and they had to apply large quantities of organic fertilizer. Moreover, the rice yield was lower than with conventional farming in the first year. Belonging to the group encouraged members to continue with organic rice. After several years, soil quality was improved and rice yields increased. The farmers had also reduced their input costs by ceasing to buy synthetic fertilizers. They learned together and could more easily obtain access to new higher-yielding technologies using less seed.

Participation in the group also created a sense of unity and enabled members to share their opinions frankly. They felt they could all share their ideas when decisions were needed and their views were regarded as equally important. Working as a group created reciprocal trust among the farmers, generating a sense of social obligation. Hence they could organize to produce and sell organic rice for a higher price than conventionally produced rice.

The farmers adopted the sufficiency philosophy in other aspects of their lives. They produced essential goods together and shared them among the members, including alternative fertilizers, plant extracts, and home goods such as washing liquid, shampoo, shower cream, and soap. As a group, it was easier for providers of external assistance to support them and knowledge was more easily transferred to members.

Ban Donmoo Community Rice Mill Group

There was also a Community Rice Mill Group in Ban Donmoo. It was formed after farmers were unsatisfied with the farm-gate price of paddy they were receiving through normal market channels, which they felt was unfair both to them and to consumers. Originally, the organic rice they produced was sold through conventional channels and farmers received the same price as rice produced with the use of chemical inputs. The farmers hoped to receive a higher price for their organic rice because at that time the rice yield was lower and they spent more time to produce it. Additionally, the farmers thought that organic rice had added value for consumers' health and so should attract a premium price. The way the existing rice mill mixed organic rice with conventional rice did not benefit the farmers or the consumers.

Consequently, seventy-four farmers formed the Community Rice Mill Group in 2005. The group started to get involved along the supply chain from farm to market so that they are now producers, buyers, processors, and sellers. The key objective was to assemble organic paddy from member farmers so they could bargain for a higher price. They also aimed to mill the organic paddy to get a premium price for the rice. The group raised about 58,500 baht (USD 1950) from the members. To construct the mill and buy paddy from farmers, the group obtained a loan of 500,000 baht (USD 17,000) from the BAAC. The group had executive members as well as purchasing, marketing, and standards teams. At the time of the study, it had increased its size to eighty-eight members, including some from nearby villages. The group bought about 50 tons of organic paddy from its members at 1 or 2 baht per kg higher than the price offered by other traders and rice mills. About 80% was sold directly to contracted buyers such as Santi Asoke and restaurants in town. Santi Asoke bought paddy while the restaurants bought milled rice. Another 20% was sold as rice directly to consumers from government offices and through the group's retail outlet in the village.

The Group's operations started from the time of planting. The standard control team frequently visited farmers to monitor and ensure that rice produce was organic. The area of members' farms was recorded and the expected yield estimated. The process of monitoring helped to prevent the problem of farmers buying non-organic paddy to sell to the Group. Separate committees worked on different aspects of the group's operations and members took part in the milling and packaging process. The

Community Rice Mill Group had a vertical relationship with the Farmer School Group. Members of the Farmer School Group could sell their organic rice at favorable prices. Thus farmers tended to continue growing organic rice. The Group also created employment in the village by hiring members to work in the milling and packing processes. They were paid the standard wage rate in the village for their duties. The Group was also seen to build human capital and farmer networks.

As a group, the farmers learned to work together and respond to their duties. The knowledge received from managing rice production and marketing was very important. The farmers normally could not undertake marketing by themselves. They were price takers and mostly sold their produce to traders. In addition, the farmers found they could manage finances and personnel. The Community Rice Mill Group confirmed that farmers were able to run a small business successfully. In addition, the linkages formed with the organic rice production group and with farmers in other villages showed them the possibilities for building a stronger and wider agricultural community.

Ban Bua Teang Agro-Tourism Group

The Ban Bua Teang Agro-Tourism Group was formed because the village had distinctive agricultural activities such as floriculture and received a push from the District and Provincial Agricultural Extension Offices (DAE and PAE). Farmers attended agro-tourism training in 2006, supported by the DAE and PAE. After the training, forty-five farmers formed the Group and selected a committee. Significant progress began when Mr. Kittipotch Seansing undertook a research project on community-based agro-tourism practice in Ban Bua Teang for his master's thesis (Seansing 2009). The project was supported by the Thailand Research Fund. The villagers involved in the project learned how to conduct effective agro-tourism. Initially, the attractions were only individual flower farms but villagers subsequently found that Ban Bua Teang had various suitable sites, including integrated farms, orchards, the community forest, and the landscape along the Mun River, where visitors could experience how villagers fished and processed local fish. Traditional farming technologies, local foods, and traditional beliefs and practices were also highlighted.

Information about the village was communicated by mass media, by government officers, and by word of mouth. As the number of visitors increased, the Group set up three more committees to handle facilities, food and beverages, and speakers. At the time of research, there were about 150 members from almost every household in the village and the group was catering for about 1000 visitors a year. The Group had also been asked to provide training organized by the BAAC for indebted farmers under the Debt Suspension and Debt Burden Reduction Project; hence, in total there were about 3000 people coming to the village each year.

The main objective of the Agro-Tourism Group was to increase villagers' income. The progress of the Group in this respect was more than the members had expected. When a large number of visitors came, all participating members received fair pay for their work which included organizing the groundwork, preparing food and beverages, entertaining visitors, and acting as resource persons. They also obtained income from selling their agricultural products and providing homestay services. The 3000 visitors and trainees coming to the village in 2013 brought in about 1.8 million baht, mostly from the BAAC for training activities. The Group received 350 baht per person for one-day training and 1250 baht per person for training over four days and three nights. Expenses for food, homestays, on-ground preparations, use of training room and facilities, and a contribution to the village temple were met from this income. The remaining income was saved in the Group's account and later distributed to members based on their time contribution.

Participants reported that the Group generated trust among the members as they worked together for the success of the Group. Members learned to take on different roles and responsibilities. Moreover, the Group gave the villagers a greater sense of confidence when visitors showed their interests in the village and its activities.

Ban Donmoo Micro Finance Institute

The Donmoo Micro Finance Institute (DMFI) had been established in Ban Donmoo to provide an accessible fund for poor households. It catered for the needs of the many different occupation groups within the village. DMFI had its origins in a village saving group—the Community Saving Fund for Pig Farming (CSFP)—formed in 1991 with the support of an NGO which contributed an initial fund of 3000 baht. Each member had to save 50 baht per month and could borrow money from the fund. With increased savings in the CSFP and by other occupational groups in the village, a local official suggested starting the DMFI in 2005. The committee members undertook training programs, especially in accounting and financial management, and visited other microfinance funds to learn how they operated. The DMFI was formally established in 2007 with support from the BAAC and working capital of 1 million baht. After ten years it continued to operate successfully, with working capital of 10 million baht, even though most of the committee members had only primary education.

At the time of the research, DMFI had fourteen committee members, elected in 2010 for a four-year term. All were residents of Ban Donmoo who understood the villagers and the issues they faced. DMFI's Members paid an enrolment fee of 50 baht. Membership was also open to residents of other villages. DMFI provided loans, paid dividends to members, and received saving deposits. The saving account was limited to 500,000 baht to control the total interest paid out; otherwise, saving would exceed borrowing and the Institute would struggle to cover its interest costs. The Institute paid a 1% annual interest rate for regular savings, 5% for fixed deposits of twelve months, and 7% for fixed deposit of twenty-four months. The rates were higher than offered by commercial banks in order to increase membership and generate more capital to circulate and use. The DMFI had regulations for membership and borrowing to minimize bad debts and maintain profitability. However, community members who struggled to meet their basic needs could borrow from the fund. In cases of overdue payments, the committee would extend the repayment period or provide an additional loan where repayment was feasible.

The DMFI was open for members on Mondays, Wednesdays, and Fridays. It operated like a bank. All members had books for their saving accounts. For withdrawals of 50,000 baht or more, members needed to give three days' notice as only 100,000 baht cash was kept in the office. However, in an emergency, a member could withdraw the money with the approval of three committee members. DMFI had also set up a welfare fund for members but this was an optional program. Each member wanting to join this fund had to pay 500 baht per year. The benefits were: (a) the ability to borrow money for family needs, (b) a hospital benefit of 60 baht per night for up to five nights per year, and (c) a death benefit of 100 baht from each fund member.

DMFI appeared to be a sustainable institution for the following reasons. (a) DMFI was formed on the basis of kinship and neighborhood ties. Thus there was a strong sense of belonging, joint liability, and participatory decision making, which strengthened the management of the organization and enhanced loan performance. (b) The committee played a crucial role in the operation of DMFI. To operate the Institute required a commitment of time and energy apart from the committee member's main occupation. The committee members had displayed sacrifice, honesty, responsibility, and accountability. (c) Committee members participated in training to improve their management skills. (d) DMFI received good support from local government agencies and the BAAC in terms of training activities and advisory services. This improved the capacities of staff and committee members and increased the sources of funds for the Institute.

Ban Bua Teang Village Fund

Village funds were part of a Thai Government program introduced in 2001. They were to provide relatively cheap microfinance to poorer borrowers in ways that mirrored informal institutions. The initiative was intended to improve the supply of rural credit through two channels: (a) stimulating local economic growth and employment; (b) targeting otherwise disadvantaged groups. The program addressed the village as the smallest administrative unit, typically comprising a few hundred households.

Each Village Fund had to be formally established with its own regulations, which nevertheless had to be approved by the National Village and Urban Community Fund Office. Part of the requirements were that the villagers form a committee of about ten persons to decide on lending policies (interest rates, maximum loan amounts, and the term of loans) and to approve borrowers. Households borrowed and repaid with interest, freeing the money to be re-lent. In this sense, village funds operated more like a formal institution. However, they had no staff or permanent office, so can be regarded as somewhere between a formal and an informal institution.

The capital provided to each Village Fund totaled one million baht. The Village Fund Committee did not handle this money directly; this was done by the BAAC. The Fund and each member had a bank book with the BAAC. Government funds were transferred to the Village Fund bank account and the committee transferred loan funds to approved borrowers.

The Ban Bua Teang Village Fund was established in 2001 with initial working capital of one million baht provided by the Government and 192 members. At the time of fieldwork, there were 220 members. The Village Fund Committee had nine members. The committee members received compensation for their work equal to 10% of total interest payments. Each member had to buy one share at 100 baht and pay a membership fee of 10 baht. There was a members' meeting once a year when the committee had to report on the Fund's performance.

DISCUSSION

The apparent success of the farmer groups studied was influenced by various factors relating to leadership, membership, networks, assistance providers, and government policies.

Good leaders were vital as they could encourage members to participate in group activities and express their opinions to the group. The leaders needed to be fair, diligent, honest, and willing to volunteer their time and energy. The leaders in the groups in Ban Donmoo and Ban Bua Teang had these qualities and contributed greatly to the success of the groups.

The active participation of members was also essential. Group success needed the opinions, participation, voluntary contributions, and unity of the members. This contributed to the necessary trust between members and leaders. The extent to which members in the groups in Ban Donmoo and Ban Bua Teang volunteered their time to support the groups was a key factor in group development.

The groups benefited from being part of wider networks. Horizontal and vertical linkages to other groups and organizations within and outside the village helped to strengthen the group's capacities. For example, the internal linkage between the Community Rice Mill Group and the Farmer School Group in Ban Donmo enhanced the profitability of organic rice and helped increase incomes. In Ban Bua Teang, the external linkage of the Village Fund with the BAAC was crucial.

The groups all had outside assistance from government, non-government, and private sector organizations. Suitable support was needed, particularly in the first stage of the group's activities, including training, seed monies, and guidance.

Government policies and institutions were instrumental in creating the opportunity for the groups to form and grow. The "sufficiency economy" principle was influential in group formation in Ban Donmoo and Ban Bua Teang, including the emphasis on organic farming, agro-tourism, and finance for the poor. This was translated into financial and training support through various government and non-government providers. The government-supported BAAC provided the backbone of village groups and microfinance institutes in the case-study villages and throughout Thailand.

CONCLUSION

The development of farmer organizations and community enterprises can help to enhance the socio-economic welfare of rural communities, beyond that achievable through independent commercial smallholder agriculture. Farmer organizations can facilitate the sharing of knowledge and skills, improve access to production technologies, encourage saving and investment, and increase farmers' market competitiveness. Community enterprises can create new employment opportunities in rural areas, especially for women and disadvantaged groups, contributing to livelihood diversification while reducing the need for out-migration. This chapter explored the development of several successful organizations and enterprises in Northeast Thailand. The key factors contributing to the success of these groups were identified as strong committed leadership, involved membership, connecting with wider networks, the role of government and non-government assistance providers, and, underlying all these, supportive government policies.

NOTES

1. Farmer Group Royal Decree BE 2547, 2004, Cooperative Promotion Department, Ministry of Agriculture and Cooperatives, Thailand, p. 6.
2. Assuming one member per farm holding, this represented 10.9% of the total number of holdings recorded in the 2013 Agricultural Census.
3. The Asoke Group is a Buddhist group that follows the "sufficiency economy" introduced by the late King Bhumibol. The Asoke Group was commissioned to conduct training for indebted farmers during the Thaksin Government. Farmers came to stay at an Asoke center for five days and learned about organic farming and recycling, and were obliged to listen to sermons on the virtues of vegetarianism and a merit-based economy (*bunniyom*) (Heikkilä-Horn 2010).

REFERENCES

AsiaDHRRA and Agriterra, 2002. *Profiles of People's Organizations in Rural Asia.* Available at http://www.asiadhrra.org/downloads/april_2005/poprofile.pdf (accessed 22 February 2013).
Chumsri, Pote, 2010. Poverty alleviation in Thailand: successful stories of competent small farmers and farmer organizations in doing farming. Asia Continental Meeting, International Year of Family Farming (IYFF), New Delhi, 23–25 March 2010.

Community Enterprise Promotion Board, 2010. *Summary of Number of Registered Small and Medium Community Enterprises and Networks at 30 December 2010.* Ministry of Agriculture and Cooperatives. Available at http://sceb.doae. go.th/Documents/STC/data301253.pdf (in Thai) (accessed 4 February 2013).

Cooperative Promotion Department, 2013a. *Agricultural Group Files.* Available at http://web2.cpd.go.th/profile_agri/ (accessed 20 October 2013).

Cooperative Promotion Department, 2013b. *Statistics of Farmer Groups and Cooperatives.* Cooperative Promotion Department, Ministry of Agriculture and Cooperatives. Available at http://www.cpd.go.th/cpd/cpdinter/Information_coop55.html (accessed 20 February 2013).

Heikkilä-Horn, Marja-Leena, 2010. Santi Asoke Buddhism and the occupation of Bangkok International Airport. *Austrian Journal of South-East Asian Studies* 3(1): 31–47.

Ishihara, Hiroe, and Pascual, Unai, 2013. *Re-politicizing Social Capital: Revisiting Social and Collective Action in Common Pool Resource Management.* UNU-IAS Working Paper No. 170. Yokohama: United Nations University Institute for Advanced Study.

Kassam, L., Subasinghe, R., and Phillips, M., 2011. *Aquaculture Farmer Organizations and Cluster Management: Concepts and Experiences.* Rome: Food and Agriculture Organization of the United Nations.

Kittisataporn, Karun, 2006. *The Well-balanced Development of Regional Economy through the OVOP Movement.* Paper presented at One Village One Product International Seminar, 20 October 2006, Millennium Hall, Ritsumeikan Asia Pacific University.

Seansing, Kittipotch, 2009. *A Study of Agro-ecotourism Practice with Community Involvement: A Case of Bua-Theung Village at Tha Chang, Sawangwirawong District, Ubon Ratchathani.* Masters Thesis, Ubon Ratchathani University.

Stockbridge, M., Dorward, A., and Kydd, J., 2003. *Farmer Organizations for Market Access.* Briefing paper presented at Stakeholders Meeting on Farmer Organizations in Malawi, 18–19 June 2003, Kalikuti Hotel, Lilongwe, Malawi.

Teerakul, Nuttamon, 2011. *An Examination of Community-based Enterprise and Poverty Reduction in Rural Northern Thailand.* PhD Thesis, University of New England, Armidale.

Thai Chaipattana Foundation, 2013. *Philosophy of Sufficiency Economy.* Available at http://www.chaipat.or.th/chaipat_english/index.php?option=com_content& view=article&id=4103&Itemid=293 (accessed 20 October 2013).

Wiboonpongse, Aree, et al., 2006. *Enterprise in Rural Communities: Research Approach, Development Policy Strategies, and Case Studies.* Research Report. Chiang Mai: Multiple Cropping Centre and Faculty of Economics, Chiang Mai University (in Thai).

PART III

A Sticky Situation

CHAPTER 5

From Subsistence to Commercial Rice Production in Laos

Vongpaphane Manivong and Rob Cramb

INTRODUCTION

The rice sector in Laos is in a "sticky situation" in several senses. First, Laos is considered the centre of origin of glutinous or sticky rice (Muto et al. 2016) and this type of rice still accounts for around 90% of production (Schiller et al. 2006; Mullen et al. 2019). Second, of the four countries considered in this volume, Laos has suffered the most from variability in rice production due to the high incidence of droughts and floods (Schiller et al. 2001, 2006). Third, this variability has made it difficult to achieve a reliable rice surplus at the national level, hindering investment in the processing capacity needed to develop a viable export industry. Fourth, the continued strong preference for growing glutinous varieties for domestic consumption has limited export growth to niche markets within the region where glutinous rice is consumed. Finally, the pursuit of market

V. Manivong (✉)
Ministry of Agriculture and Forestry, Vientiane, Laos

R. Cramb
School of Agriculture and Food Sciences, University of Queensland, St Lucia, QLD, Australia
e-mail: r.cramb@uq.edu.au

© The Author(s) 2020   103
R. Cramb (ed.), *White Gold: The Commercialisation of Rice Farming in the Lower Mekong Basin*,
https://doi.org/10.1007/978-981-15-0998-8_5

liberalisation within a socialist political regime has created a certain "stickiness" in policymaking institutions—a feature shared with Vietnam (Nguyen et al. 2017). All of this means that rice farming in Laos is the least commercialised within the Lower Mekong.

Nevertheless, as in the region as a whole, there has been a remarkable transformation of rice-based farming systems and supply chains over recent decades, including both the intensification of rice production in favourable lowland areas and the diversification of rural livelihoods to combine rice with non-rice and non-farm activities (Manivong et al. 2014; Cramb and Newby 2015). These changes reflect the broad process of agricultural commercialisation as outlined in Chap. 1. In this chapter we outline the context and trends for rice farming in Laos as a whole. Subsequent chapters present case studies of commercialisation in Savannakhet and Champasak Provinces, focusing on a comparison of rainfed and irrigated systems (Chap. 6), the supply of the key inputs of seed and fertiliser (Chap. 7), the domestic and cross-border marketing of surplus paddy and rice (Chap. 8), and the economic constraints to further intensification in the main wet-season rice crop (Chap. 9).

THE CONTEXT OF RICE FARMING IN LAOS

Laos occupies an area of 236,800 km^2, stretching 1700 km from north to south and between 140 and 500 km from east to west (MINC 2000; Fig. 5.1). Officially, three administrative regions are recognised: the Northern, Central, and Southern Regions. The Northern Region comprises seven provinces, the Central Region seven provinces (including the Vientiane Capital), and the Southern Region four provinces. The country shares a border of 416 km with China in the north, 236 km with Myanmar in the northwest, 1370 km with Thailand in the west, 492 km with Cambodia in the south, and 1957 km with Vietnam in the east (MAF 2010). Eighty-five per cent of the country's surface area lies within the Lower Mekong Basin, with only the rivers of Xam Neua Province in the north-east flowing east into Vietnam. The Mekong forms the western border of Laos for most of its length, except for Xaignbouli Province in the north-west, which lies to the west of the river and Champasak Province in the south, which the river bisects before entering Cambodia.

The majority of land in Laos is classified as mountainous, covering approximately 80% of the total land area; over two-thirds of the land has slopes of greater than 30% (MAF 2010). The landscape can be divided

Fig. 5.1 Laos with provinces and provincial capitals, 2012. (Source: CartoGIS Services, College of Asia and the Pacific, The Australian National University)

into the mountainous north, the eastern mountain chain, and the plains (MINC 2000). The mountainous north is dominated by rugged mountains with an average elevation of 1500 m above sea level. The eastern mountain chain (the Annamite Range) stretches along the border with Vietnam. Three large plateaus are located in this region, namely, the

Phuan Plateau in Xiengkhuang Province, the Nakai Plateau in Khammuan Province, and the Bolaven Plateau in the southern provinces. The plains include fourteen minor plains, twelve of which are in located in inter-montane basins in the Northern Region, and seven major plains, all located along the Mekong Valley, from the Vientiane Plain in the Central Region to the Champasak Plain in the Southern Region. The majority of lowland rainfed and irrigated rice-growing areas in the country are located in these major plains, the three most important of which are the Vientiane, Savannakhet, and Champasak Plains.

Laos has a tropical savannah climate dominated by the monsoons, with about 90% of the annual rainfall falling in the wet season from May to October while some months during the dry season between November and April may have no rainfall (see Fig. 1.2 in Chap. 1). The mean annual precipitation is 1600 mm, but this varies significantly among regions, ranging from 1000 mm in much of the Northern Region to over 3500 mm in the Bolaven Plateau in the Southern Region. It is estimated that about 270,000 million m^3 of the annual rainfall in Laos runs off into the Mekong River and contributes around 35% of the river's total annual flow (ICEM 2003). Although the climate in Laos is mostly tropical, it phases into sub-tropical in the mountainous areas in the north and along the mountain chain bordering Vietnam in the east. The temperature averages 25 °C throughout the country and the day and night temperatures differ by 10 °C. The daily temperature increases to as high as 37 °C in Champasak Province in the wet season, but drops to as low as 8 °C in Huaphan Province in the dry season (NSC 2005).

The population of Laos was estimated to be 7.1 million in mid-2019, with an annual growth rate of 1.5%. With a total area of 236,800 km^2, Laos has the lowest population density in Asia—around 31 persons per km^2—though this varies widely from 10 persons per km^2 in mountainous Phongsaly Province in the extreme north to over 200 persons per km^2 in Vientiane Capital (MPI and UNDP 2009). About half the population has settled in the large plains along the Mekong Valley (NSC 2004). The population of Laos is ethnically diverse. The 2005 Population and Housing Census reported that nearly 55% of the total population is of the Lao eth-nic group, 11% is of the Khmu ethnic group, and 8% is of the Hmong ethnic group. Most of the lowland rice farmers belong to the Lao and other Tai groups such as the Phouthai.

After decades of civil war, the Lao People's Democratic Republic (PDR), a single-party socialist republic, was declared in 1975. Two year later, the Lao Government developed the first development plan for 1978–1980 with the main focus on the development of agriculture as the fundamental base for economic development of the country. Agriculture was promoted in the form of collective production or cooperatives by increasing farm areas and supporting the use of farm machinery and irrigation facilities in order to raise production and achieve self-sufficiency in rice. The number of cooperatives rose rapidly to total 3976 nationwide by 1986 (Evans 1988). However, as reported by several authors (Evans 1995; Stuart-Fox 1997), by the early 1990s most of the listed cooperatives existed in name only and in reality very few cooperatives were actually working. The unsuccessful implementation of collective production was due to top-down management, low efficiency, shortage of inputs, lack of trained staff, and farmers' reluctance to follow the strict working conditions imposed (Stuart-Fox 1996).

In 1986 the New Economic Mechanism (NEM) was introduced to transform the country from a centrally planned economy to a market-oriented economy. The principles of the NEM were to free prices based on market demand and supply and encourage private investment from both domestic and foreign investors.[1] The government also improved infrastructure, in particular transport and communication facilities, to support the transformation to a market economy and integration with regional and international markets (UNDP 2002). Since the adoption of the NEM there has been considerable social and economic development. GDP growth averaged 6% during the 1990s, 7% during the 2000s, and 8% during the 2010s (World Bank 2019). Thus GDP per capita increased from USD 324 in 2000 to USD 2457 in 2017 and the incidence of poverty has been reduced from 39% in 1997 to 23% in 2012 (World Bank 2019).

RICE-BASED FARMING SYSTEMS

Farming systems in Laos can be broadly classified based on their occurrence in lowland, upland, and plateau environments (Table 5.1). In the lowlands, rainfed and irrigated farming systems are practised. In the sloping uplands, people have relied heavily on shifting cultivation. In the plateau environment, cash crops and fruit trees are extensively grown, replacing shifting cultivation. Noticeably, apart from the cultivation of the staple food (rice), a variety of home-garden vegetables and different types of livestock appear in almost all farming categories to serve daily house-

Table 5.1   Main farming systems in Laos

Environment	Farming system	Characteristics
Lowlands	Lowland rainfed farming system	Single cropping of traditional glutinous rice varieties. Buffaloes and cattle for draught, cash income, and occasional meat, free ranging during the dry season, confined in the wet season. Pigs, poultry, fish, and non-timber forest products (NTFPs) important for food and cash income.
	Lowland irrigated farming system	Double cropping of traditional photoperiod-sensitive rice varieties, with higher use of improved varieties, fertiliser, and other inputs for the second crop which is mainly for cash. Dry-season vegetables grown near urban centres. Relatively few livestock due to shortage of grazing land, buffaloes used for ploughing, small stock for meat and cash income.
Uplands	Upland rainfed farming system	Shifting cultivation of rice intercropped with a variety of cash crops on sloping land. Fruit tree species also grown in lower altitudes. Pigs, cattle, and poultry are the principal livestock. High dependence on NTFPs for income to purchase rice, etc. Adoption of paddy cultivation is progressing where possible in small inland valleys.
	Highland farming system	Similar to upland rainfed farming system, but with high-altitude crops such as maize and (formerly) opium, sometimes intercropped with lettuce and mustard, and temperate fruit trees such as plum, peach, and local apple.
Plateaus	Plateau farming system	Coffee, tea, and cardamom have largely replaced shifting cultivation, supplemented by fruit trees and vegetables in home gardens. Cattle important as savings and enterprise, pigs and poultry also kept.

Source: Adapted from UNDP (2002: 76)

hold consumption needs and play a key role in household saving and income generation.

Rice production is the main farming activity in Laos, accounting for over 80% of the total cultivated area (Bestari et al. 2006). Rice is grown in three main farming systems, namely, the rainfed lowland, irrigated lowland, and rainfed upland systems (Table 5.2). Rice cultivation in the rainfed lowlands normally commences at the beginning of the wet season in May or June, depending on the arrival of the rains, with land preparation involving two passes of ploughing and one harrowing (Table 5.3). Rice seed is sown in a nursery and one month later the young seedlings are transplanted to the main field. The harvesting period is in October or

Table 5.2 Three major rice-based farming systems

Farming system	Characteristics
Rainfed lowland	Rice is grown in wet season in bunded fields flooded for at least part of the season; water from rainfall
Irrigated lowland	Rice is grown in wet and dry seasons in bunded fields flooded for at least part of the season; water from irrigation and rainfall
Rainfed upland	Rice is grown in wet season in unbunded fields on sloping land under shifting cultivation system; water from rainfall

Source: Adapted from Linquist et al. (2006: 29)

Table 5.3 Seasonal rice cropping calendar for different farming systems

Farming system	Month											
	Jan	Feb	Mar	Apr	May	Jun	Jul	Aug	Sep	Oct	Nov	Dec
	Dry season				Wet season						Dry season	
Rainfed lowland					LP	Sow	TP				Harvest	
Irrigated lowland	TP			Harvest								LP, Sow
Rainfed upland	Slash		Burn, fencing		Plant		Weeding		Harvest			

Source: Adapted from Linquist et al. (2006: 32)
Note: LP, land preparation; TP, transplant

November, depending on the maturity of the varieties planted. In areas with access to irrigation, rice fields are also supplemented with irrigation water during a drought period in the wet season. In these irrigated area, farmers may grow rice in the dry season as well. After harvesting the wet-season crop, rice fields are irrigated and land preparation begins. The nursery is sown in December and the seedlings are transplanted by early January. Harvesting is completed in April or May. Traditionally, the cultivation of rainfed upland rice starts with the slash-and-burn method of land preparation between January and April. Planting is done in May, weeding between June and August, and harvesting in September or October.

In the lowland rice environment in the past, land was prepared with the use of buffaloes; however, there is now an increasing trend of using hand-held tractors for land preparation. Many farmers have their own hand-held tractors or threshing machines, while those who do not have their own can access the services provided by others for a fee. Many farmers have sold their livestock, especially cattle and buffaloes, to buy their own tractors. There has also been a recent rise in direct seeding, whether broadcasting or using drill or drum seeders (Mullen et al. 2019). Threshing is also now done with threshing machines, though manual threshing either by hand or by small machines continues to be practised, especially in remote areas. Small combine harvesters are also now starting to appear in the lowland plains. Mechanisation has thus brought some significant changes into the farming systems in the lowlands but is very limited in the upland rice production system; for example, the threshing of upland rice is still done entirely manually.

TRENDS IN RICE PRODUCTION AND MARKETING

There is evidence that the Austroasiatic farmers who occupied the Khorat Plateau on both sides of the Mekong from around 4000 BCE were already cultivating lowland rainfed rice using domesticated buffalo and iron-tipped ploughs by around 500 BCE (Schiller et al. 2006; Higham 2014). The presence of canals and reservoirs in the southernmost province of Champasak implies that lowland irrigated rice was practised during the period of Khmer dominance from the fifth to eleventh centuries CE. As Tai peoples moved down the Mekong in the first millennium CE they brought with them both rainfed and irrigated wet rice techniques and progressively occupied the minor and major plains referred to above. The Tai settlements were organised into local polities (*muang*) that exercised control over the surrounding paddy lands and forest resources, progressively pushing the pre-existing populations into the uplands.[2] From about the eleventh century, some more powerful *muang* emerged, functioning as small states that controlled land and labour over a larger area (Stuart-Fox 2006). The four oldest and strongest such *muang* were centred in what are now the provinces of Luang Prabang, Xieng Khouang, Vientiane, and Champasak. All of these depended on domination over farming populations in rice-growing areas capable of producing substantial surpluses. In the fourteenth century the state of Lan Xang, initially based in Luang Prabang, asserted control over land and people throughout the Northern Highlands and the Khorat Plateau. The decision to move the capital to Vientiane in 1560 was partly based on the larger surplus-producing capacity of the more fertile and extensive Vientiane plain (Schiller et al. 2006).

Under French colonial rule (1893–1945) there was little effort to increase rice production (Schiller et al. 2006). Almost all rice was produced under rainfed conditions and subject to periodic droughts and (in the lowlands) floods. Production was mostly no more than 350,000 t annually; hence, Laos was a rice-importing country, with only the Champasak area consistently producing a surplus. In the post-war decades, efforts to increase rice production were dwarfed by the escalating conflict in Indochina. However, in the early 1970s some IRRI (International Rice Research Institute) varieties were introduced, trials were conducted, and seed multiplication was initiated at the Salakham Rice Research Station near Vientiane (Schiller et al. 2006). In the decade after 1975, when the main thrust of agricultural policy was the collectivisation program, Vietnamese advisers introduced and evaluated many improved varieties but most were non-glutinous, had poor eating quality, and were not widely adopted (except for CR203 which was useful for noodle and beer production). In 1990 about 95% of the lowland wet-season crop was still based on traditional low-yielding varieties (Inthapanya et al. 2006). Only a small number of higher-yielding glutinous varieties introduced from Thailand were being planted in the lowlands of the Central and Southern Regions (including the aromatic RD6 that helped revolutionise rice farming in Northeast Thailand, as noted in Chap. 2).

In 1991 a long-term collaborative rice research program with IRRI was initiated, with a major focus on varietal improvement (Inthapanya et al. 2006). The priority was to develop high-yielding glutinous varieties for the rainfed and irrigated lowlands. Breeding also emphasised resistance to specific pests and diseases and selecting varieties suited to the drought-prone environments of Central and Southern Laos. A total of seventeen improved varieties were released from 1993 to 2005. All were glutinous and all but two were photoperiod-insensitive, hence potentially suitable for dry-season as well as wet-season production. While some of the varieties had to be withdrawn, there was a high level of farmer acceptance and adoption due to the new varieties' higher yield potential and responsiveness to fertiliser (Fig. 5.2). Further breeding has focused on developing more resilient varieties for specific environments, including micro-environments within the paddy fields which are more or less susceptible to drought during the wet season (Mullen et al. 2019).

From the mid-1990s there has been a steady growth in the Lao rice sector in terms of area, production, and yield (Fig. 5.3). The increase in rice production made the country notionally self-sufficient in rice in 1999,

Fig. 5.2 Lao farmer showing field trial on his paddy field. (Source: Rob Cramb)

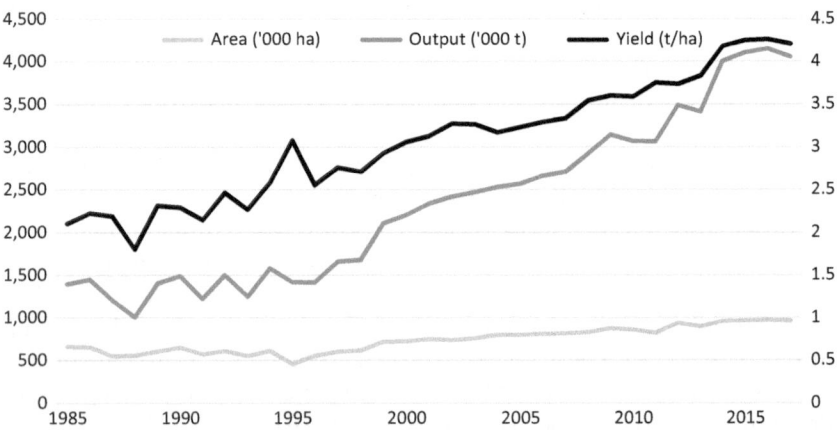

Fig. 5.3 Paddy area, output, and yield in Laos, 1985–2017. (Source: Agricultural Statistics Yearbooks (various years), Department of Planning and Finance, Ministry of Agriculture and Forestry (MAF), Vientiane)

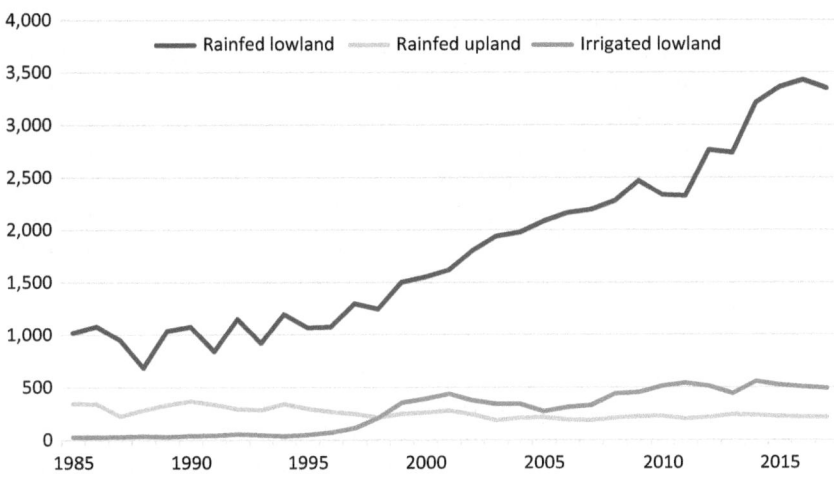

Fig. 5.4 Paddy output in Laos by production system, 1985–2017 ('000 t). (Source: Agricultural Statistics Yearbooks (various years), Department of Planning and Finance, Ministry of Agriculture and Forestry (MAF), Vientiane)

when total paddy production reached 2.1 million tonnes, compared to only 1.4 million tonnes in both 1985 and 1995. Since then the production has nearly doubled to around 4.1 million tonnes in 2017. The growth in output has been partly due to a doubling in cultivated area from 460,000 ha in 1995 to 964,000 ha in 2017 and partly to a longer-term increase in yields from 2.1 t/ha in 1985 to 3.1 t/ha in 1995 and 4.2 t/ha in 2017. Though data from field trials and farmer surveys suggest these official figures are somewhat inflated, the overall trend has been confirmed and is attributable to the widespread use of improved rice varieties and management practices, especially the use of fertilisers (Schiller 2008). By the early 2000s, improved rice varieties covered 70–80% of lowland rice-growing areas (Inthapanya et al. 2006).

Most of this growth in production has come from the rainfed lowland environment in the seven large plains,[3] though there has also been an expansion in irrigated dry-season production (Fig. 5.4). In the 1990s, investment in pump-irrigation schemes in the major plains of Central and Southern Laos increased the dry-season irrigation capacity from only 12,000 ha in 1990 to 102,000 ha in 2001, leading to a corresponding expansion in dry-season cultivated area to 13–14% of the total (Schiller et al. 2006). However, there has been little further investment in new schemes and the utilisation of existing schemes has declined, mainly due to the poor maintenance of irrigation infrastructure, such that the dry-season cultivated area fell to 97,655 ha in 2017, representing only 10% of the total. Nevertheless, dry-season yields are higher than wet-season yields, partly due to the exclusive use of improved varieties, and dry-season production has continued to hover around 500,000 t in the 2010s, contributing 12% of total paddy output in 2017 (Fig. 5.4). The area and output of upland rice have slowly declined due to a combination of restrictive government policies and diversification into cash crops (Ducourtieux et al. 2005).

Around 2000 it was estimated that only 5% of total rice production was traded (Bestari et al. 2006). Improved infrastructure, increased urbanisation, and increased regional specialisation in agricultural production (e.g., the growth of rice production in the lowlands of the Central Region and of banana and rubber production for export in the Northern Region) have led to an increase in the share of rice production entering the domestic market (Chap. 8). The marketing system involves various participants, including farmers, assemblers, millers, traders, exporters, retailers, processors, institutional buyers, and consumers. These include private-sector and state-owned enterprises, but the State Food Enterprise, with over 70% of the market, dominates and controls the rice trade (Setboonsarng et al. 2008). Milled rice flows from the provinces with high levels of rice production, such as Champasak, Savannakhet, and Vientiane, to urban centres, in particular Vientiane, and to provinces with low levels of rice production, such as Oudomxay, Luangprabang, and Huaphanh. In addition, provinces sharing borders with Thailand and Vietnam sometimes import rice to fulfil local demand, especially during periods of rice shortage in those provinces (Sengxua et al. 2009). Rice is normally sold in bulk in retail shops in fresh markets or along the streets, but is also available in mini-marts in limited quantity.

Since 2000, rice has also been exported, increasing to a value of USD 37 million or about 130,000 t in 2017, making it the sixth most valuable agricultural export but representing only 5% of total rice output (Fig. 5.5). An unrecorded quantity of paddy has also been exported, estimated to be about 248,000 t in 2016. Exports have mainly been across the borders to Vietnam, Thailand, and China. Rice exported to Thailand is mostly glutinous rice while to Vietnam and China it is largely non-glutinous rice (Bestari et al. 2006; Sengxua et al. 2009). Around 90% of the rice grown in Laos is glutinous and this limits the export opportunities to international markets, where glutinous rice accounts for less than 2% of the total traded (GDS 2005). In addition, poor milling and marketing infrastructure constrains the export competitiveness of the rice sector in Laos (Sengxua et al. 2009; Welcher and Prasertsri 2019). Hence the Government is focusing on developing niche markets for rice in which Laos has a comparative advantage, such as organic rice, black rice, or geographic indicator (Lao) rice. A small quantity of organic rice from Laos has been exported to Japan under a contract farming scheme (Setboonsarng et al. 2008). In 2015 the Xuanye (Lao) Company was approved as the sole exporter of rice to China with a quota of 8000 t, increasing to 20,000 t in 2017, including both glutinous and non-glutinous organic rice, drawing on the output of seven mills. However, the company has not been able to fill the quota, supplying only 4000–5000 t per year, largely due to issues with quality.[4]

The current government priorities for rice farming are to ensure food security and improve rural livelihoods by increasing rice productivity to achieve rice self-sufficiency and export the surplus, as well as promoting crop diversification to reduce risks and raise income (MAF 2014). As stated in the Seventh National Socio-Economic Development Plan (2011–2015), the target is to increase rice production to 4 million t with an average yield of 3.9 t/ha. The plan further sets the target to expand the irrigated area to 500,000 hectares by 2015 to increase dry-season production.

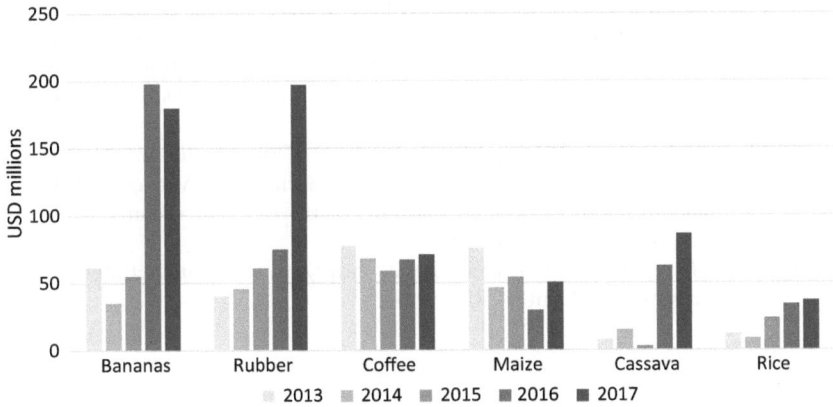

Fig. 5.5 Value of agricultural exports from Laos, 2013–2017. (Source: International Trade Centre)

Conclusion

The farming systems in Laos have been undergoing a transition from subsistence-based to market-oriented production. Rice-based farming systems are both diverse and dynamic, with households continually adapting to constraints and opportunities arising from the rapid development occurring within Laos and the wider region. Rice production is dominated by the rainfed lowland system and is still predominantly for subsistence, with only a small proportion marketed and even less exported. However, the cultivated area and especially the yield of both rainfed and irrigated rice have been increasing, contributing to the achievement of rice self-sufficiency at the national level. Moreover, rural livelihoods have become increasingly diversified as the economy of the region develops and opportunities for off-farm and non-farm employment increase.

Notes

1. The tax on rice output had already been replaced with a land tax in 1979 and official prices of rice and other crops were substantially increased in 1980. These improved incentives led to an increase in rice production of around 17% in the early 1980s (Schiller et al. 2006).
2. Fieldwork by Silinthone Sacklokham in Savannakhet Province in Laos uncovered a locally written manuscript (*Implantation des Phouthai dans La Ville de la Nam Se Pone*) recounting the history of Phouthai settlement in the upriver village of Xepon. The manuscript describes in detail how the Phouthai (a Tai group) had migrated down the Mekong to escape oppression from the Chinese in Yunnan and, having arrived in the Xepon stream, a tributary of the Banghiang River, had forced the pre-existing Mon-Khmer people (referred to as Khas, meaning "subservient peoples") into the surrounding hills.
3. Eliste and Santos (2012) observe that 70% of the production increase in 1995–2010 came from five provinces—Savannakhet, Vientiane Capital, Vientiane, Khammouane, and Saravan—with over half coming from the Vientiane and Savannakhet Plains.
4. Vientiane Times, 22 March 2017, http://www.vientianetimes.org.la/FreeContent/FreeConten_Rice.htm.

References

Bestari, N. G., Shrestha, S., and Mongcopa, C. J., 2006. *Lao PDR: An Evaluation Synthesis on Rice. A Case Study from the 2005 Sector Assistance Program Evaluation for the Agriculture and Natural Resources Sector in the Lao People's*

Democratic Republic. Vientiane: Operations Evaluation Department, Asian Development Bank.

Cramb, R. A., and Newby, J. C., 2015. Trajectories of rice-farming households in Mainland Southeast Asia. In R. A. Cramb, ed., *Trajectories of Rice-Based Farming Systems in Mainland Southeast Asia*, pp. 35–72. Canberra: Australian Centre for International Agricultural Research.

Ducourtieux, O., Laffort, J. R., and Sacklokham, S., 2005. Land policy and farming practices in Laos. *Development and Change* 36: 499–526.

Eliste, P., and Santos, N., 2012. *Lao People's Democratic Republic Rice Policy Study 2012.* Rome: Food and Agriculture Organization (FAO).

Evans, G., 1988. *Agrarian Change In Communist Laos.* Singapore: Institute of Southeast Asian Studies.

Evans, G., 1995. *Lao Peasants Under Socialism And Post-Socialism.* Chiangmai: Silkworm Books.

GDS, 2005. *Integrated Value Chain Analysis of Selected Strategic Sectors in Lao People's Democratic Republic.* Virginia: Global Development Solutions.

Higham, C., 2014. *Early Mainland Southeast Asia: From First Humans to Angkor.* Bangkok: River Books.

ICEM, 2003. *Lao PDR National Report on Protected Areas and Development. Review of Protected Areas and Development in the Lower Mekong River Region.* Brisbane: International Centre For Environment Management (ICEM).

Inthapanya, P., Boualaphanh, C., Hatsadong, and Schiller, J. M., 2006. The history of lowland rice variety improvement in Laos. In J. M. Schiller, M. B. Chanphengxay, B. A. Linquist, and S. Appa Rao, eds., *Rice in Laos.* Los Banos, Philippines: International Rice Research Institute.

Linquist, B. A., Keoboualapha, B., Sipaseuth, and Inthapanya, P., 2006. Rice production systems of Laos. In J. M. Schiller, M. B. Chanphengxay, B. A. Linquist, and S. Appa Rao, eds., *Rice in Laos.* Los Banos, Philippines: International Rice Research Institute.

MAF, 2010. *Fourth National Report to the Convention on Biological Diversity.* Vientiane: Ministry of Agriculture and Forestry (MAF).

MAF, 2014. *Strategy for Agricultural Development to 2020.* Vientiane: Ministry of Agriculture and Forestry.

Manivong, Vongpaphane, Cramb, Rob, and Newby, Jonathan 2014. Rice and remittances: crop intensification versus labour migration in Southern Laos. *Human Ecology* 42(3): 367–379.

MINC, 2000. *25 Years of Lao PDR.* Vientiane: Ministry of Information and Culture.

MPI and UNDP, 2009. *Employment and Livelihoods Lao PDR, 2009.* Fourth National Human Development Report. Vientiane: Ministry of Planning and Investment (MPI) and United Nations Development Programme (UNDP).

Mullen, J., Malcolm, B., and Farquarson, B., 2019. *Impact Assessment of ACIAR-supported Research in Lowland Rice Systems in Lao PDR.* ACIAR Impact

Assessment Series No. 97. Canberra: Australian Centre for International Agricultural Research.

Muto, Chiaki, Ishikawa, Ryuji, Olsen, Kenneth M., Kawano, Kazuaki, Bounphanousay, Chay, Matoh, Toru, and Sato, Yo-Ichiro, 2016. Genetic diversity of the *wx* flanking region in rice landraces in northern Laos. *Breeding Science* 66(4): 580–590.

Nguyen, H. T. M., Do, H., Kay, A., Kompas, T., Nguyen, C. N., and Tran, C. T., 2017. *The Political Economy of Policy Exceptionalism during Economic Transition: The Case of Rice Policy in Vietnam.* Crawford School Working Paper No. 1713. Canberra: Crawford School of Public Policy, Australian National University.

NSC, 2004. *The Households of Lao PDR: Social and Economic Indicators. Lao Expenditure and Consumption Survey 2002/03 (LECS 3).* Vientiane: National Statistics Centre.

NSC, 2005. *Statistics of Lao PDR 1975–2005.* Vientiane: National Statistics Centre.

Schiller, J. M. 2008. *Rice Production: Status and Needs Relating to Future Food Security of Lao PDR and Potential Development Options for Southern Lao PDR.* Vientiane: Asian Development Bank.

Schiller, J. M., Linquist, B., Douangsila, K., Inthapanya, P., Douang Boupha, B., Inthavong, S., and Sengxua, P. 2001. Constraints to rice production in Laos. In S. Fukai, and J. Basnayake, eds., *Increased Lowland Rice Production In the Mekong Region.* ACIAR Proceedings No. 101. Canberra: ACIAR

Schiller, J. M., Hatsadong, and Doungsila, K., 2006. A history of rice in Laos. In J. M. Schiller, M. B. Chanphengxay, B. A. Linquist, and S. Appa Rao, eds., *Rice in Laos.* Los Banos, Philippines: International Rice Research Institute.

Sengxua, P., Manivong, V., Xayachack, S., and Sayavong, S. 2009. *Rapid Assessment of Stocks and Marketing System of Paddy/Rice in Lao PDR.* Vientiane: National Agriculture and Forestry Research Institute and Food and Agriculture Organization.

Setboonsarng, S., Leung, P. S., and Stefan, A. 2008. *Rice Contract Farming in Lao PDR: Moving from Subsistence to Commercial Agriculture.* ADBI Discussion Paper 90. Tokyo: Asian Development Bank Institute.

Stuart-Fox, M. 1996. *Buddhist Kingdom, Marxist State: The Making Of Modern Laos.* Bangkok: White Lotus.

Stuart-Fox, M. 1997. *A History Of Laos.* Cambridge: Cambridge University Press.

Stuart-Fox, M., 2006. Population diversity and rice in Laos. In J. M. Schiller, M. B. Chanpengxay, B. A. Linquist, and S. Appa Rao, eds., *Rice in Laos*, pp. 1–8. Los Banos, Philippines: International Rice Research Institute.

UNDP, 2002. *National Human Development Report Lao PDR 2001: Advancing Rural Development.* Vientiane: United Nations Development Programme (UNDP).

Welcher, P., and Prasertsri, P., 2019. *Laos Rice Report MY2018–2019.* GAIN Report No. TH9016. Bangkok: USDA Foreign Agricultural Service.

World Bank, 2019. *The World Bank in Lao PDR.* Available at https://www.world-bank.org/en/country/lao (accessed 15 May 2019).

CHAPTER 6

Adapting the Green Revolution for Laos

Liana Williams and Rob Cramb

INTRODUCTION

Initial efforts to introduce new agricultural practices during Asia's 'Green Revolution' were derailed in Laos for a number of reasons: the impact of the Vietnam War; unrest associated with the seizure of power by the socialist government and proclamation of independence in 1975; an unsuccessful push to collectivise agricultural production; and limited investment in agricultural research or material support to collectives (Evans 1988). Faced with ongoing food shortages across the country, the government embraced agricultural modernisation as a central policy but lacked the resources to properly implement it.

From 1990 to 2007, the International Rice Research Institute (IRRI) and the Government of Laos built the nation's capacity in rice research and developed improved varieties suitable to Lao farming conditions.

L. Williams (✉)
Commonwealth Scientific and Industrial Research Organisation (CSIRO),
Brisbane, QLD, Australia
e-mail: liana.williams@csiro.au

R. Cramb
School of Agriculture and Food Sciences, University of Queensland,
St Lucia, QLD, Australia
e-mail: r.cramb@uq.edu.au

© The Author(s) 2020   121
R. Cramb (ed.), *White Gold: The Commercialisation of Rice Farming in the Lower Mekong Basin*,
https://doi.org/10.1007/978-981-15-0998-8_6

According to Bestari et al. (2006), the introduction of modern varieties and other inputs has been one of the key factors supporting an increase in overall rice production in Laos. IRRI credits the program with bringing the Green Revolution to Laos, supporting increases in rice production to levels of national self-sufficiency, and building national research capacity (IRRI 2006).

This chapter traces the history and processes that have seen the development, use, and spread of improved rice varieties throughout Laos, particularly in the lowlands of the Central and Southern Regions. This history represents a departure from the Green Revolution narratives of other Southeast Asian countries, where the development and use of improved varieties was predicated on access to irrigation and fertiliser and favoured yield over other qualities like taste or aroma. Instead, efforts to improve rice production in Laos emphasised plant breeding based on local conditions and preferences—low input, rainfed production of sticky rice—and built the capacity of Lao institutions and researchers to continue rice breeding after formal project efforts ceased.

The chapter begins with a brief overview of key events that shaped rural development in Laos and set the scene for the partnership between IRRI and the Lao Government. It then provides a detailed account of the practical implementation of this partnership through the Lao-IRRI Rice Research and Training Program (Lao-IRRI Project), drawing on the accounts of former Lao-IRRI staff, district officials, and farmers in Outhomphone and Champhone Districts, Savannakhet Province.[1] The chapter concludes with a reflection on the characteristics of the Lao-IRRI Project that supported its success.

SETTING THE SCENE: LAOS 1975–1990

The Political Context for Rural and Economic Development

The Lao People's Democratic Republic was declared in 1975, after almost 30 years of civil war and unrest. The threads of the conflict are complex, tied to French occupation, a growing nationalist movement, and Western concerns over the spread of Communism that saw the north of Laos used effectively as an air base (and bombing ground) by the United States (Stuart-Fox 1996; Evans 1988). When the communist Lao People's Revolutionary Party came to power and abolished the monarchy, they declared three revolutions: economic production; scientific and technical;

and cultural and ideological (Evans 1988). The policies enacted in the name of these revolutions had marked impact on the processes of agricultural development in Laos.

The government introduced controls on trading between provinces, established state-run farms, and mandated the collectivisation of agriculture (Bourdet 1996). Collectivisation was seen as the best way to rapidly modernise agriculture and protect against food shortages, but also as a means to strengthen state control against civil unrest and revolt (Evans 1988). All ownership of land was transferred to the state (Ducourtieux et al. 2005).

Rules governing the cooperative system were complicated and inconsistently implemented by local officials (Stuart-Fox 1996). Weekly meetings were held to convince villagers that small plots were inefficient and collectivisation was the best way forward—many households feared the ramifications of not joining the 'voluntary' organisations (Evans 1988). People grew increasingly dissatisfied with the cooperative system, resenting the coercion of the district officials, the limited provision of equipment or support from government, and the uncertainty about the implications of joining, generally preferring their traditional lifestyle (Stuart-Fox 1996). Farmers began to burn crops or leave land fallow rather than be forced into collectives (Stuart-Fox 1996). Far from modernising and increasing production, collectivisation led to a drop in production and was suspended after less than a year, with officials citing a need to provide better training and improved conditions (Stuart-Fox 1996).

Collective production was not the only program that failed to bring intended benefits; hence, the Party endorsed the New Economic Mechanism (NEM) in 1986.[2] Under this policy, the economy has been progressively restructured, for example, re-establishing private property rights, easing restrictions on trade, and deregulating commodity prices to support economic growth (Bourdet 1996). The reforms introduced under the NEM were enthusiastically received by organisations like the World Bank, which described the ensuing economic progress of Laos as 'unparalleled' (Rigg 2005: 22).

While the NEM may have supported growth in industry and services, growth in agriculture remained stunted (Bourdet 1996). Rice production was not enough to keep pace with population growth and Laos was dependent on food imports (Evans 1988). Bourdet (1996) suggests slow growth in agriculture was due partly to the vulnerability of farming to drought and flood and partly to the largely subsistence nature of production, lack of modern inputs, and poor infrastructure, exacerbated by an urban-centric political elite.

Economic reform was implemented within an unchanged political structure; the Lao People's Revolutionary Party remains the only legitimate political party. Party administration extends to all levels and areas—villages, district, province, ministries, and mass organisations (Stuart-Fox 2011). Appointment to local positions, though theoretically by election, is often through lines of patronage and controlled by the Lao People's Revolutionary Party (Stuart-Fox 2011). Opposition is not tolerated, with imprisonment or 're-education' the response to public expressions of discontent (Bourdet 1996). The second Five-Year National Socio-Economic Development Plan (1986–1990) brought decentralisation of government, with the central government providing 'guidelines' while the provinces were charged with the administration and implementation of programs (Hopkins 1995). In practice, information (and misinformation) flows up to the central government and decisions flow down (Stuart-Fox 2011).

Early Efforts in Rice Research

The first commitment to rice research in Laos was in 1955 with the establishment of the Salakham Rice Research Station in Hatsaiphong District, near Vientiane. Despite the centrality of rice in Lao agriculture, formal research efforts prior to this had focused on fruit trees and coffee—which were of more interest to the French (Inthapanya et al. 2006). Research at the Salakham Station during the 1960s focused on evaluation of improved varieties brought from IRRI (e.g., IR8), Thailand (e.g., *Niaw Sanpatong*), and the Philippines (e.g., C4-63-1). Early releases of improved varieties from other countries and selected Lao traditional varieties were distributed through agricultural development programs funded largely by the United States (Inthapanya et al. 2006). The Lao rejected the new varieties, preferring the taste and quality of traditional sticky rice. The seed production capacity of the station was low—far below what would have been required for wide-scale use of the new varieties (Interview 6). In any case, these early experiments with improved varieties were disrupted with the establishment of the Lao PDR.

When Laos was declared an independent state in 1975, research capacity was low: a significant proportion of educated Lao had fled the country, while physical resources and infrastructure were damaged or depleted (Stuart-Fox 1996). Though it was common for officials to undertake graduate and diploma training in Soviet-bloc countries, as one interviewee pointed out, the Russians 'weren't very competitive at rice' and thus there were limited skills in rice agronomy (Interview 7).

Agricultural cooperatives were used as a basis for agricultural extension. For the most part, resources were limited and information rather than material resources was all that could be provided (Evans 1988). The process was top-down and focused on agricultural intensification (Interview 7). Recognising the preference of the majority of the population to consume glutinous rice, researchers worked to cross IRRI lines with Lao varieties to improve yields but retain eating quality (Inthapanya et al. 2006) but none was released (Schiller et al. 2006).

From 1979 to 1982, the UN Food and Agriculture Organization (FAO) sought to build on this early work supporting rice intensification research. Researchers worked with 'farmer seed growers' to produce certified seed to distribute to other farmers the following year (Hatsadong 2013). With close relationships between farmer seed growers and extension workers, farmers were 'partners in the process … [This system] helps the farmer to understand the idea of improved seed and production, distribution, and pricing for themselves' (Interviewee 6).

From 1983 to 1988, attempts were made to build a formal seed distribution system with the establishment of Phone Ngam, Thasano, and Naphok seed multiplication centres to connect the provinces with Salakham station (Hatsadong 2013; see Fig. 6.1). Reservations have been expressed about the suitability of this kind of system compared to more locally based farmer seed production groups (Hatsadong 2013). Further support was provided through the United Nations Development Programme and FAO to strengthen linkages across the regional research stations (Hatsadong 2013). There was still no national rice research program at this time and research was relatively limited in scope and geographic reach.

With modest research and distribution capacity, the use of improved varieties was limited. Estimates vary between 2 to 5 per cent (Bestari et al. 2006) and 5 to 10 per cent (Inthapanya et al. 2006) of overall seed use. Use of improved varieties had been limited to pockets along the Mekong where the combination of access to irrigation and improved varieties enabled dry-season rice production for sale (Lao-IRRI Project 1993; Inthapanya et al. 2006). Farmers around Vientiane brought varieties (including improved varieties) over from Thailand and incorporated the new seeds alongside traditional varieties, deciding which to plant based on water availability and suitability to field conditions (Tanaka 1993).

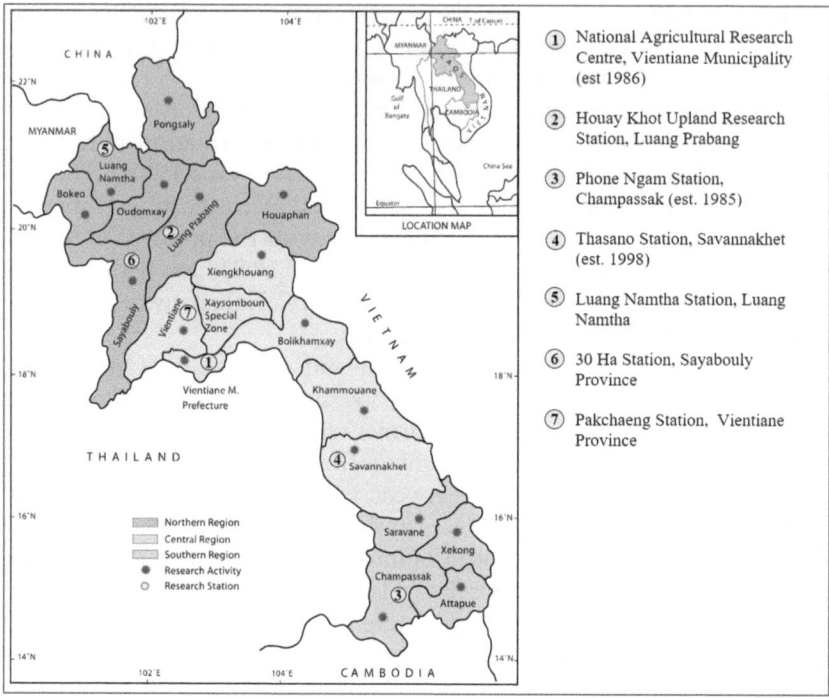

Fig. 6.1 Rice research centres and locations of Lao-IRRI research activities. (Source: Modified from Shrestha et al. 2006: 38)

Inthapanya et al. (2006) suggest three reasons for the limited uptake of improved varieties during this time. First was the absence of an effective mechanism to distribute seeds, restricting awareness of and access to improved varieties. Where improved varieties were used, this was most likely due to cross-border, farmer-to-farmer exchange rather than formal distribution programs (Tanaka 1993). Second, even though the improved varieties outperformed local varieties in terms of yield, households preferred to maintain traditional varieties where the main purpose of the crop was home consumption (i.e., eating quality was valued over yield). Third, Lao farming systems were subsistence-oriented and used no or minimal purchased inputs (Lao-IRRI Project 1993). In low-input conditions, traditional varieties were better adapted than the early modern varieties, which performed best with fertiliser and irrigation (Inthapanya et al. 2006).

Traditional practices of seed selection and multiplication contributed to a significant diversity of rice varieties in Laos, especially glutinous varieties (Bestari et al. 2006). Farmers would trial and observe 'new' varieties (received from family or friends) on small areas of their field. Decisions on which seed to retain were based on the specific agronomic conditions, yield stability, taste, and other social and cultural preferences (Appa Rao et al. 2006a). Each household would plant four or five different varieties each season to spread labour requirements and pest and disease risk by staggering the stage of crop maturation (Linquist et al. 2006). In addition to fostering new traits within their own seed stock, farmers looked for plants with desirable characteristics, often observing and swapping seed with their neighbours, or when visiting family in other districts, provinces, or countries (notably Thailand and Vietnam) (Interview 1, Appa Rao et al. 2006a).

In the late 1980s and early 1990s, rice production, and agriculture in general, remained low-input and followed traditional practices. In 1995, 83 per cent of the labour force was engaged in agriculture and fisheries, mainly for subsistence (Ministry of Agriculture and Forestry 2000). Household production relied on family or shared (exchange) labour, draught animal power, and limited or no use of chemical inputs (Lao-IRRI Project 1993). Rice has traditionally played a central role in Lao social and spiritual life. As the staple food, it is also intertwined with ideas of culture and family (Bestari et al. 2006). Key activities in the crop calendar were marked by ceremonies aimed at appeasing spirits and ensuring a good harvest (Simmalavong 2011).

Yields were low compared to similar countries and far below domestic food requirements (Worner 1996). Rice crops were regularly affected by drought and flood, and food security was often precarious at both household and national levels, especially in the North. By 1990, limited progress had been made towards the government's goal of self-sufficiency in rice (Hopkins 1995). In 1988 and 1989, severe drought cut rice production by one third and triggered emergency food aid to avert widespread food shortages and famine (Schiller et al. 2006).

THE LAO-IRRI RICE RESEARCH AND TRAINING PROGRAM, 1990–2007

Against the backdrop of ongoing food shortages, the Government of Laos and IRRI signed a Memorandum of Understanding (MoU) in 1987. The MoU articulated a commitment to developing research capabilities in

Laos and improving rice production to achieve national rice self-sufficiency. The goals listed in the MoU were implemented through the Lao-IRRI Project. The program agreement was finalised in 1989 and the program commenced in August 1990, with Swiss Development Cooperation (SDC) committing USD 16 million to the program over several phases (Shrestha et al. 2006).

The Lao-IRRI Project represented the first long-term, coordinated effort to support rice research in Laos. The main objectives reflected national policy goals to build the capacity of the Lao rice research system and increase rice production (Shrestha et al. 2006). As the project progressed and national self-sufficiency was achieved and maintained, government priorities shifted to emphasise diversification and modernisation of the agricultural sector more broadly and included consideration of sustainability and improving livelihoods. Research priorities of the Lao-IRRI Project also adjusted to reflect revised government priorities and as skills and knowledge in rice production grew (Lao-IRRI Project 2005).

The program was structured around several broad areas: improving and building research infrastructure; providing training for Lao researchers; development of a national rice research program covering varietal improvement, crop establishment, and soil and pest management; and developing a national seed collection to record and preserve traditional rice varieties (Shrestha et al. 2006).

Government policy objectives were tailored for the different rice-growing environments in Laos, and these in turn guided the research focus and emphasis within the Lao-IRRI Project. In the rainfed lowland areas, the government's priority in the early 1990s was to increase yield per hectare and expand the total area under production. Increasing irrigation access was also a priority to reduce the impact of a variable climate on rice yield and increase dry-season rice production. In contrast, in the upland areas the focus was to stop shifting agriculture, 'stabilise' production systems, and diversify crops to reduce dependency on rice (Lao-IRRI Project 1993).

IRRI oversaw the program and placed three full-time international staff in Laos: a project leader and a lowland systems specialist, both based in Vientiane, and an upland systems specialist based in Luang Prabang in the Northern Region. Close collaboration was sought with the Lao Department of Agriculture and Extension and the Provincial and District Agricultural Offices (PAFO and DAFO).

The (Political) Will to Succeed

It is important to note the value of political and institutional support in enabling the Lao-IRRI project to operate. The project had a mandate to contribute to the development of a Lao rice research system. Hence, ties between the Government of Laos and the Lao-IRRI Project were strong by necessity. The government was tightly controlled by the Lao People's Revolutionary Party and the project required government approval for basic project activities, such as field visits to the provinces, which had to be lodged for approval a month in advance (Interview 1).

Until the National Agriculture and Forestry Research Institute (NAFRI) was established in 1999, the project sat directly under the Ministry of Agriculture and Forestry. Though having strong government support allowed the project to achieve significant change, the project was not immune from the risks or tensions of working in the country. Schiller (n.d.) speculates that the initial IRRI-appointed lowland agronomy specialist's contract was ended due to his Thai nationality and the poor relationship between Laos and Thailand at the time. The support of key high-level government officials was instrumental in allowing the project to go ahead.

> [The Vice Minister for Agriculture and Forestry] would often come down for coffee just to check 'how is everything?' and [if there were] any areas where we needed support, and then he would—where appropriate—he would then make sure the support was given. Because at that time it was potentially difficult to work in Laos for a number of reasons. (Interview 1)

Schiller (quoted in Gorsuch 2002: 6) remarked that 'Lao-IRRI has been more fortunate than other projects in Laos because of political support from [the Minister]'. The Minister had studied in Russia and had strong connections to IRRI, which culminated in tenure as an IRRI board member from 1996 to 2001 (Shrestha et al. 2006; IRRI 2004). As a result, he 'was very, very conscious of the need to develop a national research capacity within Laos' (Interview 1). The Minister's background gave him a familiarity and understanding of the value of the project and what it was trying to achieve.

In addition to strong connections, the Lao-IRRI Project was directly responding to the requests of the Government of Laos, in particular the Ministry of Agriculture and Forestry, to establish a network of research stations across Laos and to achieve national self-sufficiency in rice. With government support and involvement, the project was able to contribute to the institutional architecture and agricultural research capacity of Laos. With the establishment of NAFRI came the National Rice Research Program. The program has continued to coordinate the development of the rice sector in Laos through the network of research stations and provincial and district agriculture and forestry offices. The structures and research areas established by the Lao-IRRI Project were thus effectively institutionalised. It would seem that a careful process of building research capacity, demonstrating impacts, and ensuring that local ownership and leadership within the program was developed contributed to this outcome.

Developing Research Infrastructure and Capabilities

The first phase of the Lao-IRRI Project focused on building the capabilities within Laos to conduct rice research. One aspect of this was to expand and upgrade research facilities across the country (Fig. 6.1). The National Agricultural Research Centre (NARC) near Vientiane became the principal research centre, coordinating rice research across the regions, overseeing the germplasm bank, and crossing and evaluating varieties for lowland rainfed areas (Shrestha et al. 2006). Regional seed multiplication stations were established or upgraded to support varietal improvement and testing of varieties in specific agroecological zones. Infrastructure was built to support the operations of the research network, including roads, seed storage, drying facilities, and administration and training buildings (Shrestha et al. 2006). The network of regional centres provided a connection between the project, district agricultural offices, and farmers. These connections enabled testing of improved varieties in a range of agroecological conditions (Interview 1).

In addition to provision of physical infrastructure, the Lao-IRRI Project developed the capabilities of Lao researchers. Training was provided in rice breeding and production, disease control, and cropping and farming systems, as well as English language and project management (Lao-IRRI Project 2005; Gorsuch 2002). Training—which included degree and non-degree programs, workshops, conferences, study tours, and other skill-building activities—was provided to staff from a broad spectrum of organisations including development planning, research, and extension agencies (Shrestha et al. 2006).

Training was provided to staff in national and regional centres, and especially to those in provinces responsible for field activities—57 per cent of trainees were staff from PAFOs and DAFOs who were in charge of local field trials (Shrestha et al. 2006: 28). In addition to formal training opportunities, IRRI publications and factsheets were developed in the Lao language to make information more accessible (Interview 1).

The Lao-IRRI Project sought to foster a sense of ownership of project activities (Shrestha et al. 2006). Annual meetings brought together representatives from all provinces to agree on work plans for the coming year, including which trials would be conducted in which provinces (Interview 1). Bringing the teams together in this way fostered collaborative links between the central and regional research stations and the provincial and district offices. While these links helped the internal functioning of the research program, external links and relationships were built between NAFRI and other international research agencies such as the University of Queensland, CSIRO, FAO, and ACIAR, providing access to additional funding and an ongoing portfolio of research for NAFRI (Shrestha et al. 2006).

When the project started, only five junior agricultural technicians were conducting field studies; by 1998, the National Rice Research Program employed 130 people and had activities in all provinces (Gorsuch 2002). Infrastructure—roads, buildings, dryers, and seed storage facilities—provided the foundations for developing the research network. Building the technical and administrative capacity of Lao researchers was central to embedding the ideas and approaches of the project in government institutions, while building an international network of collaborators has supported ongoing funding and research opportunities since the Lao-IRRI Project finished (Shrestha et al. 2006).

Many of the Lao researchers who were part of the Lao-IRRI Project went on to have senior positions within the National Rice Research Centres, the Ministry of Agriculture, and NAFRI, and continued to bring the experiences, perspectives, and networks gained through the project to these positions. The Lao-IRRI Project built on the training opportunities that had been provided to many Lao people in the Soviet Union and other Eastern Bloc countries through the 1970s and 1980s. This provided a pool of researchers with basic skills in remote provinces of the country (Interview 1). Formal evaluation of the Project concluded it had 'clearly played a key role in building the capacity of research and related agricultural organisations to develop and implement various programs effectively' (Shrestha et al. 2006: 40).

One interviewee suggested the capacity built through the Lao-IRRI Project has eroded as experienced people have moved to other organisations or retired (Interview 5b). A 2007 study highlighted the limited growth in the number of staff with master's or PhD degrees in the NARC and Multiplication Centres as an ongoing limitation to plant breeding and seed development (Thepphavong and Sipaseuth 2007).

Varietal Improvement and Management Practices

In the rainfed lowlands, the project developed a range of short- and medium-duration improved varieties for households to choose from. The primary focus was development of glutinous varieties for subsistence production, with secondary consideration of non-glutinous varieties for sale (Lao-IRRI Project 1993). In contrast to varieties released as part of the first phase of the Asian Green Revolution, the Lao-IRRI Project developed varieties that had high yield potential despite low input use and adapted to a range of agronomic conditions. Evaluation of varieties considered eating quality and duration (Lao-IRRI Project 1993). As these goals were met, varietal development shifted from breeding for crop duration and productivity to tolerance of specific conditions like drought or pests (Interview 10). Three types of varieties were released during the Lao-IRRI Project: Lao improved varieties that were developed specifically for Lao conditions by the Lao-IRRI Project; other improved varieties developed in other countries but suitable in some areas of Laos (e.g., IR66, RD23); and traditional Lao varieties that were found to be suitable for use in 'new' areas (Shrestha et al. 2006). The first of the Lao improved varieties was released in 1993 and, by 2005, 17 had been released (Table 6.1).

The Lao-IRRI Project collected and preserved over 13,000 seed samples from across the country (Appa Rao et al. 2006b). While establishing an important record of the biodiversity of rice in Laos, this also enabled the preservation of wild and traditional varieties before the introduction of new varieties (Interview 7). This collection is used to identify traditional varieties that may be suitable in 'new' areas of Laos and as part of improvement programs.

Improved varieties were part of a technical package including recommendations for planting times, plant spacing and density, and

Table 6.1   Release of Lao improved varieties from 1993 to 2005

Year	Varieties released	Total
1993	TDK1, TDK2, PNG1	3
1995	PNG2	1
1997	TDK3	1
1998	TDK4, TSN1, NTN1	3
2000	TDK5	1
2003	TDK6, TDK7	2
2004	TSN2, TSN3, TSN4	3
2005	PNG3, PNG5, PNG6	3
Total		17

Source: Inthapanya et al. (2006: 240)
Notes: Naming indicates the research station where breeding lines were developed—National Agricultural Research Centre in Thadokkham Village (TDK); Phone Ngam Rice Research and Seed Multiplication Centre (PNG); Thasano Rice Research and Seed Multiplication Centre (TSN); 30-ha Rice Research and Seed Multiplication Centre, Namthane (NTN). Not all varieties are still recommended, including PNG2 and TDK7, due to susceptibility to disease (Inthapanya et al. 2006)

fertiliser use (Lao-IRRI Project 1998). Additional practices in pest management and crop establishment were also explored but less broadly recommended.

Until 1997, crosses were carried out by the IRRI rice breeding division and Thai-IRRI program on behalf of the Lao-IRRI Project, using genetic material from traditional Lao varieties, varieties from Northeast Thailand, and other accessions sourced from the International Rice Germplasm Centre and IRRI. Progeny of F2 crosses[3] were transferred to the national research station for further evaluation and development in Laos (Lao-IRRI Project 1993, 1996). From 1997, the NARC had the capability to cross varieties in-house (Lao-IRRI Project 1998). Varieties that were generally adaptable, producing relatively stable yields in a range of areas, would be listed as promising lines (Lao-IRRI Project 1996).

Once evaluated, promising lines were sent to regional research centres to test for yield, adaptability, and suitability in different environments. By 1996 the project was conducting trials in every province of Laos (Lao-IRRI Project 1997).

Demonstration plots were used to conduct final assessment and evaluation of varieties with farmers, and also served as a mechanism to promote the benefits of using improved varieties (Lao-IRRI Project 1997). After harvest, farmers would be able to take seed from the demonstration plots for their own use. A number of 'collaborator farmers' were also involved in testing the varieties and management practices on their own farms. They chose which varieties they would trial and were provided with seed and other inputs, especially fertiliser (Lao-IRRI Project 1997).

The Lao-IRRI project strengthened the network of research centres and created strong links with the PAFOs and DAFOs, establishing a presence across the country. Though initial steps in the varietal development process were centralised at the NARC, attention was given to developing breeding material suited to the different regions that could be tested and adapted to local conditions. One interviewee noted this was central to the success of the varietal development activities in the project, with each region receiving breeding material tailored to local conditions (Interview 5a).

Distribution and Use of Seeds by Households

The spread of varieties across the country was described as autonomous and rapid, fuelled by demonstrable and observable results. The use of seeds 'just went boom' after they were released (Interview 3), one informant remarking that 'if it is a good thing, it spreads by itself' (Interview 7). Early releases, such as TDK-1, were most suited to areas with good soil and access to irrigation, so were readily used in these areas (Interview 3). The expansion of irrigation facilities in the lowland areas along the Mekong River supported further use by enabling dry-season rice cropping (Interview 1, Interview 3). However, in rainfed and remote areas uptake was more limited.

The project used village meetings, demonstration plots, and farmer-to-farmer communication, supported by collaborating farmers, to promote the seeds and other practices. One of the biggest constraints to this process was providing an adequate supply of seed to farmers when and where it was needed (Interview 5a, Interview 1).

Formal project efforts to promote the research were complemented by a careful process of observation, trial, and seed exchange among farmers. 'Lao farmers tend to want others to try new things so they can observe; if they see the benefits, they will consider doing it' (Interview 2). Farmer practices of seed exchange are still prevalent in Laos. A survey of farmers in Savannakhet in 2012 found 40 per cent of farmers sourced seed from other farmers, compared to 18 per cent from the seed multiplication centres, PAFO, and DAFO, and 20 per cent from international projects (see Chap. 8). Traditional practices of seed exchange initially supported the dissemination of improved varieties, accounting for most of their use. However, it is a slow process and can take many years for seeds to be distributed over a large area, by which time seeds need to be replenished (Manivong et al. 2008).

Over successive harvests, the quality of improved seed declines and yields fall. Rather than saving seeds from each harvest for planting the next year, farmers need to replace seeds every two to three years if they are to maintain yields. As farmers began to use improved varieties, two needs emerged: to adapt traditional practices to regularly replenish seed; and to establish an effective system for seed multiplication and distribution. There had never been a formal seed distribution system in Laos, although there had been some attempts through cooperatives to establish seed producer groups. The need for a more organised national approach to seed production to ensure supply was recognised by the project (Lao-IRRI Project 2003) but was not within the project's mandate (Interview 1).

Seed had to be produced in sufficient quantities (but not excessively, to avoid waste) and distributed to farmers who were accustomed to saving seed from the harvest rather than purchasing seed. Weaknesses in the nascent extension system and low technical skills of extension staff limited the adoption of improved varieties and other practices (Lefroy-Braun and Winch 2004). In the absence of an established and well-functioning extension system, collaboration with other projects became (and remains) a key facilitator for seed production and distribution.

The Lao-IRRI Project maintained formal collaboration with many international research projects. The types of projects varied from those with their own research purpose, for which improved seeds developed under the Lao-IRRI Project were one component of research, to those whose aim was explicitly to encourage adoption of new varieties. For example, the Savannakhet Integrated Rural Development Project (Phalanxai District) and the Improving Crop Yields Project (in Phalanxai and Outhomphone Districts, Savannakhet) both aimed to increase rice production through the use of 'proven, low-input and sustainable technologies for rice-based agricultural systems' (Lefroy-Braun and Winch 2004: 2). Both projects were conduits to promote and support households to access and use improved varieties. The Improving Crop Yields Project supported production of 10 t of improved seed, which was distributed to 1659 farmers (Manivong et al. 2008: 9). These farmers further distributed the seed to other households as part of normal seed exchange practices, mostly to farmers within the same village.

ACHIEVEMENTS AND LEGACY OF THE LAO-IRRI PROJECT

Evidence of Project Impacts

From 1995 to 1998, the Lao-IRRI Project conducted household-level impact studies in two villages—one each in Vientiane and Champassak Provinces. The results highlighted the potential for rapid spread of the modern varieties, with almost 100 per cent of farms in each village incorporating at least part of the recommended package (mostly use of improved varieties) within the three-year period (Lao-IRRI Project 1999). Households that applied all recommended practices had higher yields, earned higher returns, and consumed more rice, yet consumed proportionately less of their harvest (Lao-IRRI Project 1999). Partial application of the package (using improved varieties without other recommendations) meant the full yield potential was not reached, which significantly limited the potential benefit (Lao-IRRI Project 1998: 81).

A separate study in Champasak and Saravan Provinces found significant variation in the proportion of land planted with improved varieties between households and villages (Pandey 2001). Though 60 per cent of households surveyed used the varieties, they were planted on only 21 per cent of the land (Pandey 2001). Use of improved varieties and fertiliser was higher in villages with road access. The results are consistent with findings in other countries—larger farmers with better access to markets and fertiliser are more likely to use improved varieties as they are able to get the most yield benefits.

In 2004, a study surveyed villages in Outhoumphone and Phalanxai Districts to compare conditions in a village involved with the Lao-IRRI Project until 1999 with a village that had no prior involvement with development projects. Households surveyed in the former project village experienced a higher degree of self-sufficiency in rice, 'disproportionately' higher incomes, and significantly higher yields (an average of 5.4 t/ha compared to 1.4 t/ha) (Lefroy-Braun and Winch 2004).[4] In considering these results, it is important to note that the survey was conducted as a benchmark for a research project and was deliberately targeting areas with high levels of poverty and food insecurity.

External review of the Lao-IRRI Project in 2000 found the project had been 'highly successful' as indicated by the increase in rice production and self-sufficiency; the rapid adoption of modern varieties; the increase in double cropping in irrigated areas; and income and food security benefits at a household level (Shrestha in Lao-IRRI Project 2003: 175).

Table 6.2   Area of paddy land planted by seed type (%)

Region	1998/99		2010/11	
	Traditional	Improved	Traditional	Improved
Northern	93.1	7.0	87.7	12.3
Central	58.0	42.0	46.3	53.7
Southern	69.6	30.4	35.0	65.0
National	70.9	29.1	54.5	45.5

Source: Agricultural Census Office (2000, 2012)

In 1990, an estimated 90 per cent of rice production in the lowlands was from traditional varieties (Appa Rao et al. 2006b: 123). By 1998/99 an estimated 29 per cent of land area was planted with improved varieties, just six years after the first releases from the Lao-IRRI Project (Table 6.2). The following year, Laos had produced enough rice to meet national consumption needs and has been able to maintain overall self-sufficiency since, though at regional and household levels there are still production deficits (Schiller et al. 2013). By 2010/11, the area planted with improved rice varieties had increased to 45 per cent nationally and as high as 65 per cent in the Southern Region. The difference in regions shown in Table 6.2 reflects the focus of the rice improvement program on the rainfed lowland environment and the generally more suitable conditions in the lowlands. Most varieties were not suitable for upland areas due to pests and lack of water (Interview 8a).

Nationally, rice production in Laos more than doubled from around 1.5 million t in 1990 to 3.5 million t in 2012, largely following the upward trend in yield per hectare (see Fig. 5.3 in Chap. 5). Eliste et al. (2012: 63) conclude that the increase in rice production was supported by expansion of cropped area and irrigated area, but the increased use of Lao modern varieties was the 'single most important factor' to achieve these increases (Eliste et al. 2012: 63). The Rice Research Program has continued under NAFRI, with a further 13 Lao modern varieties released between 2005 and 2013 (Inthapanya et al. 2013).

At the national level, the success of the Lao-IRRI Project in enabling Laos to become self-sufficient has allowed for a policy shift away from national rice self-sufficiency. Nevertheless, the government still places significant emphasis on increasing rice production. Production and yield targets for the lowlands, once linked mainly to food security, are now framed by emphasis on the commercialisation of production (Ministry of Agriculture and Forestry 2010) and the development of rice export markets (Schiller et al. 2013).

Use of Improved Varieties in Savannakhet

A key test for any agricultural research program is whether outputs and findings are integrated and adapted into the daily life of end users over time, and particularly after formal support and funding are withdrawn. Savannakhet Province is one of the main rice-production regions of Laos and home to the Thasano Crop Research and Multiplication Centre.[5] It provides a suitable setting in which to examine the ongoing influence of the Lao-IRRI Project. Interviews and small group discussions were held with 19 farmers in four villages in Champhone and Outhoumphone Districts during October 2014 (Fig. 6.2 and Table 6.3). Villages were within 2.5 hours' drive of the provincial capital, the city of Savannakhet. Discussions considered how and when farming practices had changed with the introduction of new technologies, including improved varieties.

Villages 1 and 2 are only 9 kilometres from Savannakhet City and located just off a major road. They were relocated from another district in the 1960s and have similar agroecological conditions. The process of rice intensification in Village 1 began just one or two years before Village 2. Village 1 has had a longer history of involvement with international research projects.

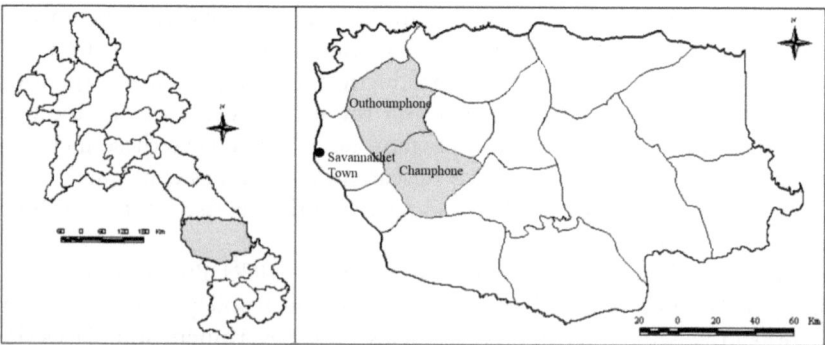

Fig. 6.2 Savannakhet Province showing Outhoumphone and Champhone districts. (Source: Modified from Manivong et al. 2008: 1)

Table 6.3   Village characteristics

Village	District	Research involvement	Access	Rice system
1	Outhoumphone	High exposure and participation	Sealed road access	Rainfed
2	Outhoumphone	High exposure and participation	Sealed road access	Rainfed
3	Outhoumphone	Low exposure and participation	Most remote of four villages. Dirt road access. Inaccessible during wet season	Rainfed
4	Champhone	High exposure and participation	Reasonable access (dirt road)	Irrigated

Village 1 was involved in the Lao-IRRI Project in the late 1990s and all households were said to use improved varieties. Since the Lao-IRRI Project, interviewees remembered at least four international agricultural research projects working in the village on different aspects of agricultural production such as crop establishment and climate adaptation. Projects facilitated access to fertiliser and other inputs that could otherwise be difficult for households to purchase. Though the farmers tried to maintain practices once projects finished, usually they were adapted to reflect the low levels of inputs they were able to access without project support.

Likewise, in Village 2, all households were said to be using improved varieties. They were first introduced in the mid-2000s by staff from Thasano, though farmers also received some improved seed from neighbouring villages. Traditional varieties were still used by some farmers interviewed. Three large international agricultural research projects had worked in this village in the last 15 years (one was ongoing at the time of the interview), each aiming to improve rice production in some way. However, as with Village 1, villagers here noted they found it difficult to continue using the practices after projects had finished because they could not afford or easily access the required inputs.

Of the four villages visited, Village 3 was the most remote (though only two to three hours from Savannakhet), connected by a narrow, bumpy, dirt road. Village 3 had the least connection to Thasano and the least exposure to international research projects. Households were still using traditional varieties but had started using improved varieties around 2010, after they were introduced by an international research project concerned with improving food production and marketing systems. Some of those interviewed stated that they wanted to maintain diversity and continue to use both traditional and improved seed.

Village 4 had irrigation access and had been involved with several international agricultural research projects. One was ongoing at the time of the visit, trialling strategies to support adaptation to climate variability, such as use of a mechanised drill-seeder. Regular field schools were held to discuss progress and challenges in the farming season and the farmers received detailed weather information to guide timing of activities and crop choices. As part of the project activity, DAFO officers visited twice each month and a PAFO officer visited once a month. Researchers from Thasano visited as part of trialling transplanting machines and other new techniques. Some farmers had started using improved varieties from Thailand in the early 1990s, while others had started to use them only in 2009.

There had been widespread use of improved varieties in each village but, as was the case during the Lao-IRRI Project, the adoption of the other practices to support yield improvement, particularly fertiliser use, remained low. Households spoke about the benefits of improved varieties in conjunction with other changes, such as mechanised land preparation. Improved varieties gave higher yields, for some farmers up to 50 per cent higher, while mechanisation helped save labour. Increased yields supported improved livelihoods but there were increased costs in terms of inputs (seed, fertiliser, machine maintenance) and pest and disease problems. Households interviewed appreciated the yield increases, but their aim was to increase the efficiency of production to meet household needs and to free labour and other resources for other (often non-farm) activities. This is consistent with other studies (Newby et al. 2013; Manivong et al. 2014) which show low returns for rice discourage farmers from investing in inputs to the 'recommended' levels (see Chap. 10).

Farmers in each village were using Lao improved varieties, Thai varieties, and traditional varieties. The diversity of varieties may have declined, but households still selected for traits that were appropriate for household needs, labour, risk, and local conditions. Some households noted that they found it difficult to know which of the suite of available improved varieties were most suitable for their land.

The persistent role of traditional practices for seed saving across seasons and farmer-to-farmer seed exchange was common across villages. However, the use of improved seed had resulted in some changes to this practice. Households expressed annoyance at having to pay for a resource they had previously been able to manage and reuse for free. However, it was clear that the benefits of increased yields outweighed the costs, as farmers continued to use improved varieties and replenish seeds when required. The dissatisfaction reflects a process of adjustment in household expectations.

International research projects continued to play a significant role in the supply of Lao improved varieties. Most, if not all, projects, source seed stock from Thasano. Once projects finish, farmers access seed from Thasano directly or from other farmers. Thasano was at the centre of farmer networks to replenish seed. It should be noted that all the villages visited had reasonably good access to Thasano relative to the rest of Savanakhet Province.

Thasano is 40 minutes' drive from Savannakhet City and multiplies seed for sale to research or development projects and to farmers. Systems for seed multiplication and distribution were still not well established. Limited farmer demand means seed stocks are kept relatively low to avoid oversupply and spoiling (Interview 2). At the same time, international research projects—which have played an essential role in distribution of seed and supported their use by farmers—have 'sudden and significant' demands which can strain under-resourced centres (Schröder 2003).

The role of the Thasano Director was crucial in building a strong profile and reputation for the Centre with farmers, across the different levels of government, and with international researchers. An evaluation of seed production activities in 2003 concluded, 'seed rice production activities are mainly left to the personal initiative of the research station manager and the Thasano Research Station in Savannakhet can be regarded as an outstanding success story' (Schröder 2003: 177). The Director's efforts extended to helping farmers in seed selection when they came to the station for seed. The Director did not leave the station during the month the farmers came because she wanted to talk to them and ask them about their fields and cropping history so she could recommend a variety and teach them how to use it (Interview 2).

Efforts to address the limitations in the seed multiplication and distribution networks are ongoing. Supported by the World Bank's Rice Productivity Improvement Project (RPIP), Thasano has collaborated with farmer groups to produce registered and certified seeds—in effect supporting the development of a decentralised seed production system (Interview 2). RPIP has funded equipment, initial seed stock, training, and technical support to farmers (World Bank 2012). Farmers multiplied seed in compliance with strict guidelines to preserve purity and quality, and either sold the seed back to PAFO or Thasano for a premium price or sold to other farmers (see Chap. 8).

A key constraint for the seed multiplication centres like Thasano has been a lack of operational funds—salaries for staff are funded by the government but centres are encouraged to cover operational costs through commercial seed production—which forces a reliance on commercial arrangements with international projects (World Bank 2012). At the household level, an absence of commercial seed markets limited the ability of households in a seed-producing farmer group to sell high-quality seeds above the price of paddy rice (World Bank 2014).

REFLECTIONS ON THE SUCCESS OF THE LAO-IRRI PROJECT

The conditions in Laos at the commencement of the project were dire: a country trying to rebuild after decades of war and revolution; a failed restructuring for collective production; significant and successive crop losses due to drought and flood, leading to severe food shortages. Interviews with key project staff conveyed the sense that this project had to succeed. Prior to the Lao-IRRI Project, there was no specific or coordinating research entity in Laos, no national rice research program, and a relatively empty landscape in terms of international research projects (Interview 1). The open space into which the Lao-IRRI Project stepped helped assure the necessary political support and gave the room to develop a national network of rice research centres. By design (and direct instruction from the government), the project was able to put in place the architecture and connections to implement project activities at a national scale, with links across the provinces and down into the districts.

Such direct access to high-level government officials and scope to build up a research program starting from the basics is rare and mostly seen in post-conflict states, where physical infrastructure, formal and informal institutions, and skills and capacity have been weakened or completely destroyed (Erskine and Nesbitt 2009). IRRI established a similar program in Cambodia in 1986 as that country struggled to rebuild after the destruction brought about by the Khmer Rouge (Nesbitt 2003). Similarly, the Seeds of Life Program in Timor-Leste supported the development of a national policy and research capability for a range of seed crops after independence from Indonesia, fostered by close relationships with the emerging Government of Timor-Leste (Borges et al. 2009).

According to one of the former Lao-IRRI Project leaders, the basis of success for these kinds of programs is 'the political will of the countries to make the programs work' (Schiller, quoted in Gorsuch 2002: 5). In this case, the combination of history and circumstance aligned the goals of the Government of Laos with the goals of IRRI. However, it is more than just an alignment of intent that supports successful project outcomes. In 'adapting the Green Revolution for Laos,' IRRI responded to criticism that its first releases in Asia were developed without consideration of farmers' circumstances or needs; hence, it shifted to more participatory modes of research, such as involving farmers in varietal selection (Douthwaite et al. 2001).

The task of the Lao-IRRI Project began with a focus on developing the capacity of the institutions and individuals within Laos to establish and take ownership of a rice research program. Horton (2002) highlights the importance of mentoring, beyond one-off training events, to effectively build capacity. This was a feature of the Lao-IRRI Project. The extent of impacts on the ground, in terms of the number of varietal releases suitable to different environments and their use across the country, would not have been possible without the scale of capacity-building that occurred. While the national program has continued and releases of improved varieties are ongoing, since the Lao-IRRI Project finished concerns have been raised about whether the research capacity has been or can be sustained (Thepphavong and Sipaseuth 2007; Clarke et al. 2015). This study does not directly affirm this concern, though it does suggest that current research capacity is dependent on key individuals. One interviewee suggested that a combination of lack of specialist agricultural skills at the district level, a lack of connection between extension and research, and government pressure to release new varieties was transferring higher risk to farmers as varieties are released without adequate testing (Interview 3).

In the absence of government operational support for breeding and with a still-nascent extension system, the Lao-IRRI Project depended on promotion of the varieties in project sites and subsequent farmer exchange to spread the varieties. Farmer observation of new practices in other farmers' fields is a long-used way for innovations to spread (Appa Rao et al. 2006a). However, the capacity of the project to provide and distribute seed also depended heavily on other international research projects that brought seeds to additional areas and supported farmers with information on their use. International projects have played an important role in expanding the use of new varieties, with many villages first gaining access to the varieties with the arrival of a project. This has led to a government preference for projects to fund seed multiplication, rather than itself ensuring basic availability as a public good (Schröder 2003). Recent efforts to encourage commercial production have been limited by a lack of households willing to pay premiums for good-quality seed (World Bank 2012).

One of the key constraints to a commercial market for seed in Laos is the long-held practices of selecting, multiplying, and exchanging seeds at the household and village levels. As a result, farmers typically had a range of varieties highly suited to their conditions and preferences. The introduction of improved varieties has shifted this knowledge from farming households to research and extension services. New varieties are developed by scientists, albeit with the involvement of some farmers, whether through participatory varietal selection or other studies that aim to understand what traits farmers value. These participatory approaches were strong themes within the Lao-IRRI Project and remain good practice in rice varietal development. However, in contrast to past farmer practices, where each farmer would be connected to seed selection and varietal development through their own processes of exchange and experimentation, most farmers are removed from the process of developing improved varieties. Participatory varietal selection directly involves only a sample of farmers.

Farmers in the village discussions reflected that they were now less certain about selecting varieties suitable for their land and soil types. Kousonsavath and Sacklokham (see Chap. 8) likewise found farmers wanted more varieties for specific environments and better information regarding suitability of varieties in different conditions. Disconnecting varietal development from farmers has undermined their familiarity with the suitability of seeds for different areas. This is observable in other aspects of production, with ritual and ceremonies traditionally used to inform key decisions such as the timing of planting now replaced by scientific knowledge and recommendations (Hatsadong et al. 2006; Simmalavong 2011). Efforts by seed centres to collaborate with farmer groups to multiply seed could be one mechanism to build farmer understanding of the range of improved varieties.

Impacts that emerge over time are more profound than an increase in rice yield at the farm level and point to fundamental transitions in production and markets. Varietal improvement does not stop with the release of a variety but is a continuous process of adaptation. The release of a relatively simple technology such as improved seed into the system likewise triggers a series of social and institutional adjustments—as households re-interpret recommendations to suit their resources; as new and old practices of resource management are adjusted to accommodate each other; or as knowledge and understanding of varieties are re-housed to sit with breeders rather than farmers.

NOTES

1. The research involved 28 individual interviews and 2 small group discussions in 2013–2014. For a full account of research methods, see Williams (2018).
2. Reforms in Laos followed those implemented by Vietnam and China, which de-collectivised agricultural production and encouraged foreign investment but retained strong protections for state-owned industry (Stuart-Fox 2011).
3. Filial generations indicate the number of generations after making a cross; F2 is the second filial generation.
4. In 2009, average yield for lowland irrigated rice was under 5 t/ha, compared to under 4 t/ha for lowland rainfed rice (see Chap. 7).
5. The Centre was formerly called the Thasano Research Station and is referred to here simply as Thasano.

REFERENCES

Agricultural Census Office, 2000. *Lao Census of Agriculture 1998/99: Highlights.* Vientiane: Ministry of Agriculture and Forestry.

Agricultural Census Office, 2012. *Lao Census of Agriculture 2010/11: Highlights.* Vientiane: Ministry of Agriculture and Forestry.

Appa Rao, S., Schiller, J., Bounphanousay, C., and Jackson, M. T., 2006a. Development of traditional rice varieties and on-farm management of varietal diversity in Laos. In J. M. Schiller, M. B. Champhengxay, B. Lindquist, and S. Appa Rao, eds., *Rice in Laos*, pp. 187–196. Los Baños, Philippines: IRRI.

Appa Rao, S., Schiller, J., Bounphanousay, C., and Jackson, M. T., 2006b. Diversity within the traditional rice varieties of Laos. In J. M. Schiller, M. B. Champhengxay, B. Lindquist, and S. Appa Rao, eds., *Rice in Laos*, pp. 123–140. Los Baños, Philippines: IRRI.

Bestari, N. G., Shrestha, S., and Mongcopa, C. J., 2006. *Lao PDR: An Evaluation Synthesis on Rice.* Manila: Asian Development Bank.

Borges, L. F., de Rosario Ferreira, A., Da Silva, D., Williams, R., Andersen, R., Dalley, A., Monaghan, B., Nesbitt, H., and Erskine, W., 2009. Improving food security through agricultural research and development in Timor-Leste: a country emerging from conflict. *Food Security* 1: 403.

Bourdet, Y., 1996. Laos in 1995: reform policy, out of breath? *Asian Survey* 36: 89–94.

Clarke, L., Grünbühel, C. M., Souvannachak, C., Keoka, K., and Phakdisoth, L., 2015. *Research Capacity and Science to Policy Processes in Lao PDR: An Initial Study*. Vientiane: Laos-Australia Development Learning Facility.

Douthwaite, B., Keatinge, J. D. H., and Park, J. R., 2001. Why promising technologies fail: the neglected role of user innovation during adoption. *Research Policy* 30: 819–836.

Ducourtieux, O., Laffort, J. R., and Sacklokham, S., 2005. Land policy and farming practices in Laos. *Development and Change* 36: 499–526.

Eliste, P., Santos, N., and Pravongviengkham, P. P., 2012. *Lao People's Democratic Republic Rice Policy Study*. Washington, DC: World Bank.

Erskine, W., and Nesbitt, H., 2009. How can agriculture research make a difference in countries emerging from conflict? *Experimental Agriculture* 45: 313–321.

Evans, G., 1988. *Agrarian Change in Communist Laos*. Singapore: Institute of Southeast Asian Studies.

Gorsuch, J., 2002. *Rice, the Fabric of Life in Laos*. Los Baños, Philippines: IRRI.

Hatsadong, 2013. The early history of lowland rice research and findings in the Lao PDR, 1975–1990. Unpublished report.

Hatsadong, Doungsila, K., and Gibson, P., 2006. Rice-based traditions and rituals in the Mekong River Valley. In J. M Schiller, M. B. Champhengxay, B. Lindquist, and S. Appa Rao, eds., *Rice in Laos*, pp. 65–79. Los Baños, Philippines: IRRI.

Hopkins, S., 1995. The economy. In Andrea Matles Savada, ed., *Laos: A Country Study*, pp. 149–202. Washington, DC: Federal Research Division, Library of Congress.

Horton, D., 2002. *Planning, Implementing, and Evaluating Capacity Development*. The Hague: ISNAR.

Inthapanya, P., Boulaphanh, C., Hatsadong, and Schiller, J. M., 2006. The history of lowland rice variety improvement in Laos. In J. M Schiller, M. B. Champhengxay, B. Lindquist, and S. Appa Rao, eds., *Rice in Laos*, pp. 325–348. Los Baños, Philippines: IRRI.

Inthapanya, P., Boualaphanh, C., Xangxayasane, P., Bounphanouxay, C., Kanhyavong, K., Latvilavong, P., and Sihathep, V., 2013. Recommended lowland rice varieties for Lao PDR. *The Lao Journal of Agriculture and Forestry* 28: 154–178.

IRRI, 2004. *Rice Today*, July-September. Available at http://ricetoday.irri.org/rice-today-vol-3-no-3-july-september-2004/ (viewed 1 October 2017)

IRRI, 2006. *The Green Revolution Comes to Laos*. Available at http://www.eurekalert.org/pub_release/2006-03/irri-tgr031506.php (viewed 9 January 2015)

Lao-IRRI Project, 1993. *Annual Technical Report 1992*. Vientiane: IRRI.

Lao-IRRI Project, 1996. *Annual Technical Report 1995*. Vientiane: IRRI.

Lao-IRRI Project, 1997. *Annual Technical Report 1996*. Vientiane: IRRI.
Lao-IRRI Project, 1998. *Annual Technical Report 1997*. Vientiane: IRRI.
Lao-IRRI Project, 1999. *Annual Technical Report 1998*. Vientiane: IRRI.
Lao-IRRI Project, 2003. *Annual Technical Report 2001–2002*. Vientiane: IRRI.
Lao-IRRI Project, 2005. *Annual Technical Report 2005*. Vientiane: IRRI and NAFRI.
Lefroy-Braun, R., and Winch, J., 2004. *Baseline Survey—Improving Crop Yields Project*. Vientiane.
Linquist, B. A., Keoboualapha, B., Sipaseuth, and Inthapanya, P. 2006. Rice production systems in Laos. In J. M. Schiller, M. B. Champhengxay, B. Lindquist, and S. Appa Rao, eds., *Rice in Laos*, pp. 29–46. Los Baños, Philippines: IRRI.
Manivong, K, Manivong, V., and Phengvichith, V., 2008. *Improving Crop Yields (ICY) Project, Savannakhet Province, Lao PDR*. Project Impact Survey Report. Vientiane, Lao PDR.
Manivong, V., Cramb, R., and Newby, J., 2014. Rice and remittances: crop intensification versus labour migration in Southern Laos. *Human Ecology* 42: 367–379.
Ministry of Agriculture and Forestry, 2000. *Agricultural Statistics, 25 Years, 1975–2000*. Vientiane: Department of Planning.
Ministry of Agriculture and Forestry, 2010. *Strategy for Agricultural Development 2011 to 2020. Sector Framework, Vision and Goals: Agriculture and Forestry for Sustainable Development, Food and Income Security*. Final Draft, September, 2010. Vientiane.
Nesbitt, H., 2003. Developing sustainable rice production systems in Cambodia: an Australian contribution. Proceedings of the 11th Australian Agronomy Conference 2003, Geelong, 2–6 February 2003.
Newby, J. C., Manivong, V., and Cramb, R., 2013. *Intensification of Lowland Rice-based Farming Systems in Laos in the Context of Diversified Rural Livelihoods*. Paper presented at 57th AARES Annual Conference, Sydney, 5–8 February 2013.
Pandey, S., 2001. Economics of lowland rice production in Laos: opportunities and challenges. In S. Fukai and J. Basnayake, eds. *Increased Lowland Rice Production in the Mekong Region*, pp. 20–30, Canberra, Australia: ACIAR.
Rigg, J., 2005. *Living with Transition in Laos: Market Integration in Southeast Asia*. London: Routledge.
Schiller, J., n.d. History of John Michael Augustine Schiller. Unpublished personal memoir.
Schiller, J., Hatsadong, and Doungsila, K., 2006. A history of rice in Laos. In J. M. Schiller, M. B. Champhengxay, B. Lindquist, and S. Appa Rao, eds., *Rice in Laos*, pp. 9–28. Los Baños, Philippines: IRRI.
Schiller, J., Manivong, V., and Phengvichith, V., 2013. Lao PDR—rice production and food security. *The Lao Journal of Agriculture and Forestry* 28: 117–134.

Schröder, P., 2003. *Seed Rice Production in the Lao PDR. Consultancy Study for Lao-IRRI.* The Lao-IRRI Rice Research and Training Project, Annual Technical Report 2001–2002. Vientiane: Lao-IRRI Project.

Shrestha, S., Boupha, T., and Khamphoukeo, K., 2006. *Sowing Seeds in Lab and Field: Socio-economic Impact of the Lao-IRRI Rice Research and Training Project.* Los Baños, Philippines: IRRI.

Simmalavong, P., 2011. *Rice, Rituals and Modernisation: A Case Study of Laos.* New Delhi: Palm Leaf.

Stuart-Fox, M., 1996. *Buddhist Kingdom, Marxist State: The Making of Modern Laos.* Studies in Asian History No. 2. Bangkok: White Lotus.

Stuart-Fox, M., 2011. *Countries at the Crossroads 2011: Laos.* New York: Freedom House.

Tanaka, K., 1993. Farmers' perceptions of rice-growing techniques in Laos: 'Primitive' or *'Thammasat'? Southeast Asian Studies* 31: 132–140.

Thepphavong, B., and Sipaseuth, 2007. *Plant Breeding and Related Biotechnology Capacity.* Vientiane: Lao People's Democratic Republic.

Williams, L. J., 2018. *Critical Reflections on 'Going to Scale' in Agricultural Research for Development: Case Studies from Southeast Asia.* Thesis, The University of Queensland, School of Agriculture and Food Sciences.

World Bank, 2012. *Laos—Rice Productivity Improvement Project: Report No. ICR2579, Implementation Completion and Results Report.* Washington, DC: World Bank. Available at http://documents.worldbank.org/curated/en/2012/12/17124491/laos-rice-productivity-improvement-project-laos-rice-productivity-improvement-project (viewed 25 August 2014)

World Bank, 2014. *Rice Productivity Improvement Project: Report No ICRR14121 Implementation Completion Report Review.* Washington, DC: World Bank. Available at http://documents.worldbank.org/curated/en/7428214745 93084547/pdf/000020051-20140625235547.pdf (viewed 3 October 2017)

Worner, W. E., 1996. Lao agriculture in transition. In Mya Than and Loong-Hoe Tan, eds., *Laos' Dilemmas and Options: The Challenge of Economic Transition in the 1990s.* Singapore: Institute of Southeast Asian Studies.

CHAPTER 7

Rainfed and Irrigated Rice Farming on the Savannakhet Plain

Silinthone Sacklokham, Lytoua Chialue, and Fue Yang

The aim of this study was to characterise rice production in the Savannakhet Plain, which has long been a major rice bowl for Laos. As this is one of the most productive and commercialised rice-growing regions in the country, an understanding of farmers' circumstances and strategies can give a good indication of how rice policy is working out in practice. If rice farmers in this region face substantial constraints on production, those in other settings will be even less able to meet government policy targets.

S. Sacklokham (✉)
SEAMEO Regional Centre for Community Education Development, Vientiane, Laos
e-mail: s.sacklokham@nuol.edu.la

L. Chialue
Faculty of Agriculture, Department of Rural Economics and Food Technology, National University of Laos, Vientiane, Laos
e-mail: lytoua.chialue@cirad.fr

F. Yang
Faculty of Agriculture, National University of Laos, Vientiane, Laos
e-mail: F.yan@nuol.edu.la

© The Author(s) 2020
R. Cramb (ed.), *White Gold: The Commercialisation of Rice Farming in the Lower Mekong Basin*,
https://doi.org/10.1007/978-981-15-0998-8_7

THE STUDY AREA

Savannakhet Province is the largest in Laos, covering 21,774 km², bordered by the Mekong River in the west and the Annamite Range in the east (see Fig. 5.1 in Chap. 5). The Province is drained by the Banghiang River, which originates in the mountains of Vietnam and empties into the Mekong about 90 km south of Savannakhet City. The river system has a comparatively steep fall and is subject to flash flooding in the upper catchment and longer-term flooding in the lower catchment, where several irrigation schemes have been established. The major rice-growing areas are found along the alluvial plain adjacent to the Mekong, with secondary areas on the residual terraces in the central part of the Province. The Province is traversed by three national roads—Route 13, which runs north-south along the Mekong corridor; Route 9, which runs on an east-west trajectory from Savannakhet City to the Vietnam border; and Route 1, which runs north-south along the eastern border range. Most of the provincial roads connecting district towns with major villages are unpaved, and most local roads are in poor condition and unusable during the wet season.

In 2011–2012 Savannakhet Province accounted for 23% of the country's rice production and 25% of irrigated rice production. Within the province, rainfed wet-season (WS) rice accounted for 78% of total production and irrigated dry season (DS) rice for 22%.[1] The yield of rice in Savannakhet averaged 3.4 t/ha for WS rice and 4.1 t/ha for DS rice, above the national average. Among the 15 districts of the Province, by far the largest rice producers were the five districts in the Mekong corridor, which together accounted for 60% of the total rice area in the Province and 86% of the irrigated area. The average WS yield in these five districts was 3.7 /ha and the average DS yield was 4.5 t/ha, somewhat higher than the provincial average.

The survey was conducted in six villages in Champhone District, the second largest rice producer in the Province, accounting for 16% of total production (Fig. 7.1). The District lies just to the east of Route 13 and the south of Route 9 and spans the middle reaches of the Champhone River, a major right-bank tributary of the Banghiang. Several irrigation schemes have been constructed along the Champhone River to service rice farmers in the district. Given its irrigation infrastructure of reservoirs and canals,

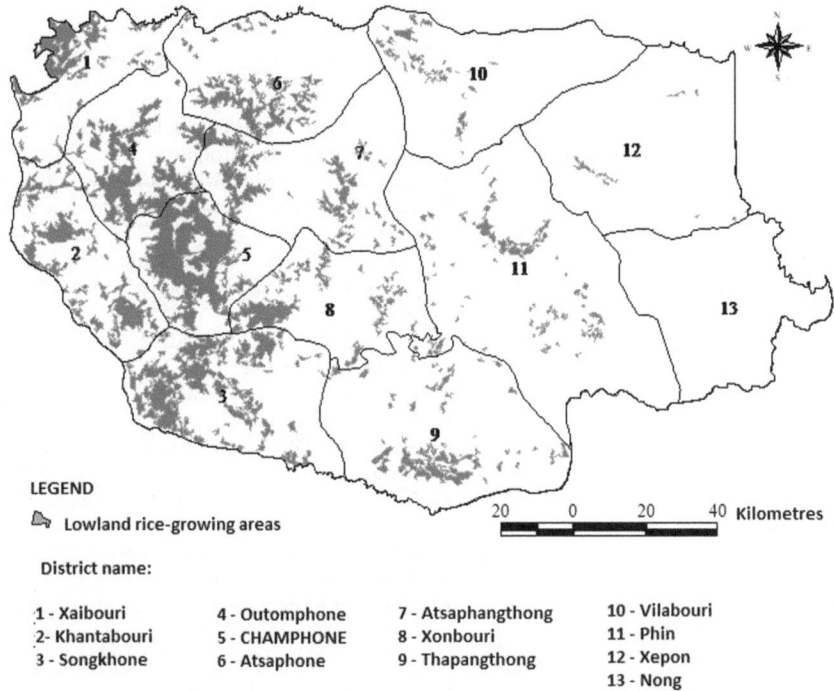

Fig. 7.1 Savannakhet Province showing Champhone and other districts and lowland rice-growing areas. (Source: Thavone Inthavong)

Champhone produced more DS rice than any other district, accounting for 41% of the Province's irrigated rice output in 2011–2012.

The villages were selected based on being located within this important rice-producing area and having potential to produce rice for the market. The characteristics of the villages and sampling details are presented in Table 7.1. The survey households were selected randomly from a list of all households in each village. The survey questionnaire focused on rice production in each season, including the area cultivated, the working calendar, input costs, production, sales, constraints, and potential. The survey was conducted in March 2012 by staff of the Faculty of Agriculture at the National University of Laos.

Table 7.1   Characteristics of survey villages

Village	No. of households	No. inter-viewed	Characteristics of village
Phalaeng	142	38	• Most farmers produced rice for market • Rice Seed Farmers Group • Irrigated DS rice in lowland areas near canal • Easy access in both seasons (1 km from main road)
Phiaka	66	22	• Irrigated DS rice in lowlands and along river • Some low-lying paddies flooded in WS in some years • Farmers with large area and DS rice produced for market • Access to village very difficult in WS
Beukthong	187	42	• Most farmers grew WS rice • Some grew DS rice near reservoir or small stream • Some farmers sold rice surplus after harvest • Rice Seed Farmers Group • Can access village in both seasons
Dondaeng	154	50	• Most farmers grew WS rice • Some grew DS rice near stream (using pump or tractor) • Some farmers sold surplus rice • Can access village in both seasons
Khaokad	127	32	• Most farmers grew WS rice, a few grow irrigated DS rice near natural ponds • Very few farmers sell rice after harvest • Can access village in both seasons
Khamsida	178	44	• Farmers grow WS rice only • Low yield due to infertile sandy soil in upper paddies • Most farmers produce rice for home consumption; some have insufficient rice in DS • Can access village in both seasons
Total	854	228	

VILLAGE AND HOUSEHOLD PROFILES

The six villages were representative of the range of conditions in Champhone District. Ban Phalaeng was located in the Champhone Village Cluster about 10 km from Champhone Town and 1 km east of Route 13. Phalaeng was established in 1809 by two groups that migrated from other villages in the region. In 2011, the village had 142 households and 192 families, with a total population of 971. All but a few households owned land and those without land rented fields for cultivation. Phalaeng had a total area of 825 ha including about 500 ha of cultivated land. Rice farming was the main source of income. WS rice was cultivated on about 440 ha and around 150 ha were used for irrigated rice and vegetables in the dry season. Water for irrigation was sourced from the Sou and Champhone reservoirs. The livestock in the village included an estimated 118 cattle, 57 water buffaloes, 83 pigs, 169 goats, and about 1800 ducks and chickens. The primary land use in the village was rainfed and irrigated rice cultivation, some cash crop cultivation, fishing, and livestock production (cattle and buffaloes). In the wet season, rice and fish culture were the main activities. Irrigated rice was grown in the dry season in the fertile floodplain near the reservoirs and along the canals. Vegetables were grown in the houseyards and in irrigable paddy fields after harvesting the WS rice crop. As the village was close to the main road and Champhone town, the villagers were quite commercialised. They could take their surplus rice and other produce to sell in Kengkok Market in Champhone every day. Moreover, local Lao and Vietnamese traders came to the village to provide fertiliser on credit. The villagers had set up a farmers group to produce rice seed for other villages with the assistance of a government agency. More than half the farmers had joined this programme.

Ban Phiaka was established more than 200 years ago about 15 km north of Champhone. The village had 66 households and a total population of 566. The village area was 520 ha, supporting 218 ha of rainfed rice in the wet season and 92 ha of irrigated rice in the dry season, as well as 30 ha of vegetable gardens and fruit tree orchards. Livestock included 319 cattle, 167 water buffaloes, 69 pigs, 75 goats, and around 10,000 poultry. More than half the villagers had their own hand tractor and there were two rice mills, six threshing machines, and three water pumps. The village had a diversity of rice ecosystems. Water for DS irrigation and WS supplementary irrigation was pumped from the adjacent Champhone River. With this source of irrigation, rice farming was the main source of income in both seasons. However, WS rice was affected by flooding in some years, though these flooded areas had fertile soil and were suitable for irrigated DS rice. Phika had also established a farmers group to produce rice seed, which provided a good income for the farmers involved. Fishing and other agricultural activities also contributed to household income. Local and Vietnamese traders came to the village to buy rice and other products.

Vietnamese traders often provided fertiliser to farmers on credit early in the season.

Ban Beukthong was located 16 km from Champhone. The village had 187 households and a population of 1306. The village area was 3997 ha, including 436 ha of rainfed lowland rice. Livestock included 312 cattle, 218 buffaloes, 211 pigs, and about 2000 poultry. The landscape in Beukthong ranged from middle-level lowlands to floodplain. DS rice could be cultivated in parts of the floodplain area with irrigation from natural ponds or a small stream. Rice farming was the main source of income. A seed production group had also been established in the village. As with the above villages, Vietnamese traders came to Beukthong early in the season to provide fertiliser on credit, with the cost being repaid after harvest, including interests of 20%.

Ban Dondaeng was established in 1937 and about 8 km east of Champhone. It was the result of a merger of smaller villages to comply with government policy. At the time of the survey the village had 154 households and a total population of 1328. The village territory was 4100 ha, with 710 ha of rainfed rice in the wet season and 147 ha of irrigated rice in the dry season. The landscape comprised two zones—a middle-level lowland area and a floodplain area. In the former, farmers used land for rainfed rice and animal raising in the wet season. In the latter, flooding prevented some areas from being used for rice in the wet season but they could be used for irrigated rice in the dry season. The source of water for irrigation and supplementary irrigation was the Talong reservoir; some farmers used water from natural ponds and small streams for their DS rice. Dondaeng had good road access to both the Kengkok markets in Champhone and the market in Xounabouly to the south. Farmers sold surplus rice as their main source of income. However, some farmers had low yields due to water shortage in the wet season and low soil fertility in the middle-level lowlands.

Ban Khaokad was established more than 300 years ago. It was located 7 km north-west of Champhone. There were 127 households and a total population of 802. The village territory was 591 ha, including 268 ha of rainfed paddies. The landscape was similar to the other villages, ranging from upper-level lowlands to lowlands, but with no irrigated rice. Rice production was mainly for household food security. Some farmers grew watermelon and vegetables in the paddy fields after the rice harvest. These activities enabled villagers to earn income to contribute to village develop-

Table 7.2 Distribution of survey households by farm size

Farm size (ha)	No. of households	% of households
<0.5	14	6.1
0.51–1.0	39	17.1
1.01–2.0	87	38.2
2.01–3.0	54	23.7
>3.0	34	14.9
Total	228	100.0

ment. A village fund had been established to lend money to villagers to develop a business or buy agricultural inputs.

Ban Khamsida was established about 200 years ago. There were 178 households and a total population of 1284. The territory was 590 ha, with 300 ha of rainfed rice (some of which could be irrigated), a garden area of 80 ha, and a forest area of 70 ha. The village was located in the upper-level lowlands and had poor soil. Hence some households had insufficient rice for two to ten months of the year. Only a few paddy fields had access to water in the dry season to grow irrigated rice. With poor soils and limited irrigation, farmers needed capital to buy inputs to increase the yield of rice.

Of the households interviewed, 80% were Lao or Phouthai and 20% were Khmu. The Khmu and related groups were the earlier settlers in Savannakhet, while the Lao and Phouthai had begun moving into the region from further north in the sixteenth century. The modal household size was 6–7 and the range was from 2 to 15. The modal number of workers per household was 3, ranging from 1 to 8. Households had from 1 to 5 plots of land. The mean farm size was 2.8 ha and the range was from 0.3 to 11.2 ha. The distribution of farm size is shown in Table 7.2, indicating that 38% of respondents had between 1 and 2 ha and 62% had between 1 and 3 ha. Almost all households (96.5%) reported that they worked on their own land, while four worked on their parents' land and only two rented land from other villagers.

RICE PRODUCTION IN THE WET SEASON

All survey farmers cultivated WS rice in 2011. The cropping calendar for WS rice is shown in Table 7.3. Nursery preparation began in April, preparation of the paddy field in May, transplanting in July, and harvesting in November. The mean area cultivated with WS rice in 2011 was 1.9 ha but

Table 7.3 Cropping calendar for wet-season and dry-season rice production

Activity	J	F	M	A	M	J	J	A	S	O	N	D
Fertilizing					■							▩
Sowing					■	■	■					▩
Land preparation 1	▩											
Land preparation 2							■					
Transplanting	▩											
Fertilising												
15-15-15	▩											
16-08-08		▩	▩					■				
16-20-00	▩							■	■			
46-00-00	▩							■	■			
Management	▩	▩				■	■	■				
Harvesting					▩	▩				■	■	
Post-harvest						▩					■	

Key: Wet season ■   Dry season ▩

Table 7.4 Mean area and yield of wet-season rice in survey villages, 2011

Village	Mean area cultivated (ha)	Mean yield (t/ha)
Phalaeng	1.90	2.43
Phiaka	2.27	2.45
Beukthong	2.79	2.29
Dondaeng	1.36	1.81
Kaokad	1.99	2.24
Khamsida	1.34	1.67
All villages	1.94	2.24

varied between villages, from 1.3 ha in Khamsida to 2.8 ha in Beukthong (Table 7.4). The range was from 0.3 to 7.5 ha.

Hand tractors were almost universally used for land preparation. Over two thirds (68%) of households reported that they used their own hand tractor, 18% hired a hand tractor, 10% borrowed a hand tractor from a relative, and only 4% still used a buffalo-drawn plough. This traditional practice was found among some of the poorest farmers in Dondaeng and Khamsida.

Farmers reported using 16 different rice varieties in the wet season, almost all glutinous, including 14 improved varieties that had been bred and distributed by Ministry of Agriculture and Forestry (MAF) research

Table 7.5 Rice varieties used by respondents in wet and dry seasons, 2011–2012

Variety	Wet season		Dry season	
	No. of households	% of households	No. of households	% of households
Improved varieties				
Thadokham 1	8	3.5	2	1.7
Thadokham 5	5	2.2	22	19.0
Thadokham 6	25	11.0	9	7.8
Thadokham 7	5	2.2	2	1.7
Thadokham 8	30	13.2	15	12.9
Thadokham 10	73	32.0	25	21.6
Thadokham 11	10	4.4	9	7.8
Phonengam 1	2	0.9	–	–
Phonengam 3	35	15.4	9	7.8
Phonengam 5	21	9.2	9	7.8
Phonengam 6	–	–	2	1.7
Thasano 3	2	0.9	8	6.9
Thasano 6	–	–	2	1.7
Thasano 7	4	1.8	2	1.7
Glutinous Mali	8	3.5	–	–
Non-glutinous Mali	7	3.1	–	–
Local varieties				
Dodaeng	24	10.5	–	–
Phanpae	2	0.9	–	–
Other	–	–	9	7.8
Total	228	100.0	116	100.0

stations (Table 7.5). Most farmers (86%) reported using an improved variety, with 32% using Thadokham (TDK) 10 (a recent release) and 15% using Phonengam (PNG) 3 (an IRRI cross released in 2005 that was high-yielding and relatively drought-tolerant). The major reason given for using improved varieties was that they yielded better than traditional varieties.

Most farmers (85%) used chemical fertilisers for WS rice production, including urea (46-00-00) (34%), ammonium phosphate (16-20-00) (46%), and compound fertilisers such as 10-08-08 (16%) and 15-15-15 (4%). The quantity used varied between households depending on the fertility of their land and their working capital.

The mean yield was 2.2 t/ha, well below the reported mean for Savannakhet as a whole (Table 7.4). Four villages averaged 2.3–2.4 t/ha

Table 7.6   Representative enterprise budget for one hectare of wet-season rice

Item	Quantity	Price (LAK/unit)	Value (LAK)	% of gross revenue
Gross revenue	2.24 t	2000/kg[a]	4,480,000	100.0
Seed	75 kg	3500/kg	262,500	5.9
Fertiliser	150 kg	6780/kg	1,017,000	22.7
Fuel			500,000	11.2
Threshing/hauling	50 bags	5000/bag	250,000	5.6
Land tax		35,000/ha	35,000	0.8
Other costs			100,000	2.2
Family labour	50 days	30,000/day	1,500,000	33.5
Hired labour		720,000/ha	720,000	16.1
Total paid-out costs			2,884,500	64.4
Total costs			4,384,500	97.9
Gross margin 1			1,595,500	35.6
Gross margin 2			95,500	2.1
GM1/day of family labour			31,910	

[a]Farm-gate price in 2012; USD 1 = LAK 1653 (24 April 2019)

but in Dondaeng and Khamsida the mean yield was only 1.7–1.8 t/ha due to lower soil fertility and the impact of drought where farmers did not have access to supplementary irrigation.

A representative enterprise budget for WS rice was prepared based on the survey data (Table 7.6). Given a yield of 2.2 t/ha and a farm-gate price of LAK 2000 per kg of unhusked rice, the gross revenue was LAK 4.48 million per ha. Enterprise expenses or paid-out costs (i.e., excluding the opportunity cost of family labour) totalled LAK 2.89 million per ha, or nearly two thirds of gross revenue. Fertiliser was the largest item, accounting for a third of expenses.

Subtracting paid-out costs from gross revenue gave a gross margin (GM1) of LAK 1.60 million per ha (Table 7.6). Calculated as a return to the input of family labour, this resulted in a figure of LAK 32,000 per day, roughly equal to the prevailing agricultural wage. Thus if the opportunity cost of family labour is valued at LAK 30,000/day, total enterprise costs were LAK 4.39 million, consuming almost all of the gross revenue and giving a gross margin (GM2) close to zero (LAK 95,500 per ha).

The observed farm-gate price in 2011 was LAK 2000 per kg, which was just enough for farmers to break even, given an average total cost of LAK

1950 per kg. However, the government subsequently introduced a minimum farm-gate price of LAK 2500 per kg for paddy rice. If this price is applied to the budget in Table 7.6, GM1 increases to LAK 2.72 million per ha and LAK 54,000 per day. A 25% increase in price thus results in a 70% increase in the return to the family's resources of land and labour.

However, the WS crop is traditionally seen as providing the household's own rice supply rather than as a major source of cash income (Fig. 7.2). With little or no alternative use of paddy land and farm labour during the wet season, to break even while ensuring the staple food supply would be considered a satisfactory outcome. In fact, many farmers also sold surplus rice from the WS harvest, converting otherwise unpaid family labour into a source of cash income for the household.

Farmers identified the major constraints facing their WS rice production. The most frequently mentioned constraints were biophysical, notably insect and pest infestation (27%) and drought (22%), followed by socioeconomic constraints such as lack of capital (13%) and shortage of labour (13%).

Fig. 7.2 Sun-drying paddy before storing for household consumption. (Source: Rob Cramb)

RICE PRODUCTION IN THE DRY SEASON

Just over half the survey households (51%) reported that they grew irrigated rice in the 2011–2012 dry season. The cropping calendar for DS rice is shown in Table 7.2. Nursery preparation began in December, straight after the WS rice harvest. Land preparation and transplanting occurred in January and harvesting in April-May. Hence the DS crop was on a tighter schedule than the WS crop.

The area of DS rice cultivated averaged 1.0 ha and ranged from 0.2 to 5.0 ha. However, 25% of those with DS rice cultivated less than 0.5 ha and 51% cultivated between 0.5 and 1.0 ha. Only 20% had 1–2 ha and 5% had more than 2 ha. Given the lower incidence of DS rice cultivation and the smaller area cultivated by each household, the total area cultivated was around a quarter of that in the wet season.

Farmers used only improved varieties in the dry season (Table 7.5). The most popular of these were, as in the wet season, TDK10 (22%), TDK8 (13%), TDK11 (8%), TDK6 (8%), and PNG3 (8%). However, TDK5, which only 2% of farmers used in the wet season, was also relatively popular due to its short duration, with 19% of DS farmers reporting its use.

Almost all farmers growing DS rice (96%) applied chemical fertiliser. As mentioned above, Vietnamese traders came to most villages at the beginning of the season to supply fertiliser on credit, to be repaid with interest at harvest. The same types of fertiliser were used as in the wet season, including ammonium phosphate (39%), urea (38%), and the compound fertilisers 16-08-08 (18%) and 15-15-15 (8%).

Another representative enterprise budget was prepared for DS rice, again based on the survey data (Table 7.7). With a higher yield of 3.0 t/ha but a lower farm-gate price of LAK 1800 per kg, the gross revenue was 20% higher at LAK 5.4 million per ha. Paid-out costs were 35% higher, totalling LAK 3.9 million per ha, or nearly three quarters of gross revenue. The major cost was again for fertiliser, accounting for 24% of revenue, but there was also an irrigation fee and higher post-harvest costs due to the higher yield.

Subtracting paid-out costs from gross revenue gave a gross margin (GM1) of LAK 1.74 million per ha, only marginally higher than for the WS crop (Table 7.7). The return to family labour was LAK 29,000 per day, marginally lower than in the wet season and just below the agricultural wage. This reflected the higher labour input for the DS crop. Thus costing family labour at LAK 30,000 per day meant that total enterprise

Table 7.7 Representative enterprise budget for one hectare of dry-season rice

Item	Quantity	Price (LAK/unit)	Value (LAK)	% of gross revenue
Gross revenue	3.0 t	1800/kg[a]	5,400,000	100.0
Seed	90 kg	3500/kg	315,000	5.8
Fertiliser	200 kg	6500/kg	1,300,000	24.1
Fuel			500,000	9.3
Irrigation fee		300,000/ha	300,000	5.6
Threshing/hauling	75 bags	5000/bag	315,000	5.8
Other costs			150,000	2.8
Family labour	60 days	30,000/day	1,500,000	27.8
Hired labour		720,000/ha	720,000	13.3
Total paid-out costs			3,960,000	73.3
Total costs			5,460,000	101.1
Gross margin 1			1,740,000	32.2
Gross margin 2			−60,000	
GM1/day of family labour			29,000	

[a]Farm-gate price in 2013; USD 1 = LAK 1653 (24 April 2019)

costs exceeded gross revenue, resulting in a gross margin (GM2) close to zero (−LAK 60,000 per ha).

If the government's minimum price of LAK 2500 per kg was applied, the calculated returns became more acceptable. The GM1 per ha increased to LAK 3.54 million and the GM1 per day to LAK 59,000, almost double the farm wage. The GM2 per ha was LAK 2.04 million.

The major constraints reported for DS rice were similar to those for the wet season—pest and insect infestation (34%), drought and inadequate water supply (24%), lack of capital (19%), shortage of labour (10%), and the absence of an irrigation scheme in the village (9%).

HOUSEHOLD RICE CONSUMPTION AND SALES

Of total annual rice production, over half (56%) was retained for household consumption and about a third (32%) was sold, including 1% as seed. About 6% was given to relatives and 6% kept for seed and poultry feed.

Nearly two thirds of households interviewed (62%) reported that they sold rice in one or both seasons. The average quantity of rice sold was 2.3 tons. The quantity sold varied with farm size and season (Table 7.8). For those selling only WS rice (46% of all sellers), the mean quantity varied

Table 7.8 Quantity of rice sold by farm size and season

Season	Farm size (ha)	% of those selling	Sales per household (tons)		
			Maximum	Minimum	Mean
Wet season only	<1.5	20	0.7	0.1	0.5
	1.51–2.5	15	2.8	0.4	1.4
	2.51–4.5	9	7.4	0.9	2.6
	>4.5	2	10.0	1.3	4.0
Dry season only	<1.5	3	3.0	1.4	2.7
	>1.5	1	4.0	1.8	3.4
Both seasons	<1.5	6	4.5	0.2	1.7
	1.51–2.5	20	5.2	0.4	2.7
	2.51–4.5	18	6.6	0.6	3.0
	>4.5	6	10.8	1.1	4.5

from 0.5 tons to 4.0 tons as farm size increased from less than 1.5 ha to more than 4.5 ha. Only a few households (4%) sold only DS rice, averaging around 3 tons. Half of the rice sellers sold both WS and DS rice, the mean quantities varying from 1.7 tons for those with less than 1.5 ha to 4.5 tons for those with more than 4.5 ha.

Nearly half of households selling rice did so in August and September (Fig. 7.3). Farmers sold rice at this time as they had enough rice in storage for household consumption and the price of paddy rice tended to rise to LAK 2000 per kg during these months, preceding the WS rice harvest. During 2012 the price fluctuated from LAK 1500 to LAK 2000 per kg for eating rice and from LAK 3000 to LAK 3500 per kg for rice seed.

On the other hand, about a fifth of households (19%) produced insufficient rice for their consumption needs, especially in Ban Khamsida and Ban Kaokad. These households experienced a period of rice shortage from one to eight or more months (Fig. 7.4). Of these rice-deficit households, most (55%) experienced a shortage of one to four months, but as many as 31% were short of rice for more than half the year. The main reasons given for facing a rice shortage were limited land (26%), poor soil (26%), and drought in some years (19%). Other problems affecting yield were pests and diseases, flooding, lack of water, and weeds. Ban Khamsida was especially prone to these problems, with mainly upper-level paddies with sandy soils that were more drought-prone and lacked irrigation. However, farmers in other villages with small holdings (23% had 1 ha or less) may also have struggled to meet their subsistence needs.

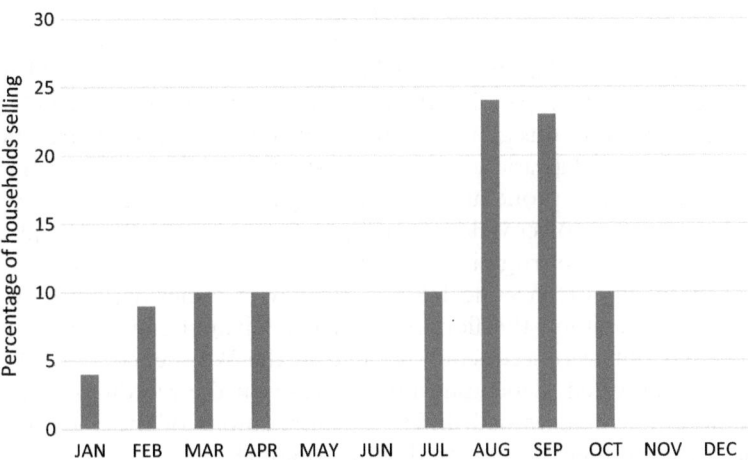

Fig. 7.3   Incidence of selling rice throughout the year (% of those selling, $n = 141$)

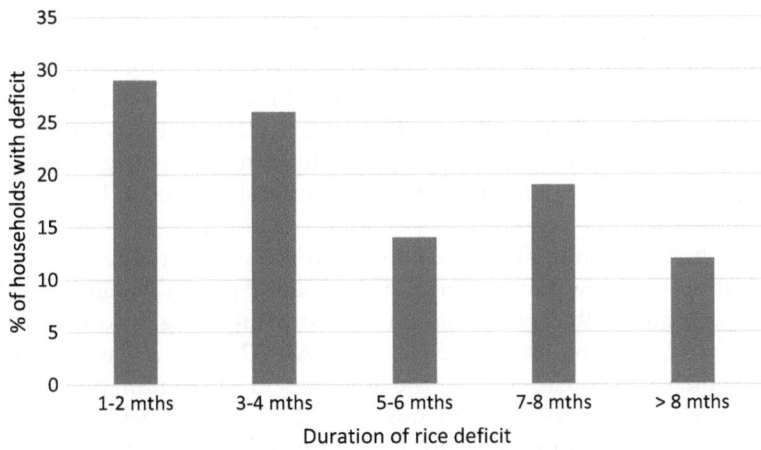

Fig. 7.4   Duration of rice deficit (% of households with deficit, $n = 43$)

CONCLUSION

The survey villages had been growing rice on the Savannakhet Plain for centuries, gradually expanding the cultivated area as population increased. Though situated in this generally favourable environment for rice, the villages encompassed a variety of agroecosystems. Upper paddies with sandy soils were drought-prone and without irrigation, hence could only support WS rice with lower yields. Lower paddies were more fertile and often had access to pump irrigation from rivers, canals, or ponds; hence, they could often support WS and DS rice crops with somewhat higher yields. Lower paddies along the floodplain of the Champhone River also had fertile soils but were frequently flooded in the WS; hence, only DS rice could be cultivated, depending for moisture on the receding floods and irrigation. The villages had different combinations of these agroecosystems, affecting their surplus-producing potential.

Farms were generally small. The mean size was 2.8 ha and 62% of households had 1–3 ha. Almost all farmers planted WS rice, cultivating about 2 ha on average, while only half of them planted DS rice, averaging about 1 ha—a function of access to reliable irrigation. Hence the total DS cultivated area was about a quarter of the WS area. Despite widespread use of improved varieties, fertiliser, and, where available, irrigation, the WS yield averaged only 2.2 t/ha and the DS yield, 3.0 t/ha—well below the Ministry of Agriculture and Forestry's target yield of 4.2 t/ha. Farmers highlighted pest infestations, drought, and insufficient irrigation as the main constraints on yield, and, less frequently, shortages of land, labour, and capital.

For the yields and prices encountered in the survey, the returns to rice cultivation were low. Given a price of LAK 2000 per kg of paddy, the WS crop gave an average gross margin (without imputing a cost to family labour) of LAK 1.60 million per ha and LAK 32,000 per day, enabling a household to just break even. At the government's minimum price of LAK 2500 per kg, the gross margin was LAK 2.72 million per ha and LAK 54,000 per day, a somewhat more attractive return. With a higher yield but a lower price of LAK 1800 per kg, the DS crop gave a gross margin of LAK 1.74 million per ha, but only LAK 29,000 per day due to the higher labour input. In this case a household would just fail to break even. Once again, at a price of LAK 2500 per kg, the gross margin jumped to LAK 3.54 million per ha and LAK 59,000 per day, making the crop somewhat more profitable.

Though more than half (56%) of total rice production was for subsistence, most of the survey farmers were highly commercialised, taking fertiliser and other inputs on credit, hiring labour and machinery, paying for irrigation, and regularly selling rice. Nearly two thirds (62%) of households sold rice; nearly half of these sold only from the WS crop and half sold from both the WS and DS crops. Overall, about a third of total production was sold. The quantity sold by each household was directly proportional to farm size—about 0.7 t/ha for those selling only WS rice and about 1 t/ha for those selling in both seasons. Farmers sold throughout the year, but half (presumably mainly DS producers) sold at the time of highest price in August-September. Farmers in some villages produced seed rice which they sold at almost double the price for eating rice.

However, even in this surplus-producing district, about a fifth of surveyed households with less-favourable resource endowments (mainly smaller, less-productive farms) were unable to meet their subsistence requirements, let alone produce a marketable surplus. In most cases (55%) the shortage was for one to four months. While the surplus producers would benefit from higher paddy prices, these net purchasers of rice would be worse off.

Overall, the survey shows that, even with low yields and low returns, rice production in the Savannakhet Plain can generate a sizeable surplus for marketing within Laos and internationally. However, farmers are going to remain poor unless they can achieve higher yields and obtain higher and more stable prices. Low incomes will increase the incentives for younger household members to migrate to Vientiane or to Thailand for employment, adding to the shortage of farm labour. Nevertheless, given its comparative advantage in rice production, the Savannakhet Plain is a good focal area for increased investment in research, extension, input supply, mechanisation, and infrastructure to boost productivity and farm incomes.

NOTE

1. A small area of upland rice is produced in the hilly interior towards the Vietnam border, though this is probably under-reported.

CHAPTER 8

The Supply of Inputs to Rice Farmers in Savannakhet

Chitpasong Kousonsavath and Silinthone Sacklokham

INTRODUCTION

The policy of rice intensification in Laos is dependent on an adequate supply of key inputs such as high-yielding seeds, good-quality fertilizers, reliable irrigation, affordable finance, and appropriate information. The study reported in this chapter focused on two crucial inputs for increased productivity in rice—seeds and fertilizers. The main objectives are to (a) map the seed and fertilizer supply chains; (b) identify the key actors in each chain and their roles; (c) identify major problems affecting the performance of each chain; and (d) provide recommendations for improvements in input supply.

Six villages in Champhone District, Savannakhet Province, were selected, as described in Chap. 7. Two villages (Phalaeng and Piakha) produced rice

C. Kousonsavath (✉)
Faculty of Agriculture, Department of Rural Economics and Food Technology, National University of Laos, Vientiane, Laos

S. Sacklokham
SEAMEO Regional Centre for Community Education Development, Vientiane, Laos
e-mail: s.sacklokham@nuol.edu.la

© The Author(s) 2020   169
R. Cramb (ed.), *White Gold: The Commercialisation of Rice Farming in the Lower Mekong Basin*,
https://doi.org/10.1007/978-981-15-0998-8_8

Table 8.1   Number of interviewees by type of actor

Type of actor	No. interviewed
Farmers	228
Vietnamese traders	1
Fertilizer import companies	2
Fertilizer shops	8
Rice millers	3
Thasano Seed Production Centre	1
Rice Production Improvement Project	1
Individual villagers	1
Individual suppliers	5

primarily for commercial purposes, two (Buekthong and Dondaeng) produced rice primarily for family consumption but regularly sold a surplus, and two (Khamsida and Khaokad) produced rice only for self-sufficiency. A preliminary survey was conducted to determine the broad picture of the seed and fertilizer supply chains in the six villages and to identify the key actors in each chain. This provided the basis for selecting interviewees in the second visit in March 2012. The types and numbers of interviewees are listed in Table 8.1. The farmers were selected randomly from the list of farmers in each village, including farmers who were members of a seed production group or involved in the government's Rice Production Improvement Project (RPIP).

THE FERTILIZER SUPPLY CHAIN

As shown in Chap. 7, most rice farmers (85%) in the Savannakhet Plain used chemical fertilizers for the wet-season (WS) crop and all used chemical fertilizers for the dry-season (DS) crop. The commonly used fertilizers were urea (46-00-00), ammonium phosphate (16-20-00), and compound fertilizers such as 16-08-08 (16%) and 15-15-15 (4%). The most common fertilizer brands used by Savannakhet farmers were Ox Brand from the Thai Central Chemical Public Company Limited, Rabbit Brand from the Chia Tai Company Limited, and Football Brand from an unidentified company in Vietnam. These were the brands with higher quality and price. Most farmers used fertilizers based on their financial capacity; only a few based their usage on technical requirements. Some farmers could not afford to apply fertilizers due to the high and fluctuating price. In many cases, the cost of applying additional fertilizers outweighed the additional return (see Chap. 10). In addition to applying chemical fertilizers, farmers in Savannakhet still applied animal manure to their rice fields before land preparation. The animal manure was sought from within the family and the village. Farmers applied as much as they could find as the number of animals had decreased and manure was increasingly scarce.

Actors in the Supply Chain

The fertilizer supply chain for Savannakhet Province is illustrated in Fig. 8.1. The fertilizers used by farmers in the province were mainly sourced from Thailand, Vietnam, and Taiwan. Most imports occurred through the border checkpoints at Savannakhet-Mukdahan (Thailand) and Dansavanh-Lao Bao (Vietnam), at either end of National Route 9 which traversed the province (see Fig. 5.1 in Chap. 5). The major types of supplier are discussed in turn.

(a) *Individual agents.* The individuals in Fig. 8.1 were villagers who acted as sales agents for a fertilizer company. They supplied a liquid organic fertilizer called *Mahalap Mahalouy*, supplied by the

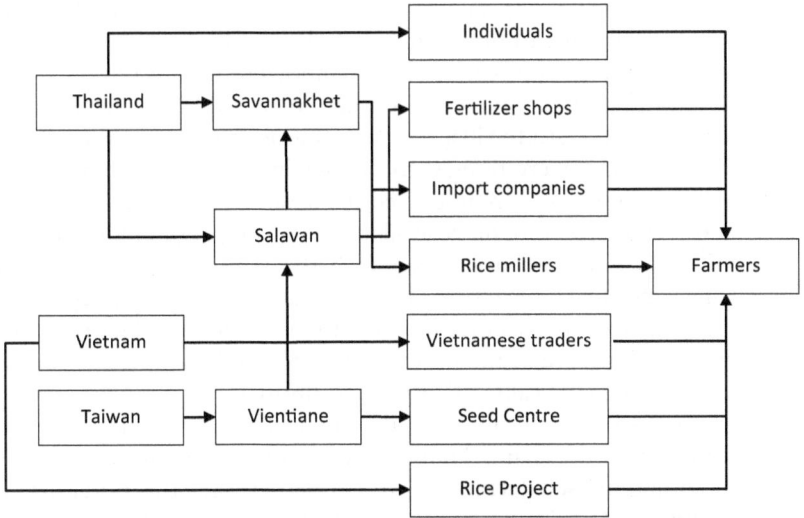

Fig. 8.1   Fertilizer supply chain for Savannakhet Province

Lifestyles Company in Thailand. The concentrated fertilizer was to be mixed with water and sprayed onto the rice leaves every seven days. Farmers who had used this fertilizer said that rice production had improved as a result, though the response was slow compared to chemical fertilizers. The fertilizers came in a set of two bottles, each costing THB 580 (around LAK 150,000).

These individual agents also supplied chemical fertilizers, buying up stocks and storing them in their houses. The types of fertilizers supplied in this way were 15-15-15, 46-00-00, 16-20-00, and 16-8-8. Farmers could purchase directly from the individual. Payment could be made in cash or the fertilizer could be taken on credit. There was no interest charged and farmers could simply repay the credit after harvest.

(b) *Import companies and fertilizer shops.* These were not solely for selling fertilizers; their main activity was selling construction materials. However, they would have a corner of the shop devoted to fertilizers during the production season, mostly imported from Thailand. The same four types of fertilizers were sold—46-0-0, 15-15-15, 16-20-0, and 16-8-8. These distributors used to provide credit to farmers but, due to the low rate of repayment, only cash sales were now made. The import companies usually imported fertilizers directly, whereas the shops were supplied by mobile vendors who visited from time to time. These vendors could not be traced in the study and it was unclear how they imported their fertilizers.

(c) *Rice millers.* There were three rice millers supplying fertilizers to farmers in the study villages. The same four types were provided. The fertilizers were bought from a fertilizer shop in Kilometre-35 Village and some were imported from Salavan Province. Both cash and credit sales were made available to farmers. For credit sales, the miller would make a contract with the farmers which stated the total amount to be repaid, the due date, the form of repayment (cash or rice—if the latter the quantity was calculated based on the current rice price at the time of drawing up the contract), and the interest rate (typically 1.0–2.5% per month).

(d) *Vietnamese traders.* These traders played a significant role in the fertilizer supply chain. Though they did not come to Savannakhet intending to sell fertilizers, in 2008 they saw the potential for supplying fertilizers to farmers in Champhone District. They imported around 30–40 tons per year from Vietnam, all in the wet season around May, June, and July. It was unclear how they brought in the fertilizers. Farmers said that at the beginning of the production season the traders came to the village with a load of fertilizers in their truck and the farmers were free to select whatever fertilizer they wanted. They brought in three main types—46-0-0, 15-15-15, and 16-20-0. Although the price of the Vietnamese fertilizers was cheaper than fertilizers from Thailand, farmers claimed they had to use almost twice as much Vietnamese fertilizers to obtain the same yield as with Thai fertilizers.

The Vietnamese traders supplied fertilizers to farmers in the village on credit; once farmers had cash or after the harvest was completed, the traders would come back to the village to collect the money. Due to the generally high price of fertilizers, this form of credit was popular with farmers who were short of working capital. Despite the poorer quality of the fertilizers, farmers were attracted by the availability of credit and the saving on the time and cost of purchasing fertilizers in town.

(e) *Thasano Seed Production Centre.* The seed production centre at Thasano, under the Ministry of Agriculture and Forestry, was located on National Road 13 just west of Champhone District. The Centre worked with village heads to organize seed production groups of 20–30 farm households to which it provided fertilizers. These groups were established because the demand for improved seeds was exceeding the Centre's own production capacity. Participating farmers had to agree with the Centre and village head to comply with the seed production techniques and standards provided by the Centre. Once a farmer group was formed, a contract was developed between the group and the Centre. The contract stated clearly the seed production techniques or standards that the farmers had to follow, that the output had to be sold to the Centre, and that the Centre would not purchase seeds from farmers who did not follow the specified procedures.

The Centre supplied two types of fertilizers to the seed producing groups—chemical and bio-fertilizers. The chemical fertilizers included 46:00:00, 16:20:00, and 15.15.15. The Centre ordered these fertilizers as required from Siam Machinery Intertrade Company Limited, based in Thailand, with importation through the Dansavanh-Lao Bao border crossing. However, the Centre had received exemption from import duty because of its public role.

The bio-fertilizers were supplied by Rfarm Company, with its head office in the capital and its factory located in Hin Hurb District, Vientiane Province. The Centre understood that the company imported fertilizers directly from Taiwan. However, further enquiries revealed that the company imports materials from Taiwan and then processes, repacks, and distributes the product in Laos.

(f) *Rice Production Improvement Project.* This project aimed to improve rice productivity for farmers and supply good rice variety for farmers who lack access to high-quality seed. The village head and the project coordinator collaborated closely in organizing farmer groups of 20 farmers each. The participants had to have at least 0.5 ha of paddy land, be hard-working farmers, and belong to a minority group that had less access to fertilizers and seeds. Once the farmer group was organized and the group committee assigned, the project supplied them with fertilizers of two types—15:15:15 and 46:00:00. The project imported fertilizers directly from Vietnam and stamped a Lao logo on the bags before supplying the farmers. One group of farmers received 50 bags of fertilizers—30 bags of 15:15:15 and 20 bags of 46:00:00. There was neither any charge to the farmer group nor any requirement to repay the cost at the end of the season. The fertilizer was only made available to farmer groups; farmers who were not members could benefit from this line of supply.

Table 8.2 shows the estimated annual volume of fertilizers imported by each of the above actors. The Thasano Seed Production Centre imported a large quantity but it was a single supplier and mainly supplied its own seed producers. The import companies and input supply shops were more

Table 8.2 Estimated annual imports of fertilizer per supplier

Supplier	Imports (tons/year)
Vietnamese trader	40
Fertilizer shop	5800
Rice miller	48
Import company	6175
Seed production centre	37,025

Source: Based on interviews with suppliers

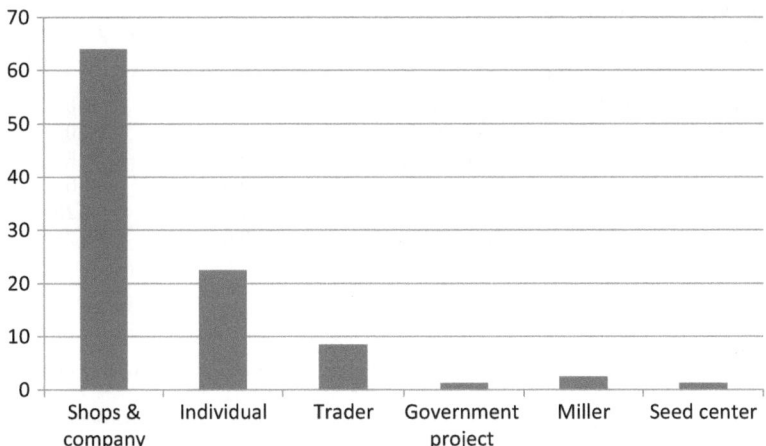

Fig. 8.2 Farmers' sources of fertilizer (% of survey respondents)

numerous and handled about **6000** tons each, hence these were the main suppliers. The Vietnamese traders and rice millers each handled a smaller quantity, but the traders were important suppliers in some villages.

About two-thirds of the farmers surveyed obtained their fertilizers from shops and import companies (Fig. 8.2). The reason given was that these suppliers had lower prices. This was confirmed in interviews with the different suppliers. Exemptions from import duty helped to lower the price. About a third of farmers purchased fertilizers from individual agents in the village and/or visiting Vietnamese traders. Millers, the seed production centre, and the Rice Production Improvement Project each supplied only a small percentage of farmers.

Fertilizer Transactions

There were two forms of payment for fertilizers—cash and credit (Table 8.3). The vast majority of farmers producing rice primarily for commercial purposes paid in cash because the price was lower than under the credit system. In the latter case, the price incorporated an implicit interest rate that varied between suppliers. Farmers using credit (34%) were those with limited capital during the production season, hence they had no choice but to pay the interest premium. These farmers stated that if they had available cash at the time of purchasing production inputs they

Table 8.3 Incidence of farmers paying cash or using credit for fertilizer purchases, by type of supplier

Supplier	Farmers paying cash (%)	Farmers using credit (%)
Shop/import company	60.0	4.0
Individual agent	5.5	17.0
Vietnamese trader	0.0	8.5
Miller	0.5	2.0
Seed centre	0.0	1.25
Rice production project	0.0	1.25

Source: Farmer survey, 2012

Table 8.4 Comparison of fertilizer prices between cash payment and credit

Type of fertilizer	Cash price (kip/50 kg bag)	Credit price (kip/50 kg bag)	Difference (kip/50 kg bag)
Thai brand (Ox Brand, Rabbit Brand)			
46-00-00	250,000	300,000	50,000
15-15-15	300,000	375,000	75,000
Vietnamese brand (Football Brand)			
46-00-00	150,000–175,000	250,000	75,000–100,000

Source: Farmer survey, 2012

would prefer to pay up-front in cash. Among the suppliers, the village agents and Vietnamese traders were the most willing to supply fertilizers on credit. These traders were flexible and willing to negotiate the time of payment and to receive whatever amount the farmer could pay.

The cash and credit prices for selected types of fertilizers are shown in Table 8.4. One point to note is that urea from Thailand was up to two-thirds more expensive than urea from Vietnam, presumably reflecting the quality differences reported by farmers. The table also shows variation in the implicit interest charge incorporated in the credit price. For Thai brand urea, supplied by shops and import companies, the premium was 20% and for Thai brand compound fertilizer (15-15-15) it was 25%. Assuming six months until payment, the annualized interest rate was 40–50%. However, for Vietnamese urea supplied by Vietnamese traders, the premium was 40–65%, representing an annualized interest rate of 80–130%. This higher rate probably reflected the greater flexibility of the Vietnamese traders in the payment time and amount.

Data were obtained on the margins between the purchasing and selling prices of different types of fertilizers for different suppliers (Table 8.5).

Table 8.5 Marketing margins for fertilizer suppliers

Supplier and type of fertilizer	Purchase price (LAK/kg)	Selling price (LAK/kg)	Margin (%)
Trader			
46-00-00	4700	5900	25.5
15-15-15	4400	4900	11.4
16-20-00	4400	4900	11.4
Shop			
46-00-00	4700	4900	4.3
15-15-15	4800	4950	3.1
16-20-00	4400	4500	2.3
Import company			
46-00-00	4100	4200	2.4
15-15-15	4300	4200	(2.4)
16-20-00	3600	4200	16.7
Miller			
46-00-00	4800	5700	18.8
15-15-15	5000	6000	20.0
16-20-00	4200	4700	11.9

The traders had the highest margins, up to 25% for urea, but this included the cost of delivery to the village. Millers also had high margins of around 20%. The shops and import companies had relatively low margins of 2–4%. The import companies purchased fertilizers 10–20% more cheaply than the other suppliers and could also distribute at a lower price. Moreover, although the purchasing price of the three main fertilizers was different, this distributor sold them at the same price—LAK 4200 per kg—which was up to 30% cheaper than other suppliers. When asked why the selling price was uniformly low, the distributor merely remarked that the price was sufficient to compensate for the purchase price.

Constraints and Problems

The survey highlighted some problems with the fertilizer supply chain, both for farmers and importers/distributors. The problems reported by farmers related to quality, capital, and price. The problems reported by suppliers related to documentation processes and bad debts.

Many farmers complained about the poor quality of the fertilizers they purchased. They said that the rice was slow to respond to some types of fertilizers. This was especially the case for Vietnamese fertilizers, while the Thai fertilizer was perceived to be of high quality. Some farmers mixed fertilizers from Vietnam with fertilizers from Thailand, which they said gave a better response. Access to financial resources was important to enable farmers to have fertilizers when they needed them most. The only financial support available was from the Agricultural Promotion Bank, but to qualify for a loan farmers had to form a group of 10–20 members and submit a production proposal to the Bank. Each member had to guarantee that every other member would repay their loan or else the group would be liable. Many farmer groups had failed to repay their loans. The only alternative for capital-scarce farmers was to take fertilizers on credit from Vietnamese traders and individual agents with much higher implicit interest rates, as shown above. Compounding these problems, the price of fertilizers was continually increasing, making it harder to purchase high-quality fertilizers and reducing the incentive to apply optimal amounts.

Many importers complained about the documentation procedure for importing fertilizers. The complexity of this process led to higher import costs, pushing up the retail price encountered by farmers. Suppliers also commented on the low rate of repayment for fertilizer credit. Many had stopped providing credit as a result. Only the Vietnamese traders and some individual agents still provided credit, with very flexible repayment times and instalments. However, they offset their risks with higher marketing margins and very high implicit interest rates.

THE SEED SUPPLY CHAIN

Origin and Uptake of Improved Varieties

Though improved rice varieties had been introduced to Laos since 1960, in 1990 about 95% of the lowland WS rice crop was still based on traditional varieties (Inthapanya et al. 2006). From the 1990s, rice breeding and seed production stations were established and a succession of high-yielding glutinous varieties were selected and disseminated to the major rice-growing areas in the Mekong Valley, with rapid uptake by farmers (see Chap. 6). At the time of this study, there were four active seed production centres, breeding and supplying improved varieties of rice throughout the country—three in Vientiane Capital, Napok, Phonengam, and Dondaeng, and one in Savannakhet, the Thasano Seed Production Centre mentioned earlier.

These centres produced a wide range of varieties with different attributes. Thasano produced ten varieties and about 50 tons of rice seeds per year. Farmers in Savannakhet also used varieties from Vientiane, including Thadokham (TDK) and Phonengam (PNG) varieties, and a variety from the Provincial Agriculture and Forestry Office (PAFO) (Homsavanh). Farmers also selected and conserved their own seeds.

In the survey, most farmers (86%) reported using an improved variety in the WS, with 32% using TDK10, a relatively recent release, and 15% using PNG 3, a high-yielding, drought-tolerant variety released in 2005. All farmers used only improved varieties in the DS, including TDK10 (22%), TDK5 (19%), TDK8 (13%), and PNG3 (8%). It was noteworthy that the use of TDK5, a short-duration variety, increased in the DS.

Actors in the Supply Chain

The structure of the seed supply chain is shown in Fig. 8.3. The principal actors were the seed production centres shown on the left, multiplying up the first and second rounds (R1 and R2) of the certified seed; the seed production groups, producing R2 and R3 seeds; the PAFO and the District Agriculture and Forestry Office (DAFO) that distributed the seed, along with millers and the Rice Production Improvement Project; and the farmers,

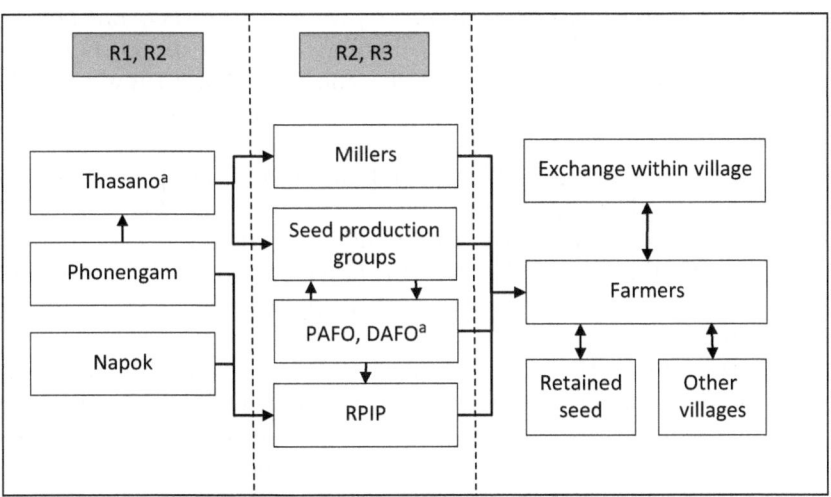

Fig. 8.3 The seed supply chain in Savannakhet. [a]Thasano Centre and PAFO/DAFO buy back seeds from the seed production groups at 10% above the market price for paddy rice

who purchased the seed, selected and retained the seed for their own use, and exchanged the seed with other farmers within and outside their village.

(a) *Seed production centres.* These centres produced both R1 and R2 seeds. However, due to the increasing demand for seeds of improved varieties, Thasano was working with groups of farmers to multiply seeds. A farmer group consisted of around 20 farmers who made an agreement with the Centre to produce and supply seeds. Centre staff visited the farmers' fields from time to time to ensure that they were meeting the required standards. If the farmers' seed production met the standards for certification, the Centre or the PAFO bought the seed at a premium price; otherwise farmers had to sell it in the market as normal eating rice.

(b) *Seed production groups.* The farmer seed production groups were organized under the supervision of the Thasano Centre and the Rice Production Improvement Project, run by the PAFO. These farmers produced either R2 or R3 seed to supply the Centre and the project. Thasano and the PAFO paid a 10% premium for seeds produced according to the requirements for certification. The farmer groups were also allowed to sell seeds directly to farmers.

(c) *Rice Production Improvement Project (RPIP), PAFO, and DAFO.* The RPIP was being implemented by the PAFO and DAFOs with financial support from the World Bank. One of its goals was to help poor and minority farmers to get access to good-quality seed. The project was working with 33 farmer groups and 615 households, including eight groups in Champhone District. The project worked with the village head to organize the farmer group. One bag of R2 seed was provided free to each household. The PAFO/DAFO then bought the seed produced by the farmer group at a 10% premium. This seed still needed further purification before selling commercially.

(d) *Rice millers.* Rice millers supplied seeds to farmers in the same way as they supplied fertilizers. The millers bought seeds from the Thasano Seed Production Centre and sold it to farmers at LAK 2000 per kg. The farmers could pay for the seeds immediately in cash or after harvest in cash or rice (calculated from the current rice price).

(e) *Farmers.* Farmers sourced seeds from many different distributors depending on their circumstances, including the seed production

Table 8.6 Sources of seed reported by survey farmers

Source	No. of respondents	% of respondents
Seed production centre	40	17.5
Rice Production Improvement Project	45	19.7
Farmers' seed production group	42	18.4
Miller	4	1.8
Within village	93	40.8
Other village	4	1.8
Total	228	100.0

Source: Household survey, March 2012

centre (18%), the RPIP (20%), and seed production groups (18%) (Table 8.6). However, the most common source was other farmers within the village (41%). Farmers reported that they observed each other's rice fields and if someone had a variety that provided higher yield and better quality, they would exchange the seed with that farmer. Observing neighbouring rice fields was a simple technique preferred by farmers to find a suitable new variety because if the variety performed well in the neighbour's field, in their experience it was likely to be well adapted to his or her own field (Fig. 8.4). This cultural practice was of long-standing and occurred throughout Laos, providing the basis for technical change. The village communities were relatively small and homogenous so that everyone knew each other, hence it was easier for farmers to observe fields and exchange rice varieties with their neighbours than to search for improved varieties independently. By this means, the improved varieties developed from the 1990s have spread rapidly in lowland areas.

The price paid by farmers for seeds from different sources is compared with the estimated cost of seed production in Table 8.7. The seed production centres had a higher cost of production and a higher selling price (LAK 5000–6000 per kg). The higher price reflected both their higher costs and significantly higher margins (66–100%). The farmer groups produced seeds at a lower cost and sold to other farmers at a lower price, with a margin of 25–50%. However, their margin in selling back to the government agencies was only 25% or less. The millers also sold seeds more cheaply and with a lower margin.

Fig. 8.4 Farmer in paddy field planted with improved variety in Savannakhet Province. (Source: Rob Cramb)

Problems and Constraints

Several problems were identified in the course of the survey. Farmers reported that the seed they bought from the main suppliers was mostly impure. This suggests that the seed production process was not properly monitored and so there was still a problem of mixing seeds of different varieties. This was compounded by the absence of a proper seed certification system to provide information on whether the seed the farmer bought was in compliance with seed production standards. Such a seed certification system would solve the problem of impure seeds and farmers would have more confidence in the quality of the seed they purchased.

Table 8.7 Production costs and selling prices of rice seeds by source

Seed producer	Production/purchase cost (LAK/kg)	Selling price (LAK/kg)	Margin (%)
Thasano Seed Production Centre	3000	5000–6000	66–100
Napok Seed Production Centre	3000	5000	66
Farmer group (selling to farmers)	2000	2500–3000	25–50
Farmer group (selling to government)	2000	2000–2500[a]	0–25
Millers	2000	2500–3000	25–50

Source: Field survey, March 2012
[a]Thasano Centre and PAFO/DAFO buy back seeds from the seed production groups at 10% above the market price for paddy rice

Farmers also felt that there was a lack of varieties for specific soil and climatic conditions (e.g., infertile sandy soils, drought, and flooding). They wanted seed that was clearly labelled regarding its suitability to specific environments (e.g., flood-tolerant). However, the four seed production centres had not yet released such site-specific varieties, focusing rather on varieties that would do reasonably well in a range of environments. Farmers also reported that the available varieties were not resistant to pests and diseases, restricting productivity in some areas.

CONCLUSION

Farmers in Champhone District had mostly adopted the seed-fertilizer technology that formed the basis of increased yields and productivity in Asian rice farming (Chaps. 1 and 6). Mechanization of land preparation through the use of hand tractors was also widespread. Many had also intensified their cropping system, using irrigation to cultivate a DS crop as well as the traditional rainfed WS crop. However, the productivity and profitability of rice farming remained low. Constraints to the supply of seeds and fertilizers can explain part of this dilemma.

Farmers used mostly improved varieties for the WS crop and entirely so for the DS crop. These were mostly glutinous varieties, incorporating introductions from the International Rice Research Institute (IRRI) and Thailand with Lao genetic material, to produce higher yields in a range of adverse environments. They had been progressively released since the 1990s and were rapidly adopted and disseminated. Just over a third of Champhone farmers (37%) sourced their seed from the formal public-sector supply chain, including seed production centres, PAFO, DAFO, and a government-implemented rice development project. The private sector played little role, apart from some millers who included seed in their advance of inputs to selected surplus-producing farmers. Most farmers (61%) obtained seed from other farmers, including 18% who bought from a seed production group, set up by the seed production centres to accelerate the multiplication of seeds, and 43% who exchanged the seed with their neighbours, after observing the performance of different varieties in the field, and then selected and retained the seed for their subsequent use. In this way, they gained access to the improved varieties, though probably with some deterioration in seed quality and hence yield (Diaz et al. 1998). Indeed, the main problems identified concerned the lack of proper seed certification, the supply of impure seeds, lack of varieties for specific soil and climatic conditions, and lack of varieties with resistance to the prevalent pests and diseases.

Farmers also used various types of fertilizers in their rice production, including chemical fertilizers from Thailand and Vietnam, organic fertilizers, and animal manure. While the increasingly limited supply of animal manure was sourced from neighbours in the village, the manufactured fertilizers were sourced from a range of mainly private-sector distributors, including import companies, input supply shops, mobile traders, individual villagers acting as agents, and rice millers. In addition, the government seed production centre and the Rice Production Improvement Project supplied fertilizers to farmers participating in their activities. Most of these suppliers provided chemical fertilizers, including urea, ammonium phosphate, and compound nitrogen-phosphorus-potassium (NPK) fertilizers; only a few provided organic fertilizers. The most important suppliers were the import companies and shops, who preferred cash payment at the time of purchase. In contrast, the traders mainly supplied fertilizers in the village on credit, to be repaid soon after harvest, with an implicit interest charge of 50–100% p.a. incorporated in the price. Farmers with limited capital were more likely to use this credit system. The major problems identified in this fertilizer supply chain were the poor quality of especially the Vietnamese product, the lack of financial resources to buy sufficient fertilizers, and the increasing price of fertilizers.

There is clearly scope for policy intervention to improve the supply and use of productive inputs for more intensive rice production. Further investment in the rice breeding and seed production centres may be needed to develop suitable varieties for the range of rice environments encountered by farmers and to improve the quality of the seed supplied. This needs to be accompanied by an official seed certification system to ensure farmers have access to high-quality seeds and information about varieties suited to their local situations. While the increasing price of fertilizers was clearly a constraint, marketing margins were quite low, implying a competitive in-country distribution system. Intervening to control or subsidize the price of fertilizers can be a costly and administratively cumbersome policy and is unlikely to be effective. The government could, however, take action to further simplify the import process, which would help reduce costs that are passed on to farmers, to increase the capacity to monitor and enforce fertilizer quality standards, and provide more site- and variety-specific information to farmers regarding optimal fertilizer use.

REFERENCES

Diaz, C., Hossain, M., Merca, S., and Mew, T., 1998. Seed quality and effect on rice yield: findings from farmer participatory experiments in Central Luzon, Philippines. *Philippines Journal of Crop Science* 23(2): 111–119.

Inthapanya, P., Boualaphanh, C., Hatsadong, and Schiller, J. M., 2006. The history of lowland rice variety improvement in Laos. In J. M. Schiller, M. B. Chanphengxay, B. Linquist, and S. Appa Rao, eds., *Rice in Laos*, pp. 325–348. Los Baños, Philippines: International Rice Research Institute.

CHAPTER 9

Rice Marketing and Cross-Border Trade in Savannakhet

Phengkhouane Manivong and Silinthone Sacklokham

INTRODUCTION

The purpose of this study was to identify the pattern of rice marketing and cross-border trade in Savannakhet Province. The province has the largest output of rice in Laos, producing a surplus for other regions in the country, and, given its location between Thailand and Vietnam, it is an important conduit for trade in the Greater Mekong Subregion (GMS). National Road 13 runs through the province along the Mekong corridor from north to south, linking the major rice-growing districts to the capital, Vientiane, and to rice-deficit provinces (see Fig. 5.1 in Chap. 5). National Road 9 traverses the province from west to east, with an international border point at either end—the Savannakhet-Mukdahan border crossing with Thailand and the Dansavanh-Lao Bao border crossing with Vietnam (Fig. 9.1). The movement of goods through these border crossings has

P. Manivong (✉)
Faculty of Agriculture, National University of Laos, Vientiane, Laos
e-mail: manivongp@afd.fr

S. Sacklokham
SEAMEO Regional Centre for Community Education Development, Vientiane, Laos
e-mail: s.sacklokham@nuol.edu.la

© The Author(s) 2020
R. Cramb (ed.), *White Gold: The Commercialisation of Rice Farming in the Lower Mekong Basin*,
https://doi.org/10.1007/978-981-15-0998-8_9

Fig. 9.1 International border points in Savannakhet Province: Savannakhet-Mukdahan (left); Dansavanh-Lao Bao (right)

accelerated since 2007, with Thailand and Vietnam now being Savannakhet's main trading partners. According to the Provincial Agriculture and Forestry Office (PAFO), about 25,000 tons of milled rice produced in Savannakhet Province was exported to Vietnam via the border point in 2010, which is double the volume exported in 2009. The province also imported rice from Thailand and about 5000 tons of Thai rice was moved to Vietnam through Savannakhet's borders.

The survey was conducted in March 2012 by a team of four staff from the Faculty of Agriculture of the National University of Laos. The survey focused on three locations. The first was Champhone District, which was part of the main rice-producing area in the Savannakhet Plain in the western part of the province and accounted for 16% of the province's rice output (Chap. 7). The second and third were the border crossings with Thailand and Vietnam. The target groups for interviews included all actors involved in the rice value chain, including producers, traders, millers, exporters, and officials at different administrative levels. A snowball sampling approach was used for the survey. Information from the first round of interviews was used to identify secondary respondents. This method was continued in both "upstream" and "downstream" directions until both the source and destination of the traded rice were identified. The study started with 144 farmers in six villages in Champhone District (Chap. 7). The sample farmers were selected from those farmers in each village who regularly sold rice from 2009 to 2011, using probability proportional to size (PPS) sampling. Data were obtained using semi-structured interviews guided by a checklist. The information provided by the producers (the initial respondents) was then used to identify the other

Table 9.1   Number of interviewees by category

Category	No. interviewed
Producer	144
Collector	2
Rice miller	11
Lao exporter	5
Retailer	23
Total	185

actors involved in the rice trade. A total of 185 interviews were conducted in this way (Table 9.1). The data are used in this chapter, first, to trace the rice market chain from farms in Champhone to domestic and international markets and, second, to examine in particular the processes of cross-border trade in rice.

THE RICE MARKET CHAIN IN SAVANNAKHET PROVINCE

A number of government policies affected rice marketing and trade in Savannakhet Province. The Ministry of Industry and Commerce (MOIC) had introduced a quota system for the export of rice in 2005 as a means to ensure domestic supply. In this instruction, rice exports had to adhere to the quota allocated by the Provincial Office of Industry and Commerce (PICO). A modest tax on rice exports was introduced in 2008, set at 5–8%. In an extreme situation, rice exports could be banned. A temporary rice export ban was imposed from November 2010 to February 2011 to safeguard domestic supply and keep domestic prices under control. In September 2011, the government issued a further policy instruction to control the price of rice. A guaranteed minimum farm-gate price of LAK 2500 (USD 0.31) per kg and a mill-gate price of LAK 3000 (USD 0.375) per kg were stipulated. The State Food Enterprise (SFE) was a major player in the domestic market, buying rice at the controlled price and holding rice stocks.

The rice market chain in Savannakhet was analysed from rice farmers in Champhone District, the main production area in the province, to buyers and sellers within Laos and in Vietnam and Thailand (Fig. 9.2). The key actors were the producers, millers, domestic retailers, and exporters. Producers in Champhone sold most of their rice surplus to a rice miller within the district. Only a few sold to a collector who then immediately

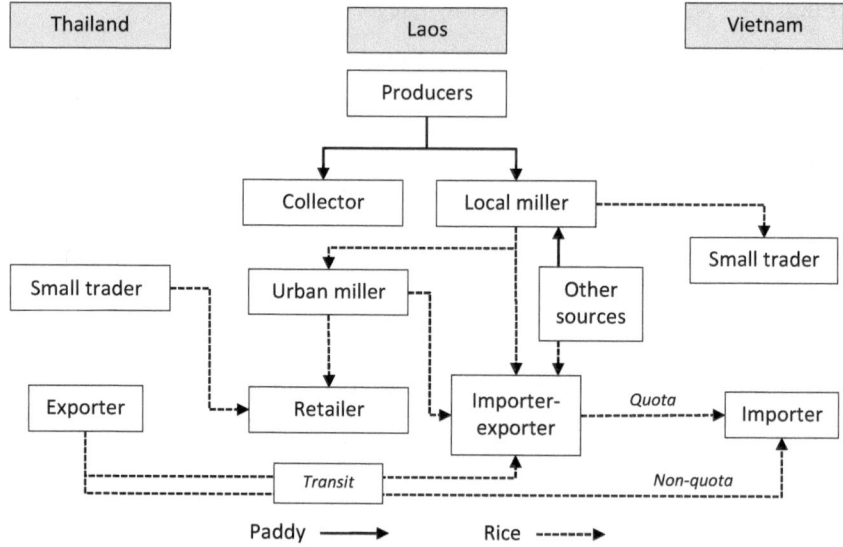

Fig. 9.2 Mapping of rice marketing and trade in Savannakhet Province

resold to a miller. The buyers typically came to the village to collect paddy rice—around 9 tons per trip for small traders and 15–20 tons for large traders.

In Champhone District, local millers were the key actors in the rice marketing system because all rice sold by producers went through a rice mill. Some of the local millers formed a trading network with their suppliers. They gave loans and farm inputs to farmers and collectors with their private funds to ensure a good quality of rice and a reliable supply. There were 57 mills registered in the district. They processed 2–4 tons of paddy rice per day, producing 60–65 kg of polished rice from each 100 kg of paddy rice. These rice mills were organized into an association at the initiative of the Trade Division of the PICO. However, purchasing and selling were still undertaken individually rather than collectively.

Farmers generally sold surplus rice twice a year. Most (90%) sold in the period from November to January following the wet-season (WS) harvest, and over a third (35%) also sold from May to July following the dry-season (DS) harvest. The millers bought 68% of their rice in the November–January period and the rest in May–July, while the collectors bought 64%

of their rice in the May–July period and the rest in November–January. Thus, more of the millers relied on collectors to make the farm-level purchases of the DS harvest.

The producer was considered a price-taker because the millers offered a farm-gate price for paddy rice based on quality criteria (e.g., average moisture content and percentage of foreign matter). The millers would set an acceptable level for each quality criterion, discount the price for each percentage point over the acceptable level, or reject the rice if damage or moisture was above a certain limit. The price was usually displayed at the front of the mill. For the polished rice, the rice mill proposed a wholesale price for the local market. However, the retail price of rice in the local market was under the control of the Trade Division of PICO in an attempt to avoid price spikes.

Farm-gate prices fluctuated seasonally in predictable ways (Table 9.2). Prices were lowest in the months following the WS (November–January) and DS (May–July) harvests, when farmers sold most of their surplus rice and supplies were abundant. Prices peaked in August–October when supplies were scarce in the lead-up to the WS harvest. Table 9.2 also shows the year-to-year fluctuation around the 2010 price spike. The peak price in 2010 was LAK 4500 per kg of paddy rice in August–October, which was 28% higher than for the same period in 2009. Exports of rice also peaked in this year. However, the price dropped by a third to LAK 2400–3000 per kg in 2011 because of the export ban introduced in November 2010 to regulate the domestic price. The price increased slightly in 2012 and had stabilized at LAK 2500–3000 per kg.

Table 9.2   Farm-gate prices of paddy rice by month, 2009–2011 (LAK/kg)

Month	2009	2010	2011
January	2800	3000	2000
March			
April			
May	2000	2500	1700
June	2000	2500	1700
July	2000	2500	1700
August	3200–3500	3800–4500	2400–3000
September	3200–3500	3800–4500	2400–3000
October	3200–3500	3800–4500	2400–3000
November	2800	3000	2000
December	2800	3000	2000

When the export of rice was surging in 2010, polished rice from the mills in Champhone District was mainly sold to export companies. Some was sold to individual traders for the informal trade. Typically the buyers came to the rice mill to make their purchases and check for product quality. However, by 2011, only 11 of the 57 mills had trading activities. Most mills, especially the small mills, had temporarily stopped trading activities due to the ban on exports imposed in late 2010. These mills bought rice and stocked it for the local market. However, they claimed to be using only 30% of their capacity as they still held stocks from the 2010 harvest.

As noted above, rice was exported from Champhone District through both trading companies and individuals (Fig. 9.2). These two channels can be classified as formal and informal trade. The formal trade was conducted by Lao import-export companies, which were classified as medium enterprises. In Savannakhet Province, there were five companies involved in the border trade in rice. Most were located in the central districts of the province but also had their representative offices in Ban Dansavanh, close to the Dansavanh-Lao Bao border checkpoint. These companies undertook four major forms of formal trading activity: imports, exports, re-exports, and transit trade. For the export of goods, the companies were required to submit all documents such as letters of request, invoices, and packaging documents to the Trade Division of PICO for their quota allocation, after which the approved documents were sent to a one-stop service centre at the border point (see below).

The informal trade was conducted by small traders at the international checkpoint. Small-scale traders often carried goods by themselves across the border and walked to local markets. It was quite common for traders from Vietnam to bring a range of consumer goods, including food items such as vegetables, garlic, fruits, noodles, and cookies, as well as clothes and plastic ware. The flows in the opposite direction commonly included rice, bananas, and non-timber forest products. These traders were usually exempt from any kind of customs or import fees, needing only to pay the border crossing fee of LAK 5000–10000 (about USD 1) per crossing.

According to the export companies interviewed, the rice from Savannakhet was sold to private companies in Vietnam who either sold the rice in their domestic market or processed the rice into starch (Fig. 9.2).

A matrix of the networks between different actors in the rice market chain is depicted in Table 9.3. In some villages, producers were organized into a farmers' group which exchanged seeds, inputs, and labour among its members. The groups were formed with the support of international

Table 9.3 Matrix of trading networks and linkages in rice market chain

Type of actor	Producer	Collector	Miller	Exporter	Importer	Retailer
Producer	Producer group	Receives credit, inputs	Receives credit, inputs	NA	NA	NA
Collector	Buys paddy rice	Managed competit-ion	Sells paddy rice	NA	NA	NA
Miller	Buys paddy rice	Buys paddy rice	Millers' group	Oral contract to supply milled rice	NA	Supplies quota of milled rice
Exporter	NA	NA	Oral contract to buy milled rice	NA	Written contract to supply milled rice	NA
Importer	NA	NA	NA	Written contract to buy milled rice	Managed competit-ion	Supplies agreed quantity
Retailer	NA	NA	Obtains quota of milled rice	NA	Orders imported milled rice	Competition

development projects. However, they did not have any marketing function and farmers made independent decisions to sell rice.

Collectors and rice millers provided services to individual farmers in their own networks by giving loans and farm inputs (seeds, fertilizers, and pesticides) at the start of the crop season. The farmers repaid their loan after harvest with interest of 3–10%. The motivation of the millers for supplying inputs on credit was to ensure the supply and a homogenous grade of rice.

Rice millers and exporters made supply agreements but without written contracts. In the opinion of the exporters, a written contract would not be respected, even by regular suppliers. Some used to make written contracts with their suppliers but they were not adhered to due to the uncertainty of the market. However, the exporters had written contracts with their foreign customers as it was a compulsory part of the export procedure.

A number of constraints to the rice marketing chain were identified, mainly around quality and grading issues. (a) Farmers used a range of varieties (e.g., long- and short-grained, glutinous and non-glutinous) which resulted in mixing of varieties when the rice arrived at the mill. (b) Much of the grain was delivered with high moisture content (15–16%). Moreover, farmers sometimes mixed dried grain with wet grain. (c) The poor processing facilities in the mills made it difficult to meet international quality standards. Lao rice is generally classed as Grade 5. (d) The fluctuation in prices, in part caused by unpredictable policies, created uncertainty for producers, traders, and processors.

Border Trade in Savannakhet Province

Laos has two types of border crossing—international and local. The international crossings are generally open to all foreign nationals in possession of a valid passport and visa, while local crossings are open only to local people on each side of the border who are able to cross back and forth using some form of border pass. Laos has 16 international border points and 20 local border points. These border points link Laos to neighbouring countries, including Thailand (via seven international border points), Vietnam (via seven international border points), China (via one international border point), and Cambodia (via one international border point). The cross-border trade is virtually synonymous with international trade due to the country's landlocked situation.

As mentioned above, Savannakhet Province has two international border points. Dansavanh-Lao Bao is located along Route 9, the east-west corridor of Mainland Southeast Asia, in Ban Dansavanh, Sepon District. This border point was opened to the public in 2002, linking Savannakhet Province to Quang Tri Province in Vietnam. Savannakhet-Mukdahan is located in Khaisonphonvihanh District, the municipal area of the province. This border point was opened to the public in 2005 and links Savannakhet Province to Mukdahan Province in Thailand via the Mekong Bridge II.

There are three major forms of formal trading activity via these border points: imports, exports, and transit trade. The main products exchanged are summarized in Table 9.4. Rice is among the main products exported from Savannakhet Province to Vietnam via the Dansavanh-Lao Bao border point, along with other unprocessed crops such as banana, cassava, and coffee. The main imported goods through this point include food products and fertilizers (as well as manufactured items such as household utensils and vehicles). On the other hand, rice is not among the main goods exported to Thailand, while coffee, cassava, and fruit are. The goods

Table 9.4 Main products exchanged via international border points of Savannakhet Province, 2010

Border point	Exported products	Imported products	Transit products
Dansavanh/Lao Bao	Rice	Garlic	Rice
	Bananas	Shallots	Fruits
	Cassava	Fertilizers	Wild orchids
	Coffee		Horticulture
	Fruits		
Savannakhet/Mukdahan	Coffee	Sugarcane	Horticulture
	Cassava	Rice	Fruits
	Fruits	Rattan	Garlic
		Rubber	Wild orchids
		Fertilizers	Shallots

Table 9.5 Value of cross-border trade in Savannakhet Province, 2010 and 2011

Border point (year)	Value of trade (USD million)		
	Exports	Imports	Transit
Dansavanh-Lao Bao (2010)	26.0	11.9	7.0
Dansavanh-Lao Bao (2011)	26.6	10.1	7.0
Savannakhet-Mukdahan (2010)	0.6	2.0	0.7
Savannakhet-Mukdahan (2011)	2.7	3.0	2.4

Source: Provincial Industry and Commerce Office, Savannakhet

imported from Thailand, however, include rice as well as other agricultural products and fertilizers. Transit products are those traded between Thailand and Vietnam through Savannakhet, including rice, fruits, and horticultural products. According to statistics provided by the PICO, the value of trade through the Dansavanh-Lao Bao border point (USD 26.6 million in 2011) was ten times more than that through the Savannakhet-Mukdahan border point (Table 9.5).

The policies and procedures for cross-border trade have evolved over the past 15 years. In 2001, the Ministry of Industry and Commerce (MOIC) issued Instruction No. 948 on Small Export Border Businesses to promote small-scale export businesses and the management of cross-border trade. In this instruction, two types of border points were distinguished—remote and non-remote. In a remote border point, import and export of all products necessary for production and consumption was allowed (within the list of permitted goods). In a non-remote border point, only inputs necessary for production were allowed to be imported; consumption goods were to be bought only from domestic markets. In October 2004, the MOIC issued Order No. 962 on importation and exportation to establish one-stop services at border points, including the services of all trade-related agencies, and to abolish export-import licences.[1] The GMS Cross-Border Trade Agreement (CBTA) took effect in 2005 with the opening of one-stop services.

For the import of goods, trading companies need to submit their plans to the Trade Division of the PICO. Approval of the import plan is based on the policy of the provincial authority with regard to the import-export balance of the province. Each year the national government sets indicative import plans for various products to manage the national trade balance. The overall plan target is then allocated to import-export companies in each province by the respective PICO. In principle, the allocation is based on the ability of a company to access the market, but in practice the allocation often does not reflect that capability.

For the export of goods, the export companies are required to submit all documents including letters of request, invoices, and packaging documents to the Trade Division of PICO, after which the approved documents are sent to the one-stop service centre at the border. Clearance procedures at the checkpoint take only 30 minutes to an hour, as long as the traders have completed the documents required.

The rice export procedure for Savannakhet Province at the time of the survey is illustrated in Fig. 9.3. After receiving an order from foreign customers, the company checked the availability of sufficient rice of suitable quality from its own trading networks. It then submitted its export plan to the Trade Division of PICO for a quota allocation. The overall export quota for the province was determined as follows: (a) the rice surplus at the provincial level was estimated as total production minus the quantity of auto-consumption (estimated at 280–320 kg/person/year); (b) the quantity required for local consumption at the provincial level was estimated and deducted; (c) the official quota was calculated as the residual surplus. For Savannakhet Province, the export quota was typically around 20,000–30,000 tons per year (10–15% of the rice surplus). This was then divided into allocations for individual export companies. After approval of its quota, the exporter had to contact the PAFO for sanitary and phytosanitary (SPS) certification. The company then went to the Taxation and Customs Unit of the Ministry

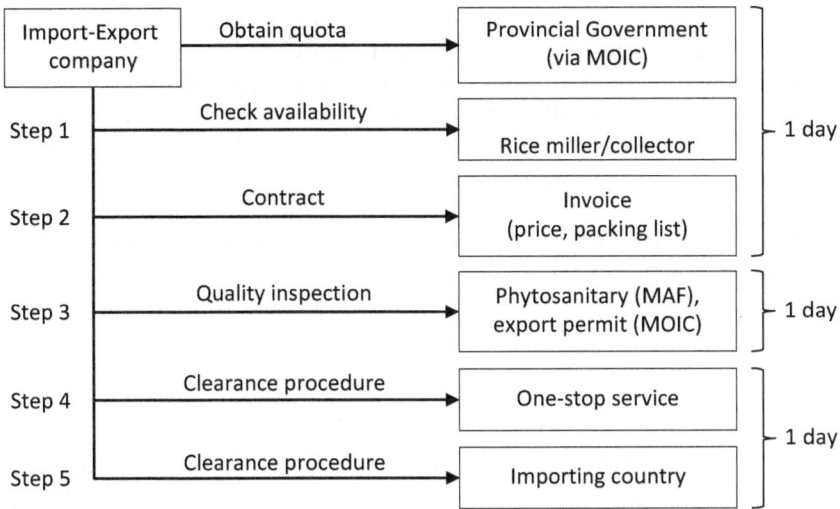

Fig. 9.3 Procedure to export rice from Savannakhet Province. (Source: Trade Division, PICO Savannakhet)

Table 9.6   Cross-border trade in rice, Savannakhet Province, 2009–2011

	2009	2010	2011
Exports (tons)	12,023	24,800	88
Imports (tons)	9498	3300	0
Transit (tons)	5137	465	1605

Source: Provincial Agriculture and Forestry Office, Savannakhet Province, Annual Reports 2007–2012

of Finance for tax and customs clearance. Finally, the company applied to the Trade Division of PICO for the export permit and licence. Each time it went through this cycle, the company paid about LAK 500,000 (USD 60) in fees.

Exports of rice have fluctuated, with a peak in 2010 of 24,800 tons, accounting for 10% of the rice surplus in Savannakhet Province and about 5% of total production (Table 9.6). The major market was Vietnam (see Chap. 20). However, the price of rice in local markets doubled between 2009 and 2010 due to a combination of factors, including official purchases of rice for southern flood victims and rice hoarding in preparation for the upcoming wet season. Despite the relatively small volume of exports, the cross-border trade was seen to have contributed to the increased price of rice at the national level. The government decided to regulate the domestic supply and keep the price under control by imposing a temporary export ban from November 2010 to February 2011. This accounts for the insignificant volume of exports from Savannakhet in 2011 (Table 9.6). The trade in rice followed seasonal fluctuations. The export of rice to Vietnam tended to be the greatest from November to January due to the high demand during the celebration of Tet, when glutinous rice is consumed. The import of rice from Thailand tended to peak from July to September due to shortage in the domestic market in the lead-up to the WS harvest.

CONCLUSION

The analysis revealed potential to expand the marketing and export of rice from Savannakhet Province. Most rice farmers in Champhone District were market-oriented, regularly producing for the market rather than only for home consumption. They used improved varieties and fertilizers and sold a substantial part of both their WS and DS crops. There was a good opportunity to increase the export of rice to neighbouring countries due to a favourable trade environment through Association of South East Asian Nations (ASEAN) and bilateral agreements. Champhone District had already been opened to the regional market and reasonably efficient trading networks were in place.

However, there were several constraints to the marketing and export of rice in Champhone District and the province as a whole. Farmers often delivered rice of mixed grades and high moisture content. The rice mills in Savannakhet had poor processing equipment, making it difficult to meet international quality standards. The export ban in 2010–2011 caused a sudden drop in prices and created market uncertainty. This reduced the willingness of farmers to produce surplus rice for the market and made it difficult for traders and millers to plan their marketing and export strategy.

These constraints could perhaps be alleviated through government policies to promote suitable varieties to ensure a homogeneous rice grade for the export trade; enhance post-harvest technology (storage and drying) to improve the rice moisture content; improve the processing infrastructure for sorting, milling, and polishing; and create a more stable policy environment for the export sector.

NOTE

1. The exceptions were gold and copper exports from Savannakhet, and vehicles, spare parts, petroleum, gas, diamonds, and other controlled goods that still required import licences from the Ministry of Commerce.

CHAPTER 10

Economic Constraints to the Intensification of Rainfed Lowland Rice in Laos

Jonathan Newby, Vongpaphane Manivong, and Rob Cramb

INTRODUCTION

Rice production in the rainfed lowlands of Laos faces a number of constraints at the farm level, including poor soil fertility, droughts and floods, and various pests and diseases (Schiller et al. 2001; Linquist and Sengxua 2001; Fukai and Ouk 2012). Furthermore, factors beyond the farm boundary such as rising input costs, fluctuating output prices, and uncertain trade policy continue to limit farmers' incentive to intensify production beyond that required to achieve household self-sufficiency. Hence, in recent years, household labour and capital have been redirected into a

J. Newby (✉)
International Centre for Tropical Agriculture (CIAT), Vientiane, Laos
e-mail: j.newby@cgiar.org

V. Manivong
Ministry of Agriculture and Forestry, Vientiane, Laos

R. Cramb
School of Agriculture and Food Sciences, University of Queensland,
St Lucia, QLD, Australia
e-mail: r.cramb@uq.edu.au

© The Author(s) 2020   201
R. Cramb (ed.), *White Gold: The Commercialisation of Rice
Farming in the Lower Mekong Basin*,
https://doi.org/10.1007/978-981-15-0998-8_10

range of other farm and non-farm activities rather than into intensifying rice production (Manivong et al. 2014). With high levels of yield- and price-risk and limited opportunities for consumption smoothing through market mechanisms (credit and insurance), households adopt income-smoothing strategies by adopting Low-Input production systems and income diversification, most notably through migration of family members to earn wages.

While the constraints are numerous, lowland rice production systems have been evolving over the past two to three decades (Chaps. 5 and 6). The traditional farming system that relied on draught animal power, traditional varieties, and organic fertilisers now accounts for a very small proportion of the country's lowland rice area, with widespread adoption of mechanised land preparation, improved varieties, and low levels of inorganic fertilisers. Despite the achievements of these "green revolution" technologies in terms of increased output, lowland rice production remains an economically marginal activity, providing limited economic incentive for farmers to intensify production beyond household consumption needs.

This poses a challenge for the Government of Laos (GOL) that seeks to keep the price of rice affordable for urban consumers (and net buyers of rice in rural areas), while providing incentives for farmers to intensify production to achieve food security (and even export) objectives. Attempts to maintain national food security, equated by policy-makers with rice self-sufficiency, have included the setting of official yield targets that are high relative to the current situation (4 t/ha for the rainfed wet season [WS] crop and 5 t/ha for the irrigated dry-season [DS] crop), as well as ad hoc trade restrictions prompted by seasonal shortfalls and price spikes. However, in many cases the strategies fail basic economic viability tests at the household level and have created further market uncertainty.

The limited intensification of lowland rice systems reflects the relative resource endowments and livelihood objectives of farm households. Induced innovation theory predicts that farming systems will respond both to changes in resource endowments and to growth in product demand, with new technologies developed and adopted that facilitate the substitution of relatively abundant and low-cost factors for those that are relatively scarce (Hayami and Ruttan 1985). In practice, this depends on the extent to which farmers' circumstances and national government policies align, and the ability of farmers to influence research and development priorities. In considering the economic and institutional constraints to improved fertility management, Pandey (1999) classifies rice production systems using a matrix of population density and the stage of economic development (as indicated by income levels). He argues that in situations with low population density and low income levels (in which he includes Laos), farms tend to be subsistence-oriented, with limited demand for improved nutrient management technologies that increase yields and returns to land. Such technologies will only be adopted if they also help save labour, the relatively scarce resource. He further argues that in order to stimulate the demand for yield-increasing technologies, policies need to focus on improving the profitability of rice production. This may include the development of export markets and improved market infrastructure, factors that lie outside the farm boundary. Nevertheless, in rainfed regions, production risk will continue to influence the demand for fertility management technologies.

In this chapter, we aim to explain farmers' decisions regarding the intensification of rainfed lowland rice systems in the context of current resource endowments, product demand, and production and market risk. We first describe the current rice production system in two major lowland provinces in central and southern Laos—Savannakhet and Champasak. We demonstrate that while the rainfed production system remains largely subsistence-oriented, farmers have selectively adopted a range of new technologies and continue to respond to changing incentives. However, to date this has largely involved the adoption of Low-Input, more labour-efficient, and more stable production systems rather than commercially oriented, High-Input, and high-yield systems. We use activity budgeting and sensitivity analysis to explore the economic performance of several input scenarios, ranging from farmers' practice to input levels required to achieve GOL policy targets. This analysis can be used to reassess aspects of rice policy for the rainfed lowlands in Laos.

Methods

Savannakhet and Champasak are two of the most important rice-producing provinces in Laos. In 2009, they accounted for around 40% of the national WS harvested area and a similar proportion of total production (Ministry of Planning and Investment 2010). A diagnosis and assessment of farming systems in these two provinces was undertaken in several phases of fieldwork, including key informant interviews with district agricultural staff, village group discussions, household surveys, and household case studies.

The fieldwork was conducted along transects reflecting different farm types, from irrigated lowlands through rainfed lowlands to uplands. However, only data from lowland villages are considered here; upland villages surveyed in the east of Savannakhet have been excluded from the analysis. Thus for present purposes the study region included six villages in Outomphone, Phalanxai, and Phin Districts in Savannakhet and six villages in Phonethong and Sukhuma Districts in Champasak (see Fig. 5.1 in Chap. 5). A household survey was carried out with 30 randomly selected households in each village, making 360 households in all. Information was sought regarding household composition and assets, cropping practices, livestock practices, off-farm and non-farm employment, migration and remittances, forest collection and hunting activities, access to water, access to credit, group membership, information sources, and rice security. More detailed case studies were conducted with 13 households in Savannakhet and 18 households in Champasak.

Survey and case-study data were supplemented with project and historical agronomic trial results in order to construct model budgets for various input scenarios. These include data from fertiliser response trials conducted by the International Rice Research Institute (IRRI) and the National Agriculture and Forestry Research Institute (NAFRI) over more than a decade (Linquist and Sengxua 2001, 2003; Haefele et al. 2010). Official yield data were not used as these tend to overestimate actual farm yields (Pandey 2001), presumably a reflection of the pressure to show progress in achieving policy targets. In 2013, the model budgets were presented to a farmer focus group for validation and updating with input and output prices relevant to the 2012 wet season. Sensitivity analysis, threshold analysis, and risk analysis (using the @Risk software package) were conducted for each scenario.

Status of Lowland Rice Farming in the Study Villages

The cultivation of rice remains an important livelihood activity for the majority of households in the lowland regions of Laos and creates the platform on which other activities and household decisions are based. Decisions regarding labour utilisation and migration, livestock management, and even religious and cultural festivals, are all made with reference to the paddy production cycle. Around 96% of surveyed households cultivated rice in WS 2010. Household access to paddy lands varied within and

Table 10.1   Status of rice-growing in surveyed villages, 2010 (n = 360)

District and village	% of hh[a] growing rice	Mean hh[a] size	Mean WS cultivated area (ha)	Mean WS yield (kg/ha)	Mean % of production sold
Outomphone	100	6.6	2.5	1466	9.7
Nagasor	100	6.1	2.1	1618	8.2
Phonegnanang	100	7.0	3.0	1314	11.2
Phalanxai	98	6.2	1.9	1572	3.8
Phanomxai	100	6.8	1.3	1987	2.1
Phontan	97	5.7	2.6	1157	5.5
Phin	88	7.2	1.2	1740	7.2
Khamsa-e	87	7.3	1.2	2545	14.1
Geangxai	90	7.0	1.1	965	0.5
Phonethong	97	7.0	2.8	1582	24.5
Phaling	97	7.3	2.4	1718	22.3
Oupalath	97	7.0	2.4	1933	27.0
None Phajao	97	6.8	3.5	1100	24.1
Soukhuma	98	6.3	1.8	1996	22.6
Boungkeo	100	6.7	1.4	2219	26.2
Khoke Nongbua	100	6.5	1.7	2109	24.1
Hieng	93	5.8	2.4	1645	17.1
Mean	*96*	*6.7*	*2.1*	*1689*	*15.3*

[a]hh = household

between villages, from less than a hectare to over 10 ha, with an average across all villages of around 2 ha (Table 10.1). There was a similar proportion of households with 1 ha or less (33%), 1–2 ha (34%), and over 2 ha (33%). Beyond farm size, other factors such as soil type, position in the toposequence, and access to water sources all affected the productivity of the land, even before any management decisions were overlaid. The stability of the livelihood platform thus varied between households and seasons.

WS 2010 was considered by farmers and researchers to be a drier than a normal year, with reported yields (calculated from farmers' estimates of cultivated area and production) somewhat lower than in previous years (Table 10.1). Droughts and floods are a common occurrence in the region, with large areas impacted by these climatic shocks. According to Schiller et al. (2006), over a period of 37 years (1966–2002) the Central Region (which includes Savannakhet) was affected by extreme events in

32 years, while the Southern Region (which includes Champasak) was affected in 22 years. These events have a profound impact on household rice self-sufficiency, given that many operate close to a subsistence threshold. Nevertheless, this means that the 2010 yields were not greatly different from the normal run of seasons. It is significant that they were below official yield data for the same season and well below the official target of 4 t/ha.

Households produced limited surplus rice for sale in WS 2010, averaging only 15% across the 12 villages (Table 10.1). Only 40% of surveyed households who were growing rice sold any rice, with the rest either producing rice exclusively for home consumption or buying rice to cover a deficit. However, sellers included some households that had access to irrigation water for the subsequent DS crop (particularly in Boungkeo and Phaling in Champasak).[1] The proportion of households selling rice, just self-sufficient, and buying rice varied significantly between the villages, as shown for the six Champasak villages in Fig. 10.1. There was also a group of households that sold rice immediately after harvest to pay off debt and re-entered the market later in the year as buyers to make up shortfalls. These households received low paddy prices when they sold their rice after

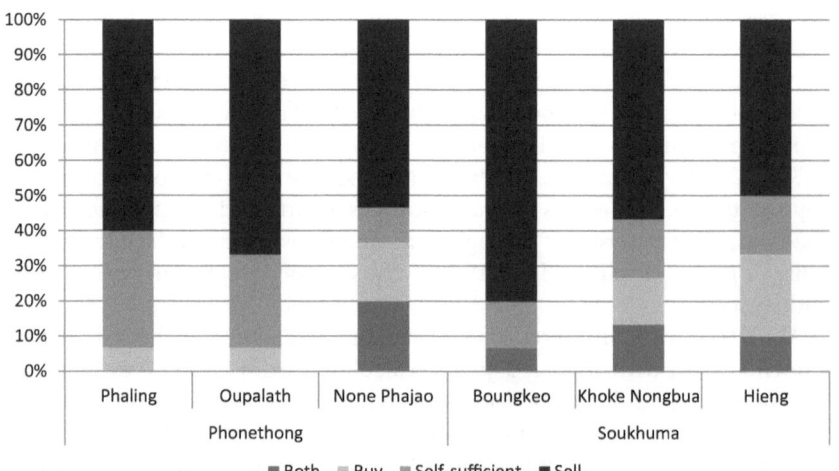

Fig. 10.1   Household rice status in Champasak for 2010, by district and village

harvest and incurred higher prices when they re-entered the market to make purchases.

The household's rice status is a function of the number of household members (or, strictly, the number of people who share the harvest); the area of paddy land available for cultivation; and the yield of the rice crop (Table 10.1). Given that yields fluctuate between years and many households are close to subsistence levels, the household's rice status is likely to change from year to year. Hence households formulate their livelihood strategy each year depending on crop performance. For example, the migration patterns of young people in some case-study households were determined by the performance of the WS rice crop and whether cash income would be required to make up shortfalls.

The average household size in the survey was 6.7 members, but this is complicated by household dynamics throughout the year. Members of the household may migrate for periods of the year and not consume from the household's rice stock. On the other hand, sometimes the rice harvest is shared beyond the immediate household, including relatives who have moved away from the village. Similarly, there are other social obligations involving sharing rice with others, including offerings to monks. Acknowledging these nuances, it is useful to take as a benchmark the national criterion for self-sufficiency, which is 350 kg of paddy (i.e., unmilled rice) per household member per year.

Figure 10.2 shows the yield required for an average household to achieve self-sufficiency for a range of paddy areas. The "self-sufficiency curve" indicates the large difference in required yield as the land size varies. For example, a household with 2 ha of paddy land only requires a yield of around 1.2 t/ha to achieve household self-sufficiency, while a household with only 1 ha would require a yield of close to 2.5 t/ha. The scatter plot presents the yield and area combinations for WS 2010. Self-sufficient households tend to track the "self-sufficiency curve", suggesting that households are trading off yield and paddy area, pursuing higher yields only when farm size is limited. As expected, most net purchasers of rice fall below the "self-sufficiency curve" in Fig. 10.2 and most net sellers are above the curve (remembering that actual family sizes vary between points). Some households remain net purchasers of rice, despite relatively large paddy areas, due to low yields, while other households achieve relatively good yields but, due to area constraints, still fail to meet household requirements.

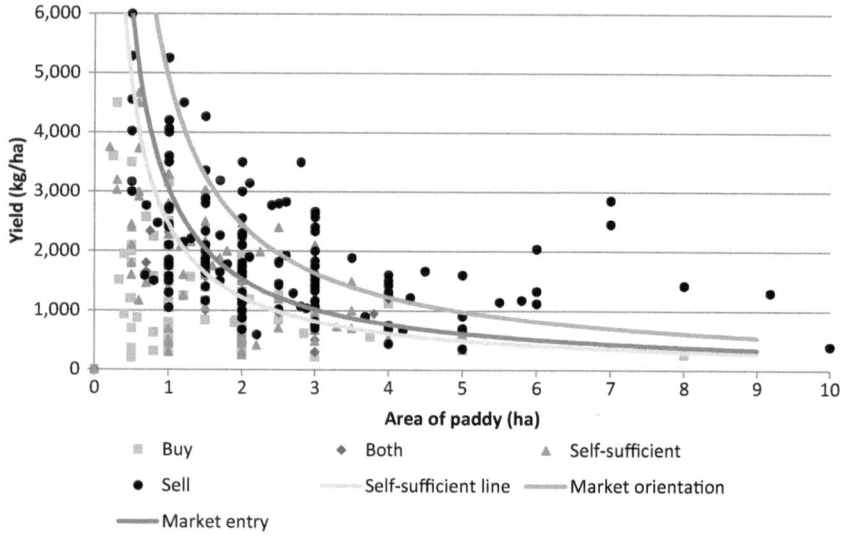

Fig. 10.2 Yield-area combinations by household rice status

The "market-oriented curve" in Fig. 10.2 shows the yield-area combinations enabling the average household to sell 50% of production, and the "market entry curve" shows the combinations for sales of 20% of production, reflecting an incipient market orientation. There were few households above the "market-oriented curve", especially in Savannakhet. As indicated in Fig. 10.2, a large proportion of households selling rice in 2010 were from Champasak, reflecting the higher average yields in 2010 in that province. Again, the scatterplot shows that the opportunity for a household to meet these market criteria varies considerably with paddy area. Households with 3 ha or more could achieve a 50% surplus with 2 t/ha or less, while the few market-oriented households with less than 2 ha were achieving yields of 3–4 t/ha.

In general, the data suggest that currently the majority of households remain largely subsistence-oriented (with respect to rice farming) and are willing to trade-off yields with paddy area to meet household requirements, limiting the incentive for intensification. Even in cases where households have access to irrigation water allowing double cropping, significant areas of the land were left fallow as rice prices fell to the extent that only 3 ha of DS rice were planted in Phaling village in 2012 compared to around 50 ha for the survey year in 2010.

ADOPTION OF MODERN TECHNOLOGY

While there are many physical and biological constraints that continue to limit rice productivity in the rainfed lowlands, the farming system has by no means remained static over the past two decades. The traditional production system that relied on draught animal power for land preparation, traditional varieties, and organic fertilisers has almost completely disappeared from the landscape. Indeed, only 11 households from the 347 households surveyed that were growing rice had not adopted any of the three main technologies—mechanised land preparation, improved varieties, or inorganic fertilisers. The current status of adoption of these technologies is summarised below.

Mechanisation

Economic growth in Laos and neighbouring countries has created considerable employment opportunities away from the farm. Migrating to Thailand is a well-established livelihood strategy for young people from lowland households; 43% of households surveyed in Champasak had at least one member working in Thailand (Manivong et al. 2014). In Outomphone, Savannakhet, 42% of households had at least one family member working in Thailand, with the incidence falling away as distance from the border increased. At the same time, employment opportunities within Laos, both in urban areas (including the construction and service sectors) and rural areas (such as working in rubber plantations) is also drawing labour away from traditional, semi-subsistence agriculture. This is not only impacting on the availability of household labour, but also increasing the cost of hiring labour, especially during peak periods such as transplanting and harvesting. Wage rates varied from LAK 25,000 to 50,000 per day depending on location, season, and activity. However, even in the remote Phin District, the wage rate for transplanting was reported to have reached LAK 50,000 per day (USD 6.25).

Mechanisation of rice production in Laos remains in its infancy, but with labour becoming increasingly scarce, changes are rapidly occurring as technology spills across the borders (Table 10.2). Around 75% of survey households utilised two-wheel tractors for land preparation rather than relying on draught animal power (mainly buffaloes). The ownership of two-wheel tractors had expanded to over 60% of households, while only 21% of households continued to use draught animal power exclusively. As

Table 10.2 Mode of land preparation by paddy area and district

	Land preparation method (% of households in each category)			
	Buffalo	Own tractor	Hired tractor	Buffalo and hired tractor
Land preparation by paddy area				
Small (*n* = 113)	21	57	16	4
Medium (*n* = 121)	19	69	7	4
Large (*n* = 113)	23	67	6	4
All (*n* = 347)	21	64	10	4
Land preparation by district				
Outomphone (*n* = 60)	18	78	2	0
Phalanxai (*n* = 59)	19	56	20	3
Phin (*n* = 53)	9	85	6	0
Phonethong (*n* = 87)	47	43	3	5
Soukhuma (*n* = 88)	6	69	16	8
All (*n* = 347)	21	64	10	4

Table 10.2 shows, the area of paddy land owned did not have a major impact on adoption. Moreover, adoption had extended into some more remote areas where rice productivity remained low and almost no surplus rice was produced. While the technology is not divisible like seed or fertiliser, the extent of adoption is not surprising given the versatility of the tractors and the extent of labour saved in both production and non-production activities, for example, transport to regional centres. However, in one village in Phonethong District (None Phajao) ownership of two-wheel tractors remained low compared to all other villages.

Other forms of mechanisation were less common, with the first transplanters, drill seeders, and harvesters only beginning to be utilised in the past few years and only in small areas. It is expected that their use will continue to expand as labour becomes increasingly expensive. Currently, in order to minimise cash outlays, households tend to extend the period of transplanting and utilise the declining household labour resource rather than hire labour or transplanters (with obvious trade-offs in terms of yield).

Improved Varieties

As shown in Chaps. 5 and 6, the adoption of improved varieties has been the single most important factor in achieving significant productivity increases since the 1990s. The first improved varieties were released in

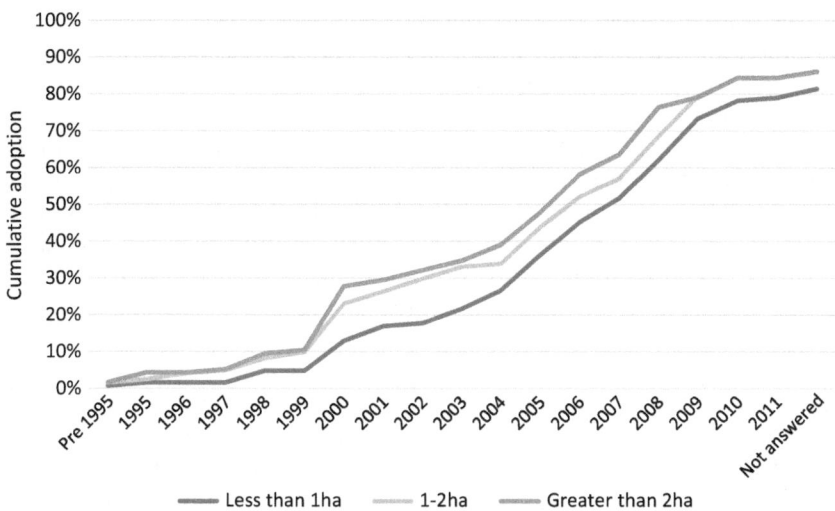

Fig. 10.3 Cumulative adoption of improved varieties by paddy area

Laos in the 1970s, and over the past two decades there has been widespread adoption. Indeed, the majority of households now grow at least one improved variety that has come out of breeding programmes in Laos or neighbouring countries,[2] with the area of traditional cultivars contracting. The adoption of improved varieties has occurred at similar rates among different farm size classes (Fig. 10.3). The impact of various projects can be seen in years (such as 2000) where significant jumps in adoption occurred.

Fertiliser Use

Soil fertility has long been recognised as one of the major constraints to rice production in Laos. The soils throughout the main lowland rice-growing areas in the central and southern plains have been described as generally infertile, highly weathered, and old alluvial deposits that comprise a series of low-level terraces with an elevation of about 200 metres above sea level (Lathvilayvong et al. 1996). Previous studies have identified nitrogen (N) as the most limiting nutrient in all regions of the country. In much of the Central and Southern Regions phosphorus (P) deficiency is also acute. Potassium (K) is the least limiting of the three tested nutrients in the central region, yet the need for K inputs is expected to increase as production is increased through double cropping or as rice yields increase through changes in management (Schiller et al. 2001).

The use of both organic and inorganic fertilisers has long been promoted in Laos. Linquist and Sengxua (2001) developed broad fertiliser recommendations based on fertility management research throughout the country. They recognised that the rainfed lowlands constitute a risky environment for crop production, hence their recommendations required relatively low investment and used nutrients with maximum efficiency rather than aiming for maximum yields. The recommendations were also based on the three fertilisers that were readily available.

For the first year of application, the recommendation is to apply 60-x-25 kg/ha of elemental nitrogen (N), phosphorus (P), and potassium (K), with the P rate varying according to soil texture. The rate of N recommended is lower than that required for maximum yields and reflects farmer risk in the rainfed environment. Higher rates of 90–120 kg/ha of N usually result in higher yields but only under good growing conditions. The recommended rate of P is 8.5 kg/ha in sandy soils, 13 kg/ha in sandy loam soils, and 19–26 kg/ha in loams and clay loams. In the second and subsequent years, the recommendation is modified to account for P that was not removed by the crop. These recommendations have been used in the scenario analysis presented in the following section.

The use of inorganic fertilisers by farmers in the lowland rainfed environment has historically been low. Surveys by Villano and Pandey (1998) for the 1996 WS crop in Champasak and Saravan Provinces found that 66% of households were using some chemical fertilisers and 48% of the area was fertilised. Of those applying fertilisers, about 54% did so to both the seedbed and the main field, 16% only to the main field, and 30% only to the seedbed.

The use of small amounts of inorganic fertilisers had expanded to around 80% of surveyed households in 2010. A range of fertility management strategies was used, including only applying fertilisers to seedlings and various combinations of basal applications and topdressing. Only around 18% of households were applying fertilisers to seedlings plus a basal application to the main field, followed by a topdressing (as recommended). Most households not using inorganic fertilisers were from the two villages in Phin District, Savannakhet. However, the reasons for not using fertilisers were very different between the two villages. The average WS yields in Khamsa-e were the highest across the Savannakhet survey, with households growing longer-duration varieties due to favourable conditions. Farmers reported that they did not use fertilisers because the land was still fertile, and hence additional (purchased) nutrients were not required. Some households reported that they had experimented with fertilisers in the past but had problems with lodging. On the other hand, Geangxai had the lowest average yields of the survey, with almost no household producing a surplus crop in 2010. Farmers in this village had frequent problems with drought as well as lower cash incomes compared to Khamsa-e. In Champasak the lowest rate of adoption was in the relatively remote village of None Phajao. Similar to Geangxai, this village had some of the lowest rice yields in the survey.

While the percentage of households using inorganic fertilisers has increased significantly, the level of use remains well below the recommended rates. The limited use of fertilisers reflects both the high cost of purchasing inputs, the limited access to credit, the high level of production risk, and market uncertainty should a surplus be produced. Physical access, counterfeit products, and limited knowledge about appropriate rates and timing contribute to the problems. Table 10.3 presents the average

Table 10.3 Average nutrient application rate by village (kg/ha)

District/village	Mean quantity of nutrient applied (kg/ha)		
	N	P_2O_5	K_2O
Outomphone	10.2	8.9	1.8
Nagasor	13.1	10.6	2.2
Phonegnanang	7.5	7.2	1.4
Phalanxai	14.4	13.0	1.1
Phanomxai	18.2	17.4	2.1
Phontan	10.9	8.9	0.2
Phin	9.5	6.9	0.0
Geangxai	10.0	6.4	0.0
Khamsa-e	7.3	9.2	0.0
Phonethong	21.1	10.5	3.2
None Phajao	5.8	5.5	1.7
Oupalath	27.4	13.6	3.1
Phaling	20.8	9.5	3.8
Soukhuma	15.9	15.3	1.7
Boungkeo	21.8	22.5	2.5
Hieng	7.1	8.1	0.1
Khoke Nongbua	17.0	13.3	2.3
All	15.3	11.8	1.9

N-P_2O_5-K_2O rates for each village. The overall average of 15-12-2 kg/ha of N-P_2O_5-K_2O converts to 15-5-1.5 kg/ha of NPK—well below the conservative recommendation developed by Linquist and Sengxua (2001) of 60-[8/26]-25 kg/ha NPK, with the P rate varying according to soil texture. It should be noted that these average amounts assume that farmers spread the fertilisers equally across their paddy fields. In practice, farmers tend to vary their application rates based on previous crop performance and perceived risk.

THE ECONOMICS OF INTENSIFYING FERTILISER USE FOR RAINFED RICE

To help understand the adoption patterns for fertiliser use, enterprise budgeting scenarios were developed for a hectare of WS rainfed rice based on household survey data and field experimental results. These representative budgets were first developed using average values for prices and yields, then sensitivity analysis was applied to allow for variability in these two key parameters. A range of indicators was used to capture farmers' decision criteria with regard to input use, including net returns to land (NR), with imputed costs for household labour deducted; net returns to household resources (NRHR), with no costing of household labour or land; and net returns to household resources per day of household labour (NRHL). When presenting the representative budgets to groups of farmers, these three indicators were assessed in terms of their usefulness for evaluating activities. Farmers preferred the NR measure to the NRHR measure as it explicitly placed a value on their own labour, but they also found the NRHL measure an easy way to compare the returns they received to the wage rate at different times of the year and for different household members.

Fertiliser-yield Scenarios

The four budget scenarios represented successively greater intensification as indicated by increasing fertiliser rates and yields.

Scenario 1 (No-Input)—Yield estimates were based largely on experimental results in which no inorganic fertiliser is added to the transplant crop. The household survey suggests that this represents around 30% of households. Both survey and experimental results show wide variation in the yields obtained where no inorganic fertiliser is used due to factors such as the indigenous soil fertility, soil-water balance properties, and other management practices. An average yield of 1.5 t/ha was assumed.

Scenario 2 (Low-Input)—This was based on the current Low-Input system that many households practice. It assumes again that households use inorganic fertilisers to establish seedlings but then apply one bag (50 kg) of 16-20-0 as a basal application, followed by a topdressing of one bag of urea. This results in a rate of 31-10-0 kg/ha of N-P$_2$O$_2$-K$_2$O. An average paddy yield of 2 t/ha was assumed.

Scenario 3 (Medium-Input)—This was developed using the current broad recommendation of 60-30-30 kg/ha of N-P$_2$O$_2$-K$_2$O (or 60-13-25 kg/ ha of NPK). This is applied through a basal application of 15-15-15 (200 kg/ha) with the remaining N coming via topdressing with urea. The yield assumption was based on adjusted experimental results (allowing for the well-known yield loss when moving from small to large plots). Again, experimental results have shown a range of responses to applied nutrients according to location. An average yield of 3 t/ha was assumed.

Scenario 4 (High-Input)—This was based on recent experimental work in the two provinces where a high rate is used in an attempt to achieve the government target yield of 4 t/ha. The recent trials had site-specific application rates with no replications and therefore it was necessary to develop an average treatment with a rate of NPK of 120-60-60 kg/ha, resulting in a yield of 3.75 t/ha, based on experimental results from the 2011 and 2012 wet seasons.

Other key assumptions are presented in Table 10.4, including the values used for sensitivity analysis. Sensitivity analysis was conducted on the

Table 10.4 Assumptions for budget scenarios

Parameter	Base assumption	Sensitivity analysis
Farm-gate price (LAK/kg)	2000	1200 and 3300
Fertiliser price (LAK/bag)		
16-20-0	230,000	250,000
46-0-0	220,000	250,000
15-15-15	250,000	300,000
Wage rate (LAK/day)	30,000	40,000

USD 1 = LAK 8000

farm-gate price of paddy based on the high 2010 price and the 2012 price in Champasak which was extremely low. The farmer focus group also considered this to be the lowest price that traders would offer before not coming to purchase rice at all. Threshold analysis was conducted on the farm-gate price of paddy to achieve various criteria. The labour required for each scenario was only varied for harvesting, threshing, and hauling, which are related to crop yield. The variation in labour for fertiliser application is minor and typically occurs during other operations.

Enterprise Budgets for the Four Scenarios

All four scenarios confirm the low profitability of rice farming in the rainfed lowlands of Laos, and the challenge facing farmers and government alike if they are to intensify the production system (Table 10.5). The gross return (GR) was calculated as the total market value of production, regardless of how much was sold. The total variable cost (VC) included all physical inputs and labour (but not land), with imputed market values used for non-cash costs. The net return (NR) was the GR minus VC, with all

Table 10.5 Economic analysis of fertiliser-input scenarios for a hectare of WS rice

	No-Input	Low-Input	Medium-Input	High-Input
Fertiliser (kg/ha, N-P$_2$O$_2$-K$_2$O)	0-0-0	31-10-0	60-30-30	120-60-60
Average yield (t/ha)	1.5	2.0	3.0	3.75
Gross returns (GR) (LAK/ha)	3,000,000	4,000,000	6,000,000	7,504,000
Variable cost (VC) (LAK/ha)	3,272,000	3,944,000	5,024,000	6,632,000
NR (LAK/ha)	−272,000	56,000	976,000	872,000
NRHR (USD/ha)	2,352,000	2,848,000	4,096,000	4,232,000
NRHL (LAK/day)	26,857	30,645	39,365	37,710
Marginal NR (USD/ha)		336,000	912,000	−112,000
Marginal rate of return (MRR)		50%	84%	D
Price of paddy rice (LAK/kg) needed for …				
NR > 0	2206	1967	1658	1757
NRHL = LAK 50,000/day	3517	2994	2388	2387
MNR > 0		1295	1121	2152
MRR > 50%		1995	1755	3316
MRR > 100%		2733	2328	4543

Note: Labour cost = LAK 30,000/day; paddy price (P$_r$) = LAK 2,000/kg; USD 1 = LAK 8,000; D = dominated scenario labour (household and hired) costed at the assumed value of LAK 30,000/day (USD 3.75).

For the No-Input scenario, the NR was negative. However, there was a positive result for the net return to household resources (NRHR), which does not involve deducting household labour costs. When NRHR was calculated as a ratio to the household labour input, the net return to household labour (NRHL) was below the wage rate of LAK 30,000/day. That is, while there were positive returns to household-owned resources (land, labour, and durable capital), these were not sufficient to provide a return greater than the opportunity cost of household labour.

The Low-Input scenario produced a positive NR and hence an NRHL slightly above the opportunity wage. Thus there was a positive marginal net return (MNR) to moving from the No-Input to the Low-Input scenario, with a marginal rate of return (MRR) of 50% on incremental investment (including household labour).

The Medium-Input scenario provided a further increase in NR and an NRHL above the opportunity wage by LAK 9000 (over USD 1). Moving from the Low- to the Medium-Input scenario provided an MRR of 84%. Thus many farmers who currently practise a Low-Input system could benefit economically from adopting the broad recommendations of the Medium-Input system, with about double the fertiliser rate and a 50% yield increase.

However, a further movement to the High-Input scenario saw the NR to land and labour both fall, although the NRHL remained just above LAK 30,000/day. Hence the MRR to this degree of intensification was negative and the scenario was deemed to be dominated (D).

Threshold and Sensitivity Analysis

Threshold analysis was conducted on the farm-gate price of paddy rice (P_r) to determine at what price (a) the NR would become positive, (b) the NRHL would be 50,000 kip/day, and (c) the MRR for moving to the next scenario would be positive, 50%, or 100%. The results, shown in the last lines of Table 10.5, indicate that, if the paddy price decreased to below LAK 1967/kg, the NR for a Low-Input system will become negative, but as long as the price is above LAK 1295/kg there is still some gain relative to applying no fertiliser at all. The threshold prices for realising positive returns to the Medium- and High-Input scenarios were in the achievable range, but the price would have to be very high indeed (>LAK 4500/kg) for the move from Medium-Input to High-Input to offer an acceptable rate of return of 100%.

In 2010 the price of fertilisers varied between locations, particularly for compound fertiliser such as 16-20-0 and 15-15-15 in more remote areas. By 2012, the price of urea had also increased across the two provinces. Furthermore, fuel prices had increased and wage rates continued to rise, adding to farmers' cash outlays. The impact of higher costs on the economic indicators is summarised in Table 10.6. The increase in input prices reduces the NR such that all scenarios produce a negative result. Increased fertiliser and fuel costs reduce the NRHL so that the Medium- and High-Input scenarios are barely above the previous opportunity wage (LAK 30,000), but are now below the new, higher opportunity wage. A move from No-Input to Low-Input still somewhat improves the NRHR, but only achieves an MRR of 30%. Similarly, a further increase to the Medium-Input scenario improves the NRHR, but again falls short of an acceptable MRR.

The incentives for intensification worsened in 2011 and 2012 when the farm-gate price fell to as low as LAK 1200/kg. At this price the NRHL would be less than half the initially assumed opportunity wage rate of LAK 30,000/day (Table 10.7). On the other hand, during the price spike in 2010 when farm-gate prices reached LAK 3300/kg in some regions, the returns to labour from intensification strategies looked much more promising. However, farmers in group interviews did not have high expecta-

Table 10.6 Sensitivity analysis of fertiliser costs and wage rates

	No-Input	Low-Input	Medium-Input	High-Input
Fertiliser (kg/ha of N-P$_2$O$_2$-K$_2$O)	0-0-0	31-10-0	60-30-30	120-60-60
Variable cost (LAK/ha)	4,184,000	4,952,000	6,336,000	8,264,000
NR (LAK/ha)	−1,184,000	−952,000	−336,000	−768,000
NRHR (LAK/ha)	2,320,000	2,768,000	3,824,000	3,728,000
NRHL (LAK/day)	26,514	29,785	36,779	33,185
MRR		30%	44%	D
Price of paddy rice (LAK/kg) needed for …				
NR > 0	2884	2525	2118	2215
NRHL = LAK 50,000/day	3539	3039	2482	2530
MRR > 50%		2335	2153	4011
MRR > 100%		3200	2856	5496

Note: Labour cost = LAK 40,000/day; paddy price (P$_r$) = LAK 2,000/kg; USD 1 = LAK 8,000; D = dominated scenario

Table 10.7 Sensitivity analysis for low and high paddy prices

	No-Input	Low-Input	Medium-Input	High-Input
Farm-gate price of paddy of LAK 1200/kg				
NR (USD/ha)	−2,248,000	−2,400,000	−2,616,000	−3,616,000
NRHR (USD/ha)	1,256,000	1,320,000	1,544,000	872,000
NRHL (LAK/day)	14,309	14,215	14,856	7,795
Farm-gate price of paddy of LAK 3300/kg				
NR (USD/ha)	552,000	1,400,000	3,368,000	3,864,000
NRHR (USD/ha)	4,056,000	5,120,000	7,528,000	8,360,000
NRHL (LAK/day)	46,349	55,086	72,404	74,443

Note: Labour cost LAK 40,000/day; input prices based on Table 10.6; USD 1 = LAK 8000

tions that prices would again be at this level in the coming season and hoped for a return to prices around LAK 2000/kg.

Optimal Farmer Strategies

Given these results, what strategy should a farm-household adopt? A move from the No-Input to Low-Input system improves the net return to land and labour; however, the NR would remain negative under 2012 conditions. Furthermore, the MRR of the change is only 50%, falling to 30% if the higher costs are assumed. Previous studies (CIMMYT 1988) have suggested an MRR of at least 100% is required before adoption is likely, although 50% may be sufficient for relatively small system changes. Assuming household self-sufficiency is an important objective, the small amount of fertiliser involved in moving to the Low-Input system may raise some households with small areas of paddy above their subsistence requirement, with returns to labour and capital treated as less important. For example, an average No-Input household with 1.2 ha could move from being 75% self-sufficient, with an output of 1800 kg, to 100% self-sufficient, with an output of 2400 kg, by adopting the Low-Input package (Fig. 10.2).

Under the 2010 price conditions, a move from the Low-Input system to the Medium-Input system provides a positive NR per hectare and an NRHL above the wage rate. This move provides an MRR of 84% (or a 71% return if moving directly from the No-Input to the Medium-Input system). The threshold analysis on paddy price suggests that this scenario is likely to provide positive NR and MNR for most price scenarios, and a small increase in the price would deliver an MRR greater than the CIMMYT (1988) rule-of-thumb. This outcome holds even allowing for an increased price of fertilisers. However, the increase in the cost of labour to LAK 40,000/day pushes this scenario into negative NR unless the paddy price is above LAK 2118/kg.

It is very unlikely that a household would adopt the High-Input scenario, given that returns to both land and labour decline compared to the Medium-Input case. Nevertheless, a land-scarce household may be forced to adopt this strategy if achieving household self-sufficiency remains the dominant objective, given that the returns to labour remain above the wage rate. However, households with acute land constraints are also less likely to have the capital to make the necessary investment.

Given that labour use does not increase much with increased fertiliser application, rising wage rates are not projected to impact greatly on WS fertility management decisions, though they will affect the overall economic performance of all scenarios. On the other hand, for households with access to irrigation that enables cultivation of a DS rice crop, the question of wage rates becomes more important, given that self-sufficiency may be achieved in the WS, allowing labour to move off-farm and earn relatively high returns in the DS. Several case-study farmers were making this decision and not growing a DS crop; rather they made their irrigable land available to households with smaller paddy areas who had not yet achieved self-sufficiency in the WS.

CONCLUSION

The survey evidence from Central and Southern Laos shows that farm-households in the rainfed lowlands continue to manage rice production systems that are largely subsistence-oriented. The adoption of new technologies, especially improved varieties, has been important in helping households meet self-sufficiency objectives and has enabled some to produce a small surplus. Despite this, rice production remains an economically marginal activity that is under increasing pressure from rising costs, particularly for labour. Rural livelihoods in the study area have become increasingly diversified, with households allocating labour to a range of alternative farm and non-farm activities. However, rice production continues to be the platform on which these other livelihood activities are based. The development and adoption of technologies that enable households to achieve self-sufficiency in a labour-efficient and cost-effective manner are important to improving household welfare in this context.

The budget models show that, given their resource endowments and the high degree of production and market risk they encounter, households in the rainfed lowlands have been rational in adopting a Low-Input system rather than intensifying rice production to achieve government yield and production targets. As the costs of labour continue to increase, technologies that improve labour productivity and enable labour to move off-farm are likely to be adopted more readily than technologies that seek to intensify production. In the same way, the development and adoption of improved varieties that are well adapted to abiotic and biotic stresses and reduce risks in specific environments can potentially improve the profitability and stability of the rainfed lowland system. Moreover, improving the efficiency of fertiliser application through site-specific recommendations may be more important than increasing absolute fertiliser rates.

While the improvements in profitability that these technologies bring may induce some intensification, we argue that the strategy of diversifying livelihoods while maintaining a largely subsistence-oriented rice production system is likely to persist, given the current economic trends. While this may not help lift rice production to reach national targets, it is likely to improve the livelihood outcomes of the numerous households living in this marginal environment.

Acknowledgements We are grateful for the comments of Ben Samson (IRRI) on the fertiliser scenarios and yield assumptions.

NOTES

1. WS rice remained largely rainfed in these villages unless subsidies were given for irrigation fees during drought years.
2. Thai varieties such as RD6 were common in lowland areas of Savannakhet.

REFERENCES

CIMMYT, 1988. *From Agronomic Data to Farmer Recommendations: An Economics Training Manual*, Rev. ed. Mexico, D.F.: CIMMYT.

Fukai, S., and Ouk, M., 2012. Increased productivity of rainfed lowland rice cropping systems of the Mekong region. *Crop and Pasture Science* 63: 944–973.

Haefele, S. M., Sipaseuth, N., Phengsouvanna, V., Dounphady, K., and Vongsouthi, S., 2010. Agro-economic evaluation of fertilizer recommendations for rainfed lowland rice. *Field Crops Research* 119: 215–224.

Hayami, Y., and Ruttan, V., 1985. *Agricultural Development: An International Perspective*, Rev. ed. Baltimore: Johns Hopkins University Press.

Lathvilayvong, P., Schiller, J. M., and Phommasack, T., 1996. Soil limitations for rainfed lowland rice in Laos. In *Breeding Strategies for Rainfed Lowland Rice in Drought Prone Environments*. ACIAR Proceedings No. 77, pp. 74–90. Canberra: Australian Centre for International Agricultural Research.

Linquist, B., and Sengxua, P., 2001. *Nutrient Management in Rainfed Lowland Rice in the Lao PDR*. Los Baños, Philippines: International Rice Research Institute.

Linquist, B., and Sengxua, P., 2003. Efficient and flexible management of nitrogen for rain-fed lowland rice. *Nutrient Cycling in Agroecosystems* 67: 107–115.

Manivong, V., Cramb, R., and Newby, J., 2014. Rice and remittances: crop intensification versus labour migration in Southern Laos. *Human Ecology* 42: 367–379.

Ministry of Planning and Investment, 2010. *Statistics Year Book 2009*. Vientiane, Lao PDR.

Pandey, S., 1999. Adoption of nutrient management technologies for rice production: economic and institutional constraints and opportunities. *Nutrient Cycling in Agroecosystems* 53: 103–111

Pandey, S., 2001. Economics of lowland rice production in Laos: opportunities and challenges. In S. Fukai and J. Basnayake, eds., *Increased Lowland Rice Production in the Mekong Region*. ACIAR Proceedings No. 101, pp. 20–30. Canberra: Australian Centre for International Agricultural Research.

Schiller, J. M., Linquist, B., Douangsila, K., Inthapanya, P., Douang Boupha, B., Inthavong, S., and Sengxua, P., 2001. Constraints to rice production in Laos. In S. Fukai and J. Basnayake, eds., *Increased Lowland Rice Production in the Mekong Region*. ACIAR Proceedings No. 101, pp. 3–19. Canberra: Australian Centre for International Agricultural Research.

Schiller, J. M., Hatsadong, and Doungsila, K., 2006. A history of rice in Laos In J. M. Schiller, M. B Chanphengxay, B. Linquist, and S. A. Rao, eds., *Rice In Laos*, pp. 3–19. Los Baños, Philippines: International Rice Research Institute.

Villano, R. A., and Pandey, S., 1998. *Technology Adoption in the Rainfed Lowland Environments of Lao PDR. Implications for Poverty Alleviation*. Los Baños, Philippines: International Rice Research Institute.

PART IV

In Pursuit of White Gold

The Commercialisation of Rice Farming in Cambodia

Rob Cramb, Chea Sareth, and Theng Vuthy

In this and the next five chapters the focus is on the commercialisation of rice farming in the Central Plain of Cambodia. Particularly since 2010, rice has come to be seen as more than merely a subsistence crop or a staple for domestic consumption but as "white gold"—a commodity with major commercial, including export potential (RGC 2010). To explore this trajectory, field studies were undertaken in Takeo Province and the lowland part of the adjacent province of Kampong Speu in the southern part of the Central Plain, embracing rainfed and irrigated lowlands (Fig. 11.1). These studies examined the economics of rice production, marketing, and trade

R. Cramb (✉)
School of Agriculture and Food Sciences, University of Queensland, St Lucia, QLD, Australia
e-mail: r.cramb@uq.edu.au

C. Sareth
Cambodian Agricultural Research and Development Institute, Phnom Penh, Cambodia
e-mail: sareth.chea@uqconnect.edu.au

T. Vuthy
Office of Food Security and Environment, USAID, Phnom Penh, Cambodia
e-mail: vtheng@usaid.gov

© The Author(s) 2020   227
R. Cramb (ed.), *White Gold: The Commercialisation of Rice Farming in the Lower Mekong Basin*,
https://doi.org/10.1007/978-981-15-0998-8_11

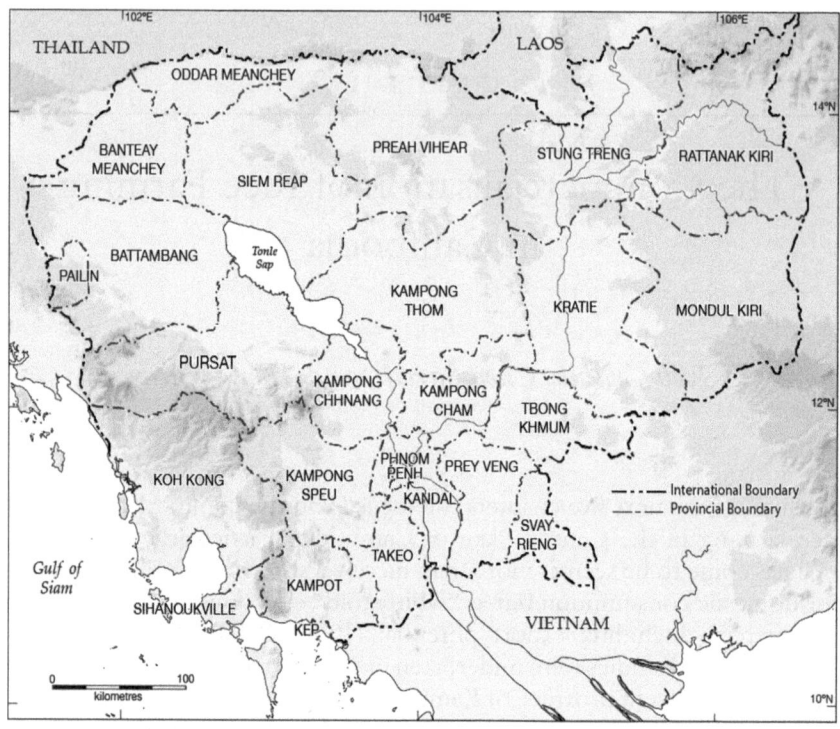

Fig. 11.1 Cambodia, showing provinces and terrain. (Source: CartoGIS, Australian National University)

in Takeo (Chap. 12), the role of the key inputs of water (Chap. 13) and fertilisers (Chap. 14) in supporting commercial rice production, the provision of credit to rice farmers by micro-finance institutions (Chap. 15), and the potential for contract farming to alleviate some of the key constraints to commercialisation (Chap. 16). This chapter sets the scene for the in-depth studies in the chapters that follow by (1) describing the rice-growing environment in Cambodia as a whole, (2) outlining the history of rice production in Cambodia, (3) examining the role of the rice sector in the rapid agricultural and economic growth in Cambodia since 1993, (4) highlighting the changes at the farm level that have underpinned this agricultural growth, and (5) providing a profile of Takeo Province within this larger context.

THE RICE-GROWING ENVIRONMENT

Cambodia encompasses a distinct physiographic region within the Mekong catchment referred to as the Tonle Sap Basin, beginning in southern Laos just above Pakse and spreading out into an extensive plain in central Cambodia, bordered on the east, north, and south-west by mountain ranges (MRC 2017). The dominant tributaries entering on the left bank are the Se Kong, Se San, and Sre Pok Rivers in Stung Treng and Rattanak Kiri Provinces in the north-east. The Tonle Sap River flows into the Mekong on its right bank at Phnom Penh, but famously reverses flow in the wet season to accommodate the floodwaters from upstream, expanding the size of the vast Tonle Sap Lake six-fold to about 25,000 km^2. Below Phnom Penh the Mekong branches into the Bassac River, its major distributary, thus forming the beginning of the Mekong Delta. Of Cambodia's total land area of 181,035 km^2, 86% lies within the Mekong Basin, forming 20% of the entire catchment. Only the coastal region to the south-west of the Cardamom and Elephant Ranges lies outside the Basin, draining into the Gulf of Siam.

Rainfall in the lowlands varies from 1250 to 1750 mm annually, with a distinct but erratic wet season (WS) from mid-April to mid-November, followed by a five-month dry season (DS) in which rice cannot be grown without some form of irrigation (Nesbitt 1997a). Hence most of the rice lands support only a single rainfed WS crop, accounting for about 87% of the annual cultivated area (MAFF 2013). In some areas around the Tonle Sap Lake and close to the Mekong floodplain which are inundated in the WS, deep-water or floating rice is grown. Some of the floodplain areas are only used for DS rice, which is planted as the floodwaters recede. Upland rice is of limited importance.

Rice soils in the lowlands are of two broad types (White et al. 1997). Those of the old alluvial and colluvial plains account for 67% of the lowland rice area and are generally light-textured soils of low fertility used for rainfed WS rice. Soils in the active floodplains around the Tonle Sap Lake and the Mekong and Bassac Rivers account for 30% of the rice area. These soils are heavy-textured and fertile, being formed from fresh alluvium deposited by annual floodwaters. They are submerged for three to five months of the year and are commonly used for deep-water rice and recessional/irrigated DS rice.

For millennia, the Cambodian population has been dependent on rice cultivation, concentrated around the Tonle Sap and the south-eastern lowlands. Rainfed lowland rice remains the mainstay of the rural economy. Cambodia's population was 16 million in 2017, of whom almost 80% resided in rural areas, most engaged in rice farming and other livelihood activities. About 91% of the population is of the Khmer ethnic group. Minorities include Vietnamese (3%), concentrated in the Delta to the south-east, Chinese (1%), and Cham, Lao, Tai, and other groups (5%). The population growth rate in 2017 was 1.6%, down from a peak of 3.9% in 1984.[1] The population density averaged 90 persons/km² but varied from 100–400 persons/km² in the Central Plain to 4–50 persons/km² in the uplands (NIS 2008).

History of Rice Production

Rice has been cultivated by Khmer farm households in these lowlands of Cambodia for perhaps 3000 years and probably longer in the uplands (Helmers 1997; Higham 2014). The more intensive lowland rice techniques developed in southern China—involving the use of the plough to prepare bunded rice fields into which seedlings are transplanted from a nursery—were introduced about 1500 years ago. The powerful kingdom of Angkor which dominated the region from the ninth to the fourteenth centuries was based on the appropriation of rice surpluses and the mass mobilisation of rural labour through corvées and slavery. The capital of Angkor located near Siem Reap to the north of the Tonle Sap Lake was surrounded by rice paddies irrigated from large reservoirs through a system of canals, permitting multiple cropping (Higham 2014: 400–403). With the decline of Angkor, the centre of population moved to the south-eastern part of present-day Cambodia, which is still the most densely populated part of the country. Rice farming in this period was probably sufficient for the needs of rural households, though it was still faced with threats from an unpredictable environment, state-imposed taxes, labour corvées, and periodic conflicts. Nevertheless, over the centuries, farmers had adapted rice-growing to the different ecosystems and selected suitable varieties for local conditions; about 2000 traditional rice varieties have been identified as unique to Cambodia (Helmers 1997).

Under the French colonial regime, little was done to improve smallholder rice production; hence yields remained at a little over 1 t/ha. The growth of production was almost entirely due to the expansion of cultivated area. From 1900 to 1950, the area cultivated increased, in line with population growth, from about 400,000 ha to 1,660,000 ha, and total production increased from 560,000 t to 1,580,000 t, but the average yield declined from 1.4 t/ha to 1.0 t/ha (Slocomb 2010: 59). From 1900 the French administration pursued a policy of promoting agricultural exports, especially of rice and cattle, to supply French agro-processing and export businesses in Saigon (Helmers 1997). French settlers were given more than 16,000 ha of land concessions to establish large rice estates on fertile soils in Battambang Province using hired labour. These estates were supported with infrastructure including irrigation works, research stations (focusing on varieties, fertilisers, and mechanisation), and a railway line to Phnom Penh. In the pre-war decades, rice exports ranged from 50,000 to 200,000 t of paddy per year, of which around 30,000 t came from the Battambang plantations and the rest from smallholders. By 1940, Cambodia was the world's third largest rice-exporting country (Helmers 1997). Smallholders did reasonably well out of these sales when prices were high, such as in the 1920s, but scaled back cultivation to subsistence levels and sought relief from the rice tax when prices fell, as in the 1930s.

Under Prince Sihanouk's Sangkum government (1953–1970), there was investment in irrigation infrastructure in some provinces and six rice research stations were established for varietal trials and seed production. The government also took control of the French rice plantations in Battambang. By 1965, paddy production had grown to 2.75 million t and exports to 500,000 t, almost entirely due to further expansion in cultivated area; yields remained around 1.1 t/ha (Helmers 1997). A state corporation was established in 1962 with a monopoly over production inputs and rice exports. By the mid-1960s, the corporation sought to forcibly collect rice at low official prices, prompting the growth of black-market trade to Vietnam and armed rebellions in Battambang and elsewhere (Kiernan and Boua 1981).

A favourable season in 1969 meant that, in early 1970, the rice crop was a record 3.8 million t. However, as the Indochina War escalated, including American carpet bombing in the east of the country, rice production was devastated. Under Lon Nol's Khmer Republic (1970–1975), total output fell by 84% (Helmers 1997; Slocomb 2010: 147–149). Exports were suspended in 1971 in an attempt to shore up domestic stocks. The Democratic Kampuchea (Khmer Rouge) regime that controlled Cambodia from 1975 to 1979 focused on developing rice production (Helmers 1997; Slocomb 2010: 205–207), not just for subsistence but to provide the surplus to fund its revolutionary programme for economic independence (or "Super Great Leap Forward"). The regime brutally forced people to work in the paddy fields and construct irrigation systems throughout Cambodia as part of its ambitious plan to achieve two or three crops a year and raise yields to 3 t/ha Himmel 2007). However, most of the irrigation schemes failed and the forced collectivisation of labour left the country's agriculture in disarray. Rice was requisitioned to supply the army and to export in exchange for arms, while locals starved. When Vietnamese forces took over in 1979, they found that the countryside was devastated and famine was widespread.

The Vietnamese-installed People's Republic of Kampuchea (PRK) (1979–1989) focused on rehabilitating rice farming, but with very limited resources (Helmers 1997; Slocomb 2010: 207–209). Farming was again organised on socialist lines, with all land collectivised and groups of 20–25 households constituted as the basic unit of production, though in practice it was common for individual households to manage their own plots within the village communal land and for the groups to merely share animals and equipment and to exchange labour. Vietnamese advisers introduced some International Rice Research Institute (IRRI) varieties such as IR36 and IR42 but there was no rice research service to test or promote these and other modern inputs. The Cambodia-IRRI-Australia Project (CIAP) was established in 1987 and began to build the country's rice research capacity, but the impact was not seen until the 1990s.

Slocomb (2010: 209) reports that, from 1980 to 1989, the area cultivated increased by only 31% from 1,441,000 ha to 1,890,000 ha, short of the PRK's target of 2.5 million ha, and total production increased by 54% from 1,670,000 t to 2,570,000 t, below the target of 3 million t. Average yields increased only slightly from 1.2 to 1.4 t/ha. Nevertheless, by the end of the decade, Cambodia was almost self-sufficient in rice. In 1989, the PRK was renamed the State of Cambodia and crucial reforms were introduced (Helmers 1997; Slocomb 2010: 225). Private land tenure was established, with the communal lands broken up and allocated to individual households based on the number of household members, and the market economy was legitimised, in recognition of its de facto reassertion in the preceding decade.

After the United Nations (UN)-supervised elections in 1993, the Royal Government of Cambodia (RGC) was installed, paving the way for increased foreign investment and aid directed to agricultural and rural development. At this point, farmers in the lowlands were still largely dependent on conventional farming practices, low-yielding traditional varieties, very low rates of inorganic fertilisers, almost no use of agrochemical inputs, and little mechanisation of land preparation or harvesting. They were subject to various pressures, including the seasonality and variability of rainfall, lack of irrigation, poor soil fertility, weed and pest problems, few farm resources, and limited access to inputs, credit, and markets. The average yield from rainfed lowland rice was only 1.5 t/ha, one of the lowest in Asia (Nesbitt 1997b; Javier 1997). Hence the majority of households were producing rice at subsistence levels. In addition, opportunities for productive employment of land and labour in the dry season were limited. Meanwhile population growth in the decade to 1993 had surged to between 2.9 and 3.9%.

As mentioned earlier, agricultural research had resumed in the late 1980s under CIAP and this began to have an impact in the 1990s. The primary objective was to improve rice production to alleviate the country's chronic rice shortage. By 2006, 37 improved varieties had been developed and released, mainly for the rainfed lowlands, with a potential yield range of 2.5–4.5 t/ha (Sakhan et al. 2007). The programme also covered rice agronomy, pest management, soil classification, and mechanisation (Nesbitt 1997b). This research effort has had a significant impact on rice yields and production in Cambodia, providing the basis for the expansion of output and exports in recent decades.

AGRICULTURAL AND ECONOMIC GROWTH SINCE 1993

After the war-time devastation of the 1970s and 1980s, the Cambodian economy has experienced more than two decades of rapid growth, averaging 7.6% over the period 1994–2015 (World Bank 2017). Gross national income (GNI) per capita reached USD 1070 in 2015, giving Cambodia the status of a lower-middle-income country. This growth has been associated with a marked reduction in poverty, from 48% in 2007 to 14% in 2014, though most families who escaped poverty remain "near-poor" and economic inequality is increasing. With the growth of the industry and service sectors (particularly garment manufacture, construction, and tourism), agriculture's share of the economy has declined. Agriculture Value Added as a proportion of gross domestic product (GDP) fell from 50% in 1995 to 28% in 2015, though agricultural workers still comprised 51% of the labour force in 2012.

While industry and services have grown faster than agriculture, the agricultural sector has also grown at a rapid rate. Gross Agricultural Production grew at 8.7% during 2004–2012 and Agricultural Value Added at 5.3% (World Bank 2015). This rapid growth was driven by crop production, mainly the rice sector, the output of which has grown at 5% from 1990 to 2017 (Fig. 11.2). About two-thirds of the reduction in poverty during this period was attributable to agricultural growth, where higher rice prices stimulated increased production and farm incomes as well as pushing up farm wages. According to a review of the agricultural sector by the World Bank (2015), Cambodian agriculture has benefited from a market-oriented policy, including (1) an open trade policy, enabling farmers to benefit from improved access to the European Union (EU) market as well as cross-border trade with Thailand and Vietnam; (2) wider availability of machinery services such as threshers and combine harvesters; (3) better access to rural finance, especially micro-finance; and (4) investment in rice milling.

The growth in rice production was due partly to an expansion of cultivated area (at a rate of 1.7% during 1990–2017) but more so to an increase in yields (at a rate of 3.5% in the same period). Moreover, the area expansion has levelled off while there is still potential for further yield growth (Fig. 11.2). The national rice yield now averages 3.5 t/ha,

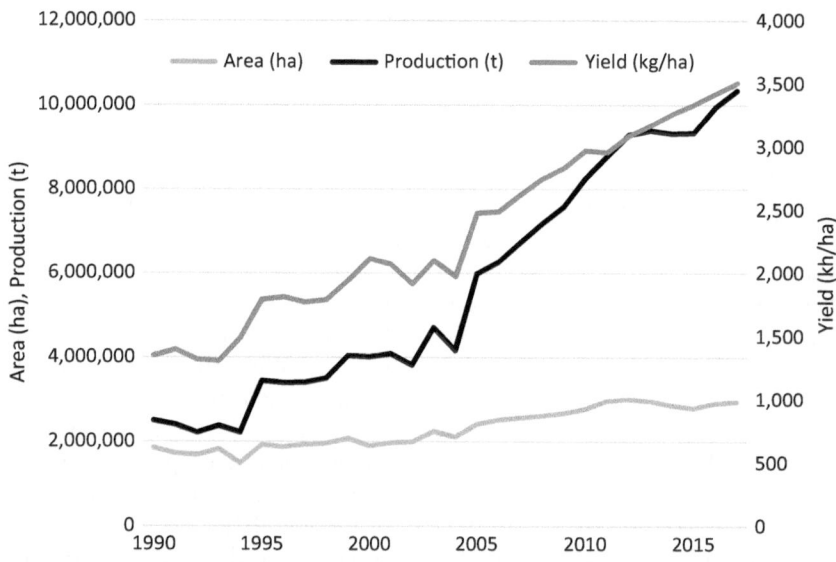

Fig. 11.2 Area, yield, and output of rice in Cambodia, 1990–2017. (Source: FAOSTAT)

Table 11.1 Rice production data for Cambodia and Takeo Province, 2017–2018

Variable	Cambodia	Takeo	Takeo %
Total area (ha)	3,335,929	302,546	9.1
Wet season area (ha)	2,739,446	199,643	7.3
Dry season area (ha)	596,483	102,903	17.3
Overall yield (t/ha)	3.35	3.90	116.4
Wet season yield (t/ha)	3.09	3.39	109.7
Dry season yield (t/ha)	4.50	4.90	108.9
Total production (t)	10,891,735	1,179,936	10.8
Wet season production (t)	8,212,893	676,051	8.2
Dry season production (t)	2,678,842	503,885	18.8

Source: Ministry of Agriculture, Forestry and Fisheries, Cambodia

compared with 1.1 t/ha in the 1960s and 1.5 t/ha in the 1990s (MAFF 2013; Helmers 1997; Nesbitt 1997b). The yield increase was mainly due to the adoption of improved varieties and increased use of fertiliser; only 8–9% of arable land is irrigated in the DS and the potential for expanding surface irrigation infrastructure is limited (Johnston et al. 2013). Nevertheless, the DS rice area increased from 13% of the total rice area in 2008 to 18% in 2017–2018 (World Bank 2015; Table 11.1). The more profitable Cambodian aromatic rice varieties, mainly grown in the western provinces such as Battambang, now account for 10% of the annual cultivated area and 30% of total production (World Bank 2015).

The growth in production has generated a rice surplus above estimated domestic requirements (World Bank 2015). The notional surplus increased from 1.44 million t of milled rice in 2008 to 2.38 million t in 2013. With increasing liberalisation of trade and explicit government encouragement—especially with the 2010 *Policy Paper on the Promotion of Paddy Production and Rice Export*, in which rice was designated as "White Gold"—the export of rice has increased sharply (ADB 2014). In 2013, formal exports of rice were 378,850 t, exported to over 50 countries, 91% of which were in the EU or Asia; half of these official exports were of high-quality fragrant rice. This was well below the target enunciated in the 2010 Policy Paper of 1 million t of milled rice exports by 2015.[2] However, in 2013, informal cross-border trade of rice and paddy (unmilled rice) to Vietnam and Thailand was estimated to be 1.5 million t (in milled rice equivalent).[3] Thus total exports of rice and paddy increased nearly 20-fold from 2008 to 2013, from just over 100,000 t to 1.9 million t (in milled rice equivalent).[4] Though drying and milling capacity remains a constraint and exports are still mainly in the form of paddy, capacity has increased from 20 t/hr in 2009, when there were only two large rice milling companies, to over 700 t/hr in 2013.[5] In 2018 official data indicated that milled rice exports had risen further to 626,225 t. There was no corresponding figure for official and unofficial paddy exports, but the evidence is that these continue to exceed the exports of milled rice by a substantial margin.

Changes in Rice Production at the Farm Level

Realising the benefits of this growth at the farm household level depends on access to land, labour, water, finance, and markets. However, many rice-growing households, especially in the south-eastern lowlands, have farms of less than one hectare. For Cambodia as a whole, the average size of rice farms increased from 1.9 ha in 2008 to 2.1 ha in 2012 (World Bank 2015), reflecting out-migration from the densely populated south-east and in-migration and area expansion in the north and north-west. However, this disguises an increase in inequality, with the average size of small farms decreasing during 2008–2012 (from 1.0 ha to 0.9 ha) and the average size of medium and large farms increasing (from 1.6 ha to 2.4 ha, and from 3.6 ha to 7.0 ha, respectively).

At the same time, the unit profitability of rice farming has increased. Gross margins (GMs) for WS rice were estimated to be about USD 250/ha and USD 5/day in 2013, and for DS rice, USD 300/ha and USD 10/day (World Bank 2015).[6] However, for small farms using modern technology, the GM in 2013 was USD 522/ha in the WS and USD 276/ha in the DS (comparable to the GMs found in the Takeo field studies reported in the next five chapters). During 2005–2013, the GM per ha for WS rice grew in real terms at 2.4% due to the increase in yield and price. In the same period, the GM per ha for DS rice grew at 2.1%. The growth in profitability as well as the growth in population has increased the demand for land, as reflected in rising prices—the purchase price for rainfed lowland plots increased by 271% from 2005 to 2013 and for irrigated land by 620% (World Bank 2015).

Rice farming has also been undergoing rapid mechanisation, following the trend in Thailand and Vietnam, with significant implications for labour requirements (World Bank 2015). Mechanisation initially took the form of power tillers (two-wheeled tractors) for land preparation and small, moveable pumps for irrigation (whether from rivers, canals, receding floodwaters, farm ponds, or tube wells). These were attractive because they were affordable, multifunctional, and saved on labour, time, and costs. Contracted services of reapers and threshers were also widely taken up but are being overtaken by combine harvesters, the number of which is rapidly increasing. However, methods to save on planting labour such as drum seeders and rice planters have not been widely adopted. As a consequence of this mechanisation process, the labour used in WS rice production decreased from 85 days/ha in 2005 to 48 days/ha in 2013, and even more in DS rice production, from 90 days/ha to 28 days/ha (World Bank 2015: 67). There is potential for these labour requirements to be reduced further if the trend in neighbouring countries is a guide.

As noted earlier, access to irrigation schemes has been limited and there are few suitable sites for further expansion of surface irrigation infrastructure. However, in recent decades there has been an increase in the use of groundwater for irrigation, especially in the south-eastern lowlands (Johnston et al. 2013). While groundwater has been used for domestic purposes for centuries, the availability of small, portable pumps has encouraged many farmers to sink tube wells in their rice fields and use this source for supplementary irrigation of WS or DS (recessional) rice, or for alternative, less-water-demanding DS crops. The sustainability of this use of groundwater is still a matter for research, with some evidence of long-term decline in Prey Veng and Svay Rieng Provinces, though in Kandal and Takeo, closer to the main channel of the Mekong and Bassac Rivers, it seems that aquifers are readily recharged during each wet season. As discussed in Chap. 12, access to this form of on-farm irrigation can have a significant impact on the productivity of rice-based cropping systems.

Another trend affecting the capacity of small-scale rice farmers to increase production and incomes is the availability of credit. Before the 1990s, farmers only had access to short-term, high-interest loans from local moneylenders. From 1993 there was a proliferation of non-government organisations (NGOs) involved in rural development, some of which offered microfinance. One of these has grown into the Association of Cambodian Local Economic Development Agencies (ACLEDA) Bank, Cambodia's largest commercial bank with branches in Myanmar and Laos. From 2000 onwards, government reforms enabled many of these NGOs to become specialised micro-finance institutions (MFIs), providing loans at commercial interest rates for use as working capital (to pay for seed, fertiliser, hired labour, and other inputs) and to purchase durable capital items such as pumps and two-wheeled tractors (and also land). By 2011 there were 29 MFIs and the ACLEDA Bank providing financial services in 24 provinces, covering almost 60,000 villages and 1.1 million borrowers with outstanding loans of USD 573 million. The number of borrowers has increased three-fold between 2005 and 2011 and the value of loans by a factor of 11 (Chap. 15). The average value of outstanding loans was USD 515 per borrower.

Apart from rice farming, a significant proportion of farm households throughout Cambodia now also depend on non-farm work opportunities for their livelihoods. The recent rapid development of construction, light industry, and the services sector in urban centres, especially in and around Phnom Penh, has provided many job opportunities. Young household members frequently migrate to urban areas to seek non-farm employment to help support their families in the villages. However, these young wage earners have few skills and little future earning potential, while their absence from the villages can severely constrain the farm labour force. Another source of non-farm employment is the provision of contract services in rural areas by households with the resources to purchase large machinery, particularly combine harvesters.

THE STUDY AREA

The location for the specific field studies reported in the next five chapters was Takeo Province and the adjacent lowland portion of Kampong Speu Province in the southern part of the Central Plain (Fig. 11.1).[7] Most of the study area was characterised by the sandy, infertile soils of the old alluvial and colluvial plains, suited to WS rice, but in the south-east corner the richer, heavier soils of the Bassac floodplain predominated, providing the opportunity for DS rice. The annual rainfall ranges from 1000 to 1500 mm, lower than the national total of 1500–2000 mm. Mean monthly rainfall records for Takeo and Kampong Speu Provinces for the 31-year period from 1982 to 2012 show that the lowest monthly rainfall occurred from December to March (5–35 mm/month) and the highest in September and October (195–230 mm/month), although there was considerable variation in WS rainfall from year to year. For example, October rainfall varied from 0 to 500 mm. Only July, August, and September avoided a complete drought over the three decades, with a minimum monthly rainfall of between 35 and 85 mm.

Takeo Province covers an area of 3563 km² and in 2010 was divided into 10 districts, 100 communes, and 1117 villages, with 199,373 registered households (NCDD 2010). The town of Takeo, at the centre of the province, is located about 80 km south of the capital (Fig. 11.1). The province's location between Phnom Penh and the Mekong Delta has been favourable for accessing new agricultural technologies and markets for farm inputs and outputs, particularly with the improvement of transport infrastructure. National Highway 2 passes through the centre of the province, extending to the Vietnam border in Kiri Vong District. National Highway 3 passes through the north-west districts on the way to Kampot Province. This infrastructure has also favoured periodic migration to take up non-farm employment in Phnom Penh.

The study area has for centuries been an important site for the concentration of rural population based on rainfed lowland rice, perhaps dating back to the ancient polity of Funan around 2000 years ago. In the 1950s, the population density was already high, between 150 and 200 persons/km², with some districts nearing 500 persons/km²; hence holdings of paddy land at that time were smaller than 2 ha, while grain yields from the single WS rice crop were low, averaging less than 1 t/ha (Delvert 1961). Apart from growing rice, farm households traditionally produced palm sugar, wove silk fabrics, and made bamboo baskets to support their livelihoods.

The total population of Takeo Province at the 2008 Census was 965,835, consisting of 186,247 households. The average household size was 4.9, similar to the nationwide average. The province had the second highest population density (276 persons/km²) after Kandal Province (364 persons/km²). In 2008, 47% of the Takeo population was aged less than 20 years and 5% was aged over 60 years, reflecting the youthfulness of the population in Cambodia as a whole (Fig. 11.3). The population pyramid showed a bulge in the 10–19 years age cohort, reflecting the rapid population growth in the 1980s and the slowing of this growth since the mid-1990s. The bulge indicates a high population momentum, despite a declining birth rate, as well as increased numbers of young people seeking employment in what is already a densely populated province, putting pressure on landholdings and spurring outmigration.

The most important source of livelihood in the province remains agriculture, with 92% of households recorded as rice-producers in 2010. WS rice accounts for 67% of arable land in the province and DS rice for 39% (mostly land that is not available for WS cultivation due to flooding). Only 4% of arable land is used for non-rice crops, including maize, soybean, mung bean, peanut, cassava, sweet potato, sesame, and vegetables, grown

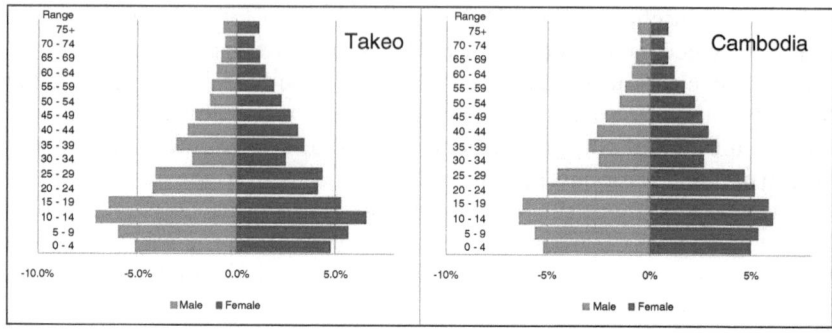

Fig. 11.3 Population pyramid for Takeo and Cambodia. (Source: Cambodian Census 2008)

mainly in Prey Kabbas District. Takeo is also a major producer of livestock, including cattle, pigs, and poultry. Surplus livestock is marketed to Phnom Penh or exported to Vietnam. Currently, the province has the smallest average farm size in the country; about 57% of households have less than 1 ha of paddy land and 5% have no paddy land. Even so, average farm size increased from 0.63 ha in 2008 to 0.85 ha in 2012 (World Bank 2015), reflecting permanent outmigration to frontier zones and to Phnom Penh. In addition, around 9% of the resident population aged between 18 and 60 years migrates periodically to seek non-farm employment, mainly as factory and construction workers in the vicinity of Phnom Penh. The incidence of non-farm employment is unevenly distributed among the ten districts, ranging from 4% to 16% (NCDD 2010).

Takeo was the third largest WS rice–producing province in Cambodia in 2012, accounting for 8% (198,768 ha) of the area under WS rice cultivation nationally and producing 681,184 t (nearly 10% of national production) (MAFF 2013). Though Takeo is relatively close to Phnom Penh, the area committed to rice cultivation has not been affected by the rapid growth of light industry in the vicinity of the capital. In fact, the area of WS rice had increased by 26% from 2008 to 2012. Almost all WS rice cultivation in Takeo is rainfed, with only a small area having access to supplementary irrigation to cope with droughts. Of this small irrigated area, 62% is irrigated from dams, 33% from natural sources of surface water, less than 5% from groundwater, and less than 1% from farm ponds. These sources of water are mostly unavailable during the DS, limiting opportunities for double cropping of rice.

Nevertheless, Takeo Province is the largest producer of DS rice in Cambodia, producing 466,010 t from 96,507 ha in 2012 (MAFF 2013). The area under DS rice in 2012 represented a 21% increase from 2008. The proportion of farm households with access to DS irrigation varies from 11% in Tram Kak District in the west to 64% in Borei Chulsa District in the floodplain area to the east. The opportunity for DS rice cultivation and the expansion of the DS rice area have been influenced by a number of factors. Areas close to the Bassac River that are subject to regular flooding in the WS are increasingly being used for growing high-yielding varieties in the form of DS recession rice. The development of dams and canals to manage the floodwaters is providing favourable conditions for irrigating these DS rice crops. The expansion of cross-border trade with Vietnam has also provided an incentive for DS rice cultivation in the province. Apart from being a source of inputs and providing a ready market, this cross-border channel means the harvested paddy rice can be conveniently transported by road to mills in Vietnam immediately after harvest.

Although Takeo is a major rice-producing province, rice cultivation is still heavily reliant on traditional practices—draught animal power, manual labour, and conventional tools. The total number of draught animals recorded in 2012 was 373,800 head (mainly cattle), a 12% increase from 2008 (MAFF 2013; NCDD 2010). The average number of large ruminants per cattle-raising household ranged from 2.1 to 2.6 head between districts. Despite the large number of draught cattle, mechanisation of land preparation is increasing, either as a result of individual household investment or through contract operations. Nearly 8500 power tillers, accounting for 5% of rice farmers, were registered in Takeo Province in 2012 (MAFF 2013). The number of threshing contractors has also increased in recent years—in 2012 there were 1300 threshers registered, an increase of 400 over 2011 (MAFF 2013).

More significant has been the increasing use of combine harvesters, especially in the DS. There were 3743 small combine harvesters in the province in 2012, accounting for almost 50% of the national total. This represented a fourfold increase over the number in 2011. Large-capacity harvesters did not become available until 2012, with 91 machines being reported (MAFF 2013). Despite the rapid increase, the number of combine harvesters is not sufficient to harvest the entire rice crop in the province. In addition to the farmers' financial constraints, field conditions and the characteristics of WS rice varieties are not always suited to mechanical harvesting.

Takeo Province has a large number of rice mills. Nevertheless, while there has been an increase in the level of mechanisation of farming operations, the number of small rice mills declined from 3800 in 2011 to 3500 in 2012. There was also a decline in the number of large rice mills from ten in 2011 to six in 2012 (MAFF 2013). As noted, much of the DS rice harvest is transported directly to more efficient mills in Vietnam which incorporate large-scale driers, capable of handling paddy that has been transported directly from the farm without the usual sun-drying.

Despite the widespread use of traditional practices, there has been sufficient adoption of modern seed-fertiliser technology (and improved water management in the DS) to see a steady increase in yields and output (MAFF 2013). The average WS yield increased from 2.9 t/ha in 2007 to 3.4 t/ha in 2012, and the average DS yield increased from 4.2 t/ha to 4.8 t/ha over the same period. These yield increases, combined with the expansion in cultivated area in both seasons mentioned above, has meant that WS production increased 35%, and DS production 52%, from 2007 to 2012. Thus total rice production in 2012 was 1.2 million t—59% from the WS harvest and 41% from the DS harvest. This represented 13% of national output. The estimated surplus for Takeo in 2012 was about 800,000 t, worth around USD 160 million,[8] representing about 17% of the national surplus. Most of the surplus produced in Takeo was exported as paddy, either legally or illegally, to Vietnam.

CONCLUSION

Cambodia has a long history and comparative advantage in rice production. With relative political stability and access to improved varieties and other inputs, farmers have been able to increase the area cultivated and especially per-hectare yields so that total production has grown at over 5% since 1990. From being a rice-deficit country in the 1980s, the country has achieved self-sufficiency and, since 2010, become a serious exporter of paddy and milled rice. Rice farmers in Takeo Province have long made an important contribution to Cambodia's rice production and currently contribute 8% of WS output and 19% of DS output, as well as a major share of exports. The next chapters consider different dimensions of the rice value chain in Takeo to provide insights into the constraints and opportunities facing industry actors in that province and the potential for policy interventions.

Notes

1. Data from UN Department of Economic and Social Affairs, Population Division, as analysed by Worldometers, http://www.worldometers.info/world-population/cambodia-population/ (viewed 13 June 2017).
2. The 2010 Policy Paper also set a target of producing a paddy surplus of 4 million t by 2015.
3. In 2013, 1.7 million t of paddy were exported to Vietnam and 250,000 t of paddy and 450,000 t of milled rice were exported to Thailand (ADB 2014).
4. Formal exports of rice increased from 1500 t in 2008 to 378,850 t in 2013. Informal exports of paddy (in milled equivalent) increased from 100,000 t to 1.54 million t (ADB 2014).
5. In 2011, there were 28,474 small rice mills scattered throughout the rice-growing areas.
6. The daily wage for labour hired in rice production in 2013 was USD 4.50 (World Bank 2015).
7. The southern part of Kampong Speu Province comprises part of the lowland plains region, while the northern part lies within the plateau and mountains zone.
8. A paddy price of KHR 800/kg or USD 200/ton has been used to estimate the value of the paddy surplus.

References

ADB, 2014. *Improving Rice Production and Commercialization in Cambodia: Findings from the Farm Investment Climate Assessment.* Metro Manila: Asian Development Bank.

Delvert, J., 1961. *Le Paysan Cambogien,* translated by Sonal Shah. Paris: Mouton and Co.

Helmers, K., 1997. Rice in the Cambodian economy: past and present. In H. J. Nesbitt, ed., *Rice Production in Cambodia,* pp. 1–14. Manila: International Rice Research Institute.

Higham, C., 2014. *Early Mainland Southeast Asia: From First Humans to Angkor.* Bangkok: River Books.

Himmel, B. J., 2007. Khmer Rouge irrigation development in Cambodia. Unpublished manuscript.

Javier, E. L., 1997. Rice ecosystems and varieties. In H. J. Nesbitt, ed., *Rice Production in Cambodia,* pp. 39–81. Manila: International Rice Research Institute.

Johnston, R., Roberts, M., Thuon Try, and De Silva, S., 2013. *Groundwater for Irrigation in Cambodia.* Colombo: International Water Management Institute.

Kiernan, B., and Boua, C., 1981. *Peasants and Politics in Kampuchea, 1942–1981.* London: Zed Books.

MAFF, 2013. *Bulletin of Agricultural Statistics and Studies.* Phnom Penh: Department of Planning and Statistics, Ministry of Agriculture, Forestry and Fisheries.

MRC, 2017. Physiography of the Mekong Basin, Mekong River Commission. Available at http://www.mrcmekong.org/mekong-basin/physiography/ (viewed 8 June 2017).

NCDD, 2010. *Provincial Database.* Phnom Penh: National Committee for Sub-national Democratic Development.

Nesbitt, H. J., 1997a. Topography, climate, and rice production. In H. J. Nesbitt, ed., *Rice Production in Cambodia*, pp. 16–19. Manila: International Rice Research Institute.

Nesbitt, H. J., 1997b. Constraints to rice production and strategies for improvement. In H. J. Nesbitt, ed., *Rice Production in Cambodia*, pp. 108–122. Manila: International Rice Research Institute.

NIS, 2008. *National Census in 2008.* Phnom Penh: National Institute of Statistics, Ministry of Planning, Cambodia.

RGC, 2010. *Policy Document on Promotion of Paddy Rice Production and Export of Milled Rice.* Phnom Penh: Royal Government of Cambodia.

Sakhan, S., Seang, L., Leng, L., and Then, R., 2007. Rice ecosystems and cultivation in Cambodia. In S. Men, ed., *Rice Crop in Cambodia*, pp. 134–155. Phnom Penh: Cambodian Agricultural Research and Development Institute.

Slocomb, M., 2010. *An Economic History of Cambodia in the Twentieth Century.* Singapore: NUS Press.

White, P. F., Oberthur, T., and Pheav, S., 1997. *Soils Used for Rice Production in Cambodia: A Manual for their Recognition and Management.* Manila: International Rice Research Institute.

World Bank, 2015. *Cambodian Agriculture in Transition: Opportunities and Risks.* Economic and Sector Work, Report No. 96308-KH. Washington, DC: World Bank.

World Bank, 2017. Cambodia Overview. Available at http://www.worldbank.org/en/country/cambodia/overview (viewed 5 June 2017).

CHAPTER 12

The Production, Marketing, and Export of Rice in Takeo

Chhim Chhun, Theng Vuthy, and Nou Keosothea

INTRODUCTION

Though the rice sector has demonstrated rapid growth in Cambodia in the past decade, many studies show that there are significant issues in production and post-harvest operations to do with the cost of production, rice quality, storage capacity, the structure and performance of the milling sector, and the management of cross-border trade with Thailand and Vietnam (ACI and CamConsult 2006; ADBI 2008; Gergely et al. 2010; RGC 2010; Sok et al. 2011). Understanding the rice value chain may help to increase the benefits that accrue to smallholder producers. Takeo is one

C. Chhun (✉)
Cambodia Development Resources Institute, Phnom Penh, Cambodia
e-mail: chhun@cdri.org.kh

T. Vuthy
Office of Food Security and Environment, USAID, Phnom Penh, Cambodia
e-mail: vtheng@usaid.gov

N. Keosothea
National Committee, Economic and Social Commission for Asia and the Pacific,
Ministry of Foreign Affairs and International Cooperation,
Phnom Penh, Cambodia

© The Author(s) 2020
R. Cramb (ed.), *White Gold: The Commercialisation of Rice Farming in the Lower Mekong Basin*,
https://doi.org/10.1007/978-981-15-0998-8_12

of the main rice-producing provinces in Cambodia and increasingly engages in cross-border trade with Vietnam. The aim of this case study was to examine the production and marketing of rice in Takeo Province with a view to identifying ways to increase the benefits accruing to rice growers. The specific objectives were to (1) map the rice value chain (from producers downstream); (2) analyse costs and margins along the value chain; (3) examine relationships, governance, and flows of information along the value chain; and (4) identify policy options to improve the value chain.

To map the rice value chain in Takeo Province, both qualitative and quantitative approaches were adopted, drawing on primary and secondary data (M4P 2008). Data collection was conducted from mid-February to mid-March 2012. Focus group discussions (FGDs) and key informant interviews (KIIs) were used to obtain information from various actors, ranging from rice producers, rice collectors, rice traders, rice mill owners, rice exporters, and institutions working with rice marketing. FGDs were conducted with farmers in three districts: Tram Kak (predominantly rainfed wet-season [WS] rice), Prey Kabbas (rainfed WS rice with supplementary irrigation), and Koh Andaet (rainfed WS rice and dry-season [DS] and/or recession rice). For the value chain actors, information was also collected in two more districts—Angkor Borei and Kiri Vong—and in Takeo town.

Economics of Rice Production

The first link in the rice value chain is on-farm production. Table 12.1 presents typical gross margin analyses for WS and DS rice based on information provided by farmers in 2012 (i.e., for the 2011–2012 farming season). Farmers averaged yields of about 2.3 t/ha in the WS and 7.2 t/ha in the DS, the latter attributable to the use of high-yielding IR varieties (derived from the International Rice Research Institute), higher fertiliser rates, irrigation, and the higher level of insolation. Farm-gate prices were around USD 250/t for WS paddy and USD 194/t for DS paddy, the IR varieties required for export to Vietnam being of lower quality. Despite the lower price, DS rice production provided nearly 2.5 times the gross revenue of WS rice and twice the gross margin per ha. Nevertheless, the gross margin per day of family labour was similar, at about USD 8 per day, compared with daily wage rates of about USD 3 in the WS and USD 3.75 in the DS.

Farmers in the study area felt they faced high production costs, accounting for 52% of gross income in the WS and 61% in the DS. The main costs

Table 12.1   Gross margin analysis for rice farming (1 ha)

Activity	Wet season		Dry season	
	USD	%	USD	%
Value of paddy produced	575	100	1396	100
Input costs				
Land preparation	40	7	90	6
Seed	19	3	90	6
Hired labour	88	15	–	0
Chemical fertiliser	61	11	321	23
Pesticide	0	0	125	9
Irrigation	75	13	160	11
Threshing	15	3	70	5
Total input cost	298	52	856	61
Gross margin (excl. family labour)	277	48	540	39
Gross margin (incl. family labour)	150	26	297	21
Gross margin per day of family labour	8.15		8.31	

Source: Farmer interviews in study villages
Note: 1 USD = 4000 riels

in the WS were hired labour, fertilisers, and irrigation, while in the DS the main costs were fertilisers and irrigation.

> Costs of production are very high due to the high price of water fees, fertilisers, and pesticides. The high price of diesel also contributes to high production costs. In addition, we bought inputs on credit and paid back at harvest. For instance, fertilizer (DAP), the current price was USD 35 per 50 kg bag, but we paid back at the price of USD 40 per bag at harvesting. (FGDs with farmers in Koh Andaet and Prey Kabbas)

Most WS production followed traditional cropping practices, with transplanting and harvesting by hand, and thus hired labour was a significant cost. However, DS rice involved direct seeding (broadcasting) and mechanical harvesting, and hence all tasks could be handled with family labour and the cost of hired labour was zero.

Purchasing chemical fertilisers was another significant production cost, especially in the DS. This is because DS varieties are more responsive to fertilisers and DS yields are more assured, given access to irrigation. Hence, there are higher returns to increased fertiliser use and less risk of making a loss in the additional investment. However, farmers had little guidance on appropriate fertiliser rates.

> *There are no government agents or extension workers who come to teach us how to use agricultural inputs properly. Nowadays, we apply inputs following our neighbours in the villages, or retailers tell us how to use both fertiliser and pesticide and the application rate. (FGD with farmers in Koh Andaet)*

Irrigation fees were one of the cost items that interviewed farmers complained about the most, especially DS farmers. WS farmers also spent a significant amount on irrigation because they faced a short drought in 2011 requiring them to hire water pumps and buy fuel.

WS farmers did not usually apply pesticides unless there was a severe pest outbreak. However, DS farmers typically incurred a high cost for pesticides. In some years, pesticide costs were as high as for fertilisers, but in the study year there had been little problem with insect pests, hence pesticides accounted for around 15% of total production costs.

Purchasing rice seeds was not common for WS farmers; they retained their own seeds by drying and storing it carefully for use in the following season. However, DS farmers purchased seeds, accounting for 11% of production costs. This was because they used high-yielding seeds provided through Vietnamese traders, and broadcasting required a higher seeding rate.

The Rice Value Chain

Various value chain actors were identified during field interviews, as shown in Fig. 12.1. There were two main pathways, one for the WS harvest and one for the DS harvest. Most WS rice was produced for home consumption, with a small surplus sold. DS rice, by contrast, was produced for commercial purposes and most was sold immediately after harvesting. As a consequence, value chain actors were less active in the WS but very busy in the DS. WS rice was mainly traded to supply domestic consumers, with the flow from farmers to village collectors, regional traders, and local and regional millers. The DS market chain led to Vietnam, with paddy being transported by barge or road transport to the border at Phnom Den for milling within Vietnam.

Village collectors and traders were key actors in buying paddy from individual farmers and selling to rice millers and regional rice traders,

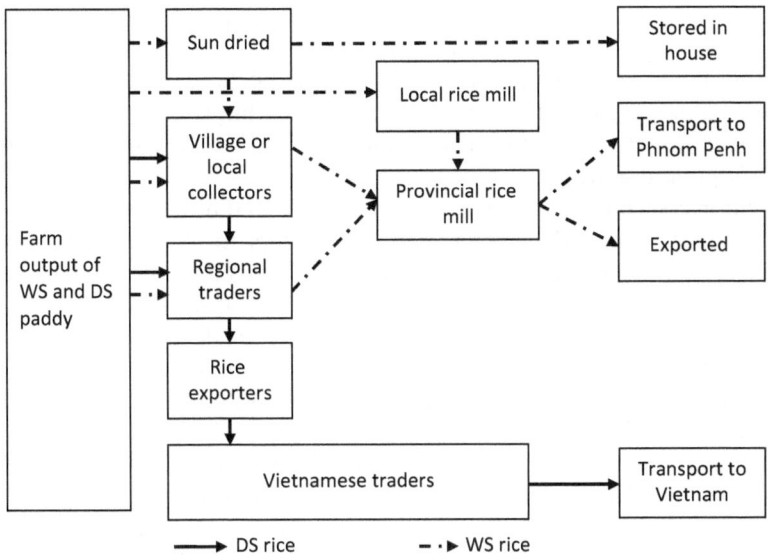

Fig. 12.1 Map of rice value chain in Takeo

mostly the latter. They were small businesses with a buying capacity of 10–30 t due to limited transport and capital. They typically loaded the paddy onto an oxcart or hand-tractor cart to transport from the village to the regional traders. They generally bought different varieties and mixed them, though they separated premium quality paddy which they sold for a higher price. Village collectors and traders were commonly farmers from the same village. Some bought paddy using capital advanced by the regional traders and transported the purchased stock to the regional traders for a commission USD 0.05 per 50 kg bag; some used their own working capital to buy paddy and sell for a profit.

Regional traders were larger businesses with the capacity to buy more than 100 t and sell to exporters. They usually had their own trucks and hired labourers. During the harvest season, given the recent progress in rural road development, regional traders had access to almost all collectors. They parked their trucks in the villages to collect paddy from the collectors and pay them immediately in cash. Sometimes they stored paddy in their warehouses for speculative reasons, but usually they transported the paddy directly to the next point in the chain. During the WS harvest, regional traders brought the collected paddy to local or provincial millers, who then sold rice to local consumers and retailers or transported rice to the Phnom Penh market or exporters. During the DS harvest, regional traders collected paddy to sell to Vietnamese traders within Cambodia but mostly they sold directly to Cambodian exporters who had a regular relationship with Vietnamese traders.

Exporters were larger businesses, usually located near a river port or the border, collecting paddy from regional traders for Vietnamese buyers. They were well connected with the traders from Vietnam and thus knew which varieties to buy in what quantities. To some extent they were financed by the Vietnamese traders, especially if the demand for paddy was high and the exporters experienced a capital constraint.

Local rice millers were also actors in the rice value chain. They bought paddy directly from farmers, milled it, and sold the rice in the local market. They also sold some paddy to provincial rice millers. Normally, they purchased and milled only WS paddy as local consumers demanded good quality local rice varieties rather than the bulk-export varieties planted in the DS.

VALUE CHAIN ANALYSIS

To estimate the value added by actors along the value chain, data were collected during the field interviews in February–March 2012 regarding buying and selling prices, handling and transportation costs, and mark-ups.

Table 12.2 Margins in value chain for wet-season paddy (USD/t)

	Village collector		Local trader		Regional trader		Provincial rice miller	
	$/t	%	$/t	%	$/t	%	$/t	%
Purchase price	250.0	100	262.5	105.0	325.0	130.0	375.0	150.0
Handling	1.0	0.4	2.0	0.8	2.0	0.8	2.0	0.8
Transportation	1.4	0.6	4.1	1.6	5.0	2.0	0.0	0.0
Materials	1.0	0.4	1.5	0.6	1.5	0.6	0.0	0.0
Informal fee	0.0	0.0	0.0	0.0	1.5	0.6	0.0	0.0
Mark-up	9.1	3.6	54.9	22.0	40.0	16.0	0.0	0.0
Total	262.5	105.0	325.0	130.0	375.0	150.0	377.0	150.8

Source: Authors' calculations from field data obtained in May 2012

Table 12.3 DS rice marketing value chain in Takeo province (USD/t)

	Village collector		Local trader		Regional trader		Exporter		Vietnamese trader	
	$/t	%	$/t	%	$/t	%	$/t	%	$/t	%
Price	192.5	100.0	202.5	105.2	217.5	113.0	237.5	123.4	262.5	136.4
Handling	1.0	0.5	2.0	1.0	2.0	1.0	2.0	1.0	2.0	1.0
Transport	1.4	0.7	4.1	2.1	5.0	2.6	5.0	2.6	0.0	0.0
Materials	1.0	0.5	1.5	0.8	1.5	0.8	1.5	0.8	0.0	0.0
Informal fees	0.0	0.0	0.0	0.0	1.5	0.8	2.0	1.0	0.0	0.0
Mark-up	6.6	3.4	7.4	3.8	10.0	5.2	14.5	7.5	0.0	0.0
Total	202.5	105.2	217.5	113.0	237.5	123.4	262.5	136.4	264.5	137.4

Source: Authors' calculations from field data obtained in May 2012

The results are presented in Table 12.2 for WS paddy and Table 12.3 for dry-season paddy.

The value chain for WS paddy began at harvesting, when farmers sold some surplus for cash income, and ended at the provincial rice millers (leaving aside the paddy used for household consumption that was taken to the village rice mill and returned to the household). The overall value added along this value chain was about USD 127/t or 50% of the farm-gate price (Table 12.2).

Most farmers sold paddy to small-scale village collectors at farm-gate prices of about USD 0.25/kg (USD 250/t). The village collectors bore the costs of loading, materials (bags, twine, and containers), and transportation, totalling around USD 3.4 or 1.4% of the farm-gate price. Most village collectors used their own means of transportation such as motorbikes, ox-carts, or hand-tractor carts, and hence had lower transport costs than local or regional traders. Their mark-up was about USD 9/t, and hence the value added by the collectors was about USD 12.5 or 5%. They accepted the market price offered by local traders from outside the village.

These local traders had somewhat higher costs than the village collectors and a significantly higher mark-up at USD 55/t, representing nearly half the value added from the farm gate to the miller. This suggests that the local traders had access to more price information and working capital than the collectors and could manipulate their buying price to a degree to increase their profits. However, some of the higher mark-up may have been due to storage costs beyond those incurred by the village collectors.

Regional traders were usually engaged in inter-provincial trade, spending relatively more on transport to their warehouses or to exporters, including "informal fees" along the way. Their mark-ups (USD 40/t) were somewhat lower than those reported for the local traders. They too were price-takers when selling to the local or provincial millers but may have been able to exercise some market power with the local traders.

In contrast to the WS, DS rice farmers produced rice solely for commercial purposes. Actors in the value chain were very active and competitive. Two additional DS actors were identified from the field interviews—exporters and Vietnamese traders (Fig. 12.1). The paddy traded in the DS mostly comprised IR varieties of lower quality than the local varieties grown in the WS. Hence the farm-gate price was lower, at around USD 193/t (Table 12.3). As for WS paddy, the value added from the farm gate onwards was in part due to the costs of handling, materials, transportation, and informal fees incurred by each actor; these expenses were similar between seasons.

However, the traders' mark-ups were significantly lower for the DS crop and did not differ greatly from the village collectors' mark-up, ranging from USD 7 to 10/t (or 3.4 to 7.5% of the farm-gate price). This indicates that the market was more competitive in the DS, squeezing the margins of all actors. The exporters, however, obtained a higher mark-up of USD 15/t, perhaps reflecting a degree of market power as the number of exporters was fewer and there was little domestic demand for the paddy. During the field interviews, it was not possible to obtain information on the transportation costs from the Cambodian port to Vietnam, only the handling cost of the Vietnamese traders at the border. Hence the remainder of the value chain and the final selling price in Vietnam was not captured (see Chap. 18 for the story from the Vietnamese side of the border).

In general, the market showed a high degree of competition, with many actors involved at each stage and prices set largely by market forces. Farmers could sell their paddy throughout the year into a highly competitive market. Paddy prices for different types and qualities were widely communicated on a daily basis (Gergely et al. 2010). However, there were obvious deficiencies in the market infrastructure, especially for export paddy. Takeo exported most of its rice surplus as paddy to Vietnam. Thus, the rice market in Takeo was highly dependent on the demand from Vietnam; if the border was closed or buying prices were reduced, there would be a major income crisis for value chain actors within Takeo,

Table 12.4 Rice prices in Cambodia, Thailand, and Vietnam, August–October 2011 (USD/t)

	Cambodia	Thailand	Vietnam	Price difference relative to	
				Thailand	Vietnam
White rice					
Farm gate	250–350	340–350	340–350	(0–90)	(0–90)
Milled rice	650	490–493	461	157–160	89
Export price[a]	680	605–610	565–575	70–75	105–115
Fragrant rice					
Farm gate	354–452	402–452	–	(0–47)	–
Milled rice	870	907–910	–	(37–40)	–
Export price[a]	900	1075–1085	675–685	(175–185)	185–225

Source: Ministry of Agriculture, Forestry and Fisheries, Cambodia
[a]FOB

especially the producers. This has occurred in the case of the cassava trade between north-western Cambodia and Thailand.

The amount of paddy exported across the border depended almost solely on the differential in paddy price between Cambodia and Vietnam. As shown in Table 12.2, the farm-gate price of paddy in Cambodia was much lower than in Vietnam (and Thailand), stimulating the flow of exports from southern Cambodia to Vietnam, including both official and unofficial exports (hence there was no official record of the amount of paddy exported). This indicates that there was a lack of storage and milling capacity within Cambodia to process and export milled rice to Vietnam or the international market. The high cost of milling in Cambodia is reflected in the relatively high prices of milled and export white rice in Table 12.4. Nevertheless, the relative prices of Cambodian fragrant rice make it potentially competitive with Thailand in this sub-sector of the market.

RELATIONSHIPS AND GOVERNANCE IN THE VALUE CHAIN

There is a two-way flow of information in the DS (export) rice value chain in Takeo. On the one hand, information about the availability of paddy in the villages is transferred along the chain from farmers to Vietnamese traders. On the other hand, information about prices and requirements for quality and quantities flows from Vietnamese traders back to farmers in the villages. This information flows through the intermediate actors in the value chain—exporters, regional traders, local traders, and village collectors. The price, quality, and quantity are set by the Vietnamese traders; the information is then passed on and manipulated by the different actors to cover their costs and obtain a margin, and finally farmers are faced with the farm-gate market price, quantity, and quality requirements. Mostly the Cambodian traders have little chance to negotiate the price and quality with the Vietnamese traders. When the demand is high, the Vietnamese traders seem not to take the quality problems so seriously, but they often take advantage of their position in the chain to downgrade the paddy and reduce the price.

There are no formal rules and regulations relating to setting the price of paddy in Takeo. Usually, the price is simply agreed between buyers and sellers, but it is ultimately limited by the price level set by the Vietnamese traders, otherwise the actors along the value chain will make a loss. Since rural roads have been markedly improved over the last decade, traders can now easily access most villages. Therefore, farmers have a degree of choice to sell their paddy to whomever can provide a better price.

> *Nowadays, we can sell our rice to someone who can give us the higher price. We don't care who they are. (FGD with farmers in all villages interviewed)*

> *There are paddy traders now; the buying price is very competitive. To get enough paddy, sometimes we have to increase prices; however, the price is not higher than the price set by the Vietnamese traders. (KII with village traders in Angkor Borei)*

There is also no formal or systematic mechanism in place to classify paddy quality at each link in the value chain; actors make judgements based on their own knowledge and experience before accepting paddy at agreed prices. The main quality criteria considered are moisture content and damaged or mouldy grain. Vietnamese traders particularly emphasise moisture content (a function of Cambodian traders buying paddy straight after harvest by combine harvesters when moisture content is still high). The Cambodian paddy exporters complain that the Vietnamese traders are too strict in setting quality standards as Cambodian farmers generally produce paddy that is not as good as the benchmark sample.

> *Normally, the Vietnamese traders give the sample of paddy quality and [associated] price to the exporters. Then the exporters pass on the quality requirements and prices to the regional traders to buy paddy for them. Most*

often, exporters are faced with quality problems because the collected paddy is usually of mixed quality or farmers grow mixed varieties, so it is difficult to distinguish them according to the quality demanded. Sometimes, Vietnamese traders downgrade the paddy, not accepting the quality of the paddy that we have collected and transported to the port. Therefore, negotiations had to take place and finally the price was decreased. (KII with rice traders and exporters in the study area)

RICE POLICY IN TAKEO

Takeo is one of the main rice producers in Cambodia, accounting for 12.5% of national production and 17.6% of the national rice surplus. Thus, Takeo is one of the key provinces contributing to the government's policy promoting rice exports, with a goal of exporting 1 million t of milled rice. About 41% of Takeo's paddy output came from DS production in 2011–2012. Though Takeo shows potential, there are many shortcomings in the rice sector, including the varieties used, low-quality seeds, limited extension services, and post-harvest issues.

Recently, the Ministry of Agriculture, Forestry and Fisheries (MAFF) promoted ten varieties, including three early-maturing IR varieties—Sen Pidor, IR66, and Chulsar—that have the potential to meet the quality standard for high-value rice exports. However, farmers continue to use more Vietnamese varieties. IR504 from Vietnam is widely used by farmers in irrigated and recession rice areas in the DS. Though this variety is not of good quality for the local market, the high yield and the demand from Vietnam has meant that farmers widely adopt it for commercial production.

As indicated above, farmers continue to use low-quality seeds. Though there are companies producing seeds, supplies are still limited in many areas, including Takeo. Hence, most farmers do not renew their seeds regularly, particularly for WS rice. Seeds are often mixed during storage and reduced in quality after being used for many years. Farmers renew their seeds only when collectors or millers demand better quality and offer higher prices.

During the field interviews, farmers complained about the difficulty of finding technical assistance to control rice pests, especially in the DS. They applied many kinds of pesticides; some were banned and very dangerous to human health and the environment. Most of the pesticides sold in the market were imported from Vietnam or Thailand, with original language labels. Furthermore, there was a dearth of information from extension services to advise farmers on fertiliser application. Farmers applied at a rate they felt they could afford or merely followed the advice of the fertiliser merchants. Some fertilisers sold on the market also had low quality, as discussed in Chap. 14.

About 69% (764,902 t) of total paddy rice production in Takeo is surplus, available for export. The milling sector has limited capacity to absorb this surplus to process and export. Thus, the export market for paddy remains vital for Takeo rice farmers. As indicated earlier, paddy in Takeo is mainly traded with Vietnam; the trade is dominated by Vietnamese traders in setting prices and the required quality. Vietnamese traders can downgrade the paddy and hence lower the export price. Measures are needed to formally grade paddy and encourage better quality so the trade is fair and beneficial to value chain actors on both sides of the border.

CONCLUSION

Rice production in Takeo provides a subsistence base for farm households, an adequate return to household labour and, for those who have access to irrigation in the DS, an important commercial activity. The returns to farmers could be improved by providing better information about and regulation of the key inputs—seeds, fertilisers, and pesticides. The rice market in Takeo is well structured with a network of collectors, traders, and exporters. Farmers can readily sell their paddy at a competitive market price. The marketed surplus is traded and milled efficiently in the domestic market, but the milling sector does not have the physical capacity or capital to handle the DS paddy surplus, which is exported directly to Vietnam. Thus, the export of paddy remains crucial for the commercial rice industry in Takeo. Though Cambodia and Vietnam have an agreement with regard to the cross-border trade in paddy, if Vietnam's rice policy changed to protect its own farmers, the rice sector in Takeo would be vulnerable. Nevertheless, while Cambodia continues to develop its rice processing and export capacity, the cross-border trade in paddy provides a viable source of income for a sub-sector of rice farmers in Takeo.

REFERENCES

ACI and CamConsult, 2006. *Cambodia Agriculture Sector: Diagnostic Report.* Prepared for AusAID. Phnom Penh: Agrifood Consulting International (ACI) and CamConsult.

ADBI, 2008. *Rice Contract Farming in Cambodia: Empowering Farmers to Move Beyond the Contract Toward Independence.* Manila: ADB Institute.

Gergely, N., Baris, P., and Meas, C., 2010. *An Economic Survey of Rice Sector in Cambodia.* Phnom Penh: Agence Francaise de Développment, Supreme National Economic Council, and G.I.G. Consultants.

M4P, 2008. *Making Value Chains Work Better for the Poor: A Toolbook for Practitioners of Value Chain Analysis,* vol. 3. Phnom Penh: Agricultural Development International.

RGC, 2010. *Policy Document on Promotion of Paddy Rice Production and Export of Milled Rice.* Phnom Penh: Royal Government of Cambodia.

Sok, S., Chap, S., and Chheang, V., 2011. *Cambodia's Agriculture: Challenges and Prospects.* Phnom Penh: Cambodian Institute for Cooperation and Peace (CICP).

CHAPTER 13

The Role of Irrigation in the Commercialisation of Rice Farming in Southern Cambodia

Chea Sareth, Rob Cramb, and Shu Fukai

INTRODUCTION

This chapter is based on a study to explore the key constraints to rice-based farming systems in the rainfed lowlands of Cambodia and the role of different sources of irrigation in alleviating some of those constraints (Chea 2015). The research was carried out in lowland districts in Takeo and Kampong Speu Provinces in the southern part of the Tonle Sap Basin, representing a major lowland rice-growing region with high population density, small farm sizes, and severe production constraints (Fig. 13.1). Three villages were selected with similar biophysical and socioeconomic environments but different degrees of access to irrigation:

C. Sareth (✉)
Cambodian Agricultural Research and Development Institute,
Phnom Penh, Cambodia
e-mail: sareth.chea@uqconnect.edu.au

R. Cramb • S. Fukai
School of Agriculture and Food Sciences, University of Queensland,
St Lucia, QLD, Australia
e-mail: r.cramb@uq.edu.au; s.fukai@uq.edu.au

© The Author(s) 2020   261
R. Cramb (ed.), *White Gold: The Commercialisation of Rice Farming in the Lower Mekong Basin*,
https://doi.org/10.1007/978-981-15-0998-8_13

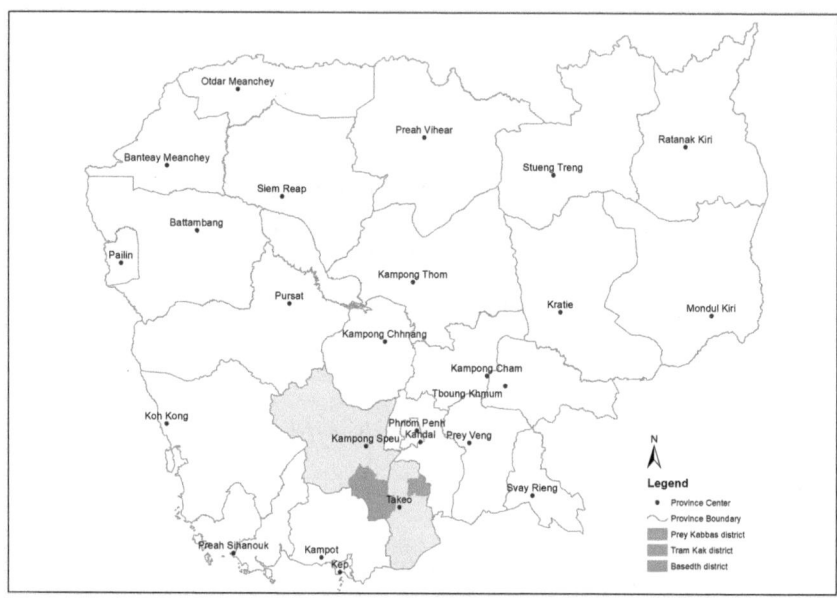

Fig. 13.1 Locations of the three study districts in Takeo and Kampong Speu Provinces. (Source: Cambodian Agricultural Research and Development Institute)

- Trapeang Run, in Tram Kak District in Takeo Province, shows the full extent of the development problem facing farm-households and villages in the rainfed lowlands, with all the constraints attributed to this zone, including very limited access to irrigation, restricted to small house-yard ponds.
- Snao, in Prey Kabbas District, also in Takeo Province, shows what options become available to farm-households with access to on-farm sources of irrigation in the form of shallow tube wells to draw on groundwater resources, in addition to farm ponds. This case also shows the potential for agricultural development with little or no intervention by government or other development agencies.
- Ta Daeng Thmei, in Basedth District in Kampong Speu Province, shows what farmers can do when they have access to a medium-scale, gravity-fed irrigation facility. Where public investment in such irrigation schemes is feasible, farming options are increased, though there are issues that must be addressed at the community level to maintain the irrigation infrastructure and manage water use.

A range of research methods were employed between 2010 and 2013 for data collection, including reconnaissance visits, household surveys (with 200 respondents across the three villages), discussions with village heads, key informant interviews, analysis of market trends, farm walks and direct observation, use of village data manuals and documents, surveys of pond-water and groundwater, analysis of rainfall data, soil surveys, and field crop experiments (Chea 2015). Each village was studied as an individual case, with cross-case comparison used to develop broader generalisations. It is this comparative analysis that is presented in this chapter.

CHARACTERISTICS OF CASE-STUDY VILLAGES

Village Settlement and Population

The main geographical and demographic characteristics of the case-study villages are shown in Table 13.1. All three villages were located 70–75 km south or south-west of Phnom Penh, but Trapeang Run was more favourably situated in terms of access to district and provincial centres for both farm transactions and non-farm employment. The settlement patterns of Snao and Ta Daeng Thmei were typical of rural Khmer communities, with houses clustered on areas of higher land which are dry year-round. However, in Trapeang Run, the houses were scattered throughout the village territory, singly or in small clusters, on or adjacent to paddy fields, giving farmers greater capacity to manage their rice and non-rice crops and livestock.

The highest population density was in Trapeang Run (700 persons per sq. km), about double that of the other two villages. However, the villages had similar areas of paddy land (90–120 ha) and there was little difference in the available paddy land per capita (around 0.1 ha). There were no major differences in the demographic characteristics of farm-households, except that the average age of household heads in Ta Daeng Thmei was 5–6 years lower than in the other two villages, consistent with a younger total village population and a high percentage aged less than 25 years. This may have been due to a lower rate of outmigration, especially when compared with Trapeang Run. Between 86% and 95% of household heads considered farming as their primary economic activity, as did their spouses. Economically active daughters (those aged 15 years and above who had

Table 13.1   Major characteristics of the case-study villages

Characteristic	Trapeang Run (rainfed)	Snao (on-farm irrigation)	Ta Daeng Thmei (fully irrigated)
Province	Takeo	Takeo	Kampong Speu
Distance to ...			
• Phnom Penh (km)	75	75	70
• Provincial capital (km)	12	>30	>30
• National road (km)	2	15	20
Access to market	Favourable	Less favourable	Less favourable
Topography	Central plain (15 masl)	Next to floodplain (3–15 masl)	Gently sloping (27–36 masl)
Flooding regime	Flash-floods	Part flooded in WS	Flash-floods
Total land (ha)	113	451	200
Paddy land (ha)	90	120	120
Irrigation source	Small ponds	Groundwater, ponds	Reservoir
Cropping pattern (WS/DS/EWS)	Rice/fallow/rice	Rice/radish-cucumber/rice[a]	Rice/peanut-rice/rice
Settlement pattern	Dispersed	Clustered	Clustered
No. of households	157	277	158
Pop. density (pers./km^2)	697	292	372
% under 25 years	35	55	62
Paddy land (ha/person)	0.11	0.09	0.16
Household size	5.4	4.9	5.0
Family workforce	4.0	3.7	3.3
Household head			
• Age (years)	46.4	47.0	41.4
• Male (%)	89	97	92
• Education (years)	6.0	6.0	5.7
• Occupation (% farming)	86	94	93
Children's occupation			
• Female (% non-farm)	42	42	38
• Male (% non-farm)	35	22	23

[a]Cropping pattern for Snao is for WS paddy land only, excluding the DS paddy land to which some villagers had access which was flooded in the WS

finished studying) were twice as likely to be engaged in non-farm jobs as farming in all three villages. This was consistent with the predominant employment of young female workers in the nearby garment industry. Economically active sons, however, were equally likely to be employed in farming as in non-farm activities (typically, construction).

Land Resources

Trapeang Run occupied a level plain and experienced only very short periods of flash-flooding (Table 13.1). Snao occupied a level plain adjacent to the Tonle Bassac floodplain, and hence some of the lower paddy land was subject to wet-season (WS) flooding while the upper paddy land was subject to drought. Some households in the village also had access to floodplain land that was uncultivable in the WS due to flooding but highly suitable for a dry-season (DS) rice crop—that is, flood recession rice. Ta Daeng Thmei was located on a gently sloping plain downstream of low hills and below a dam providing gravity-fed irrigation. It was only subject to flash-flooding when excess water was discharged from the reservoir.

All three villages had access to three land types—WS paddy land (cultivable in wet and dry seasons), upland used for non-rice crops and residential upland—and (as noted above) some households in Snao had access to DS paddy land (only cultivable in the dry season). The WS paddy lands in all villages were of the Prateah Lang soil type—the infertile, sandy soils that predominate in the lowland rice-lands of Cambodia. However, the DS paddy land to which some villagers in Snao had access were highly fertile alluvial soils. Almost all survey households in the three villages owned WS paddy fields. The mean area of WS paddy land was lowest in Snao (0.6 ha), intermediate in Trapeang Run (0.9 ha), and highest in Ta Daeng Thmei (1.3 ha). However, nearly 50% of households surveyed in Snao owned on average 0.85 ha of DS paddy land in addition to their WS land.

All villages showed the spatial dispersion of paddy landholdings arising from the land reform of the late 1980s and the subsequent fragmentation of land through equal inheritance among children. However, paddy land was more dispersed in Ta Daeng Thmei, averaging 5.7 plots per household, than in Trapeang Run (3.3 plots) and Snao (2.5 plots). The more recent settlement, larger average landholding, and access to irrigation could have influenced the greater degree of land fragmentation in Ta Daeng Thmei.

Water Resources

Households in Trapeang Run had established small ponds close to the house for their domestic water supply, which were also used to a small extent for the irrigation of vegetables in the house-yard, and irrigation of field crops on small plots of paddy land adjacent to the house (Table 13.1).

Pond-water was also used to supplement the water needs of rice seedlings when rainfall was inadequate early in the WS. The minimal use of ponds for agriculture was because of their limited storage capacity, such that they could potentially become dry early in the DS. Households in Trapeang Run also accessed groundwater through open wells and tube wells, but only for domestic use. Hence in terms of water resources for agriculture, it is accurate to characterise Trapeang Run as a purely rainfed village.

Households in Snao also had access to small ponds, sometimes in the farm (Fig. 13.2). However, the village had made the important change to extracting groundwater through tube wells in the farms for irrigation, after which farmers have made little use of ponds for irrigation (Fig. 13.3). The use of groundwater was reflected in the much higher incidence of pump ownership in this village (90%). Groundwater was a highly reliable irrigation source, sufficient to fully irrigate two DS crops of radish or cucumber, as well as provide supplementary irrigation for the early-wet-season (EWS) and WS rice crops. Despite increasing extraction over the past three decades, the water table had shown no sign of a significant drawdown. Although there was sufficient groundwater for a large irrigated

Fig. 13.2 Farm pond with portable pump in Takeo. (Source: Rob Cramb)

Fig. 13.3 Farmer in Takeo with tube well and pump. (Source: Rob Cramb)

area, only part of the paddy land could be irrigated because the land was fragmented and financial constraints restricted households from installing tube wells in every plot.

Ta Daeng Thmei had a community irrigation scheme, drawing water from a large reservoir, which also supplied five neighbouring villages.[1] The water level in the reservoir decreased late in the DS due to intensive irrigation and lack of rainfall. Hence the irrigation supply could be unreliable for up to two months but gradually recovered from late May because of the large catchment area to the north. The slight slope of paddy land from north to south permitted a gravity-fed irrigation system, but some households occupied paddy lands that could not be reached in this way. Hence portable pumps were used in these cases to get water from the main canals to farmers' fields, but at a higher cost that limited the options for these less favourable plots.

Village Characteristics in Context

The characteristics of the three case-study villages can be seen in the context of the general features of the lowland plain. All the villages had high population densities, characteristic of the rice-lands of south and south-eastern Cambodia. Hence all were experiencing the long-term rural-rural (e.g., to north-east and north-west Cambodia) and rural-urban (to Phnom Penh) migration that has been a feature of the south and south-east in recent decades. That the population density of Trapeang Run was twice as high as in the other two villages implies greater pressure to migrate, explaining the low proportion of the village population aged less than 25 years. The potential for agricultural development in all lowland villages in the south needs to be seen against this backdrop of continuing out-migration.

All the case-study villages had reasonable access to Phnom Penh, the largest and fastest-growing agricultural market in the country, as well as having close proximity to Vietnam. Hence future expansion of agricultural production was unlikely to encounter a market constraint. However, Trapeang Run also had particularly good access to district and provincial centres, giving it an advantage in terms of supplying fresh produce to these markets, as well as engaging in business activities, non-farm employment, and higher education, including high school and university. This was reflected in the generally higher grades of school-age children. The greater distance from national roads and market centres seen in Snao and Ta Daeng Thmei was more typical of rainfed lowland villages. Nevertheless, the widespread improvement in transport infrastructure in the southern lowlands in the past decade has created significant new market opportunities, even for these relatively remote villages.

While the dispersed settlement pattern of Trapeang Run was also atypical, it could indicate the future pattern for lowland villages as the population grows and farming becomes more intensive and diverse. The traditional Khmer pattern of clustered housing in a village centre was already beginning to change in the other two villages as a number of young farm families had settled on their inherited paddy land rather than adjacent to the parental household.

The three villages' reliance on WS paddy land with infertile, sandy soils, and only small upland plots used for house-yards and non-rice crops, was representative of the general situation in the rainfed lowlands. The land potential of Trapeang Run was more typical in that paddy lands made up most of the village area apart from residential land, whereas the other two villages had greater access to uplands for cropping and some in Snao had access to DS paddy land beyond the village boundary (not a general feature of the rainfed lowland zone).

Establishing small ponds in house-yards has long been a practice in lowland villages, though they are mainly used for domestic purposes, as in Trapeang Run. Likewise, accessing groundwater through open wells is a traditional practice, but not generally for irrigation. However, the case of Snao, with widespread on-farm irrigation based on groundwater, reflects an emerging trend in parts of the southern and south-eastern provinces. As in Trapeang Run and Snao, there is limited potential in the lowlands for the kind of canal irrigation development seen in Ta Daeng Thmei.

The variation in ownership of WS paddy land is a feature of the lowlands and a critical determinant of economic differences between households. However, the generally small landholdings seen in the case studies, even the very low mean of 0.6 ha in Snao, are common for the southern lowlands. The fragmentation of paddy land that was seen in all three villages, influenced by the 1980s land reform and the pattern of land inheritance, was also a general phenomenon in the lowlands, potentially hindering the adoption of both mechanisation and irrigation.

The increasing engagement of household members in non-farm employment in all three villages was characteristic of the lowlands, despite varying distances from Phnom Penh. In particular, the garment industry in Phnom Penh employs around 650,000 young female workers from a wide range of rice-growing areas. In each of the study villages, young women were twice as likely to be engaged in non-farm work as in farming. While young women from more favourably located villages could commute to the factories, many others still opted to take up this employment and reside in Phnom Penh rather than focus on farming. Many young men from the lowlands also took up employment in Phnom Penh, mainly in construction, but in the study villages they were just as likely to be engaged primarily in farming. In Trapeang Run, with its better access to local markets, young men and some older household members were also engaged in local trade, business, and wage employment.

COMPARATIVE ANALYSIS OF WET-SEASON RICE PRODUCTION

WS rice was the traditional mainstay of the farming system, being cultivated by every survey household in the three villages as the main or only source of household rice supply, as well as a potential source of cash income (Table 13.2). In each village, the available paddy land was fully cultivated. The mean cultivated area was the lowest in Snao (0.6 ha), but even in Ta Daeng Thmei, where the cultivated area was more than twice this figure,

Table 13.2 Characteristics of WS rice cultivation in the case-study villages

Practices	Trapeang Run (n = 79)	Snao (n = 62)	Ta Daeng Thmei (n = 59)
Mean area (ha)	0.9	0.6	1.3
No. of traditional varieties	15	3	9
No. of modern varieties	5	3	1
Varieties/household	2.4	1.2	2.0
Land preparation	Draught animal, plough, and harrow	Draught animal, plough, and harrow	Draught animal, plough, and harrow
Establishment method	Transplanting	Transplanting	Transplanting
Main water source	Rainfed	Rainfed	Rainfed
Supplementary irrigation	Small ponds	Groundwater	Reservoir
Irrigate nursery (%)	39	77	29
Irrigate main field (%)	16	71	25
Manure nursery (%)	100	77	93
Manure main field (%)	85	66	34
Fertilise nursery (%)	22	79	54
Fertilise main field (%)	95	82	100
Weeding (%)	89	71	25
Weeding method	Manual	Manual	Manual
Harvesting method	Sickle	Sickle	Sickle
Threshing method	Manual and thresher (11%)	Manual and thresher (35%)	Manual
Transport of paddy	Oxcart and shoulder pole	Oxcart and shoulder pole	Oxcart and shoulder pole
Drying paddy	Sun drying	Sun drying	Sun drying
Storage of paddy	Rice barn and bags	Rice barn and bags	Rice barn and bags

there was adequate labour to fully utilise the available land, even without mechanisation.

As elsewhere in the lowlands, traditional rice varieties were preferred in the WS, despite low yields, because of their good grain quality and adaptability to abiotic stress (Javier 1997). Lowland farmers were still unwilling to adopt modern IR varieties (derived from the International Rice Research Institute) for the WS crop, despite their higher yield potential, because of their inferior eating quality. There were up to 15 different traditional varieties in a village, but the suite of varieties (at least, as identified by farmers) differed between villages; only the *Srau Kraham* (Red Grain) variety was reported by every village. A few modern varieties were also grown but on no more than 5% of the total cultivated area in a village.

All activities from land preparation through to storage of the paddy were very similar across the three sites. Land preparation was undertaken with a pair of draught cattle and a traditional plough and harrow, as has been the practice for centuries. The low level of mechanisation reflected the general situation in the lowlands. In Takeo Province, the ratio of cultivated rice area to two-wheeled tractors is 23.5 ha per unit and in Kampong Speu, 14.7 ha per unit (MAFF 2011–2013). In the WS, farmers had an extended window for land preparation (June–September) and in any case farms were small. Moreover, households mostly owned enough draught cattle to manage land preparation and did not want to outlay the money to buy a tractor, or even to hire one from the few tractor-owners in each village.

The traditional labour-intensive transplanting method was used in all villages. Direct seeding by dry-seed broadcasting has been practised in north-western provinces such as Battambang and Banteay Meanchey, with larger farms, more fertile soils, and distant field locations, but there was no apparent trend to direct seeding in the south and south-east. This was presumably because the population density was higher, farm sizes smaller, and the household labour supply not yet limiting.

Supplementary irrigation was used for the seedling nursery and the transplanted crop. The incidence was much higher in Snao (over 70%) because of the ease of irrigating from tube wells. There was a low incidence of manual weeding in Ta Daeng Thmei (25%), reflecting a greater ability to maintain an adequate level of standing water in the paddy field. The incidence of weeding in the other two villages (70–90%) was high compared to other rainfed lowland areas (Rickman et al. 1997). Both farmyard manure and mineral fertiliser were widely applied in all villages.

In Snao, there were also probably carryover effects from the heavy application of nutrients to the DS crops grown on the same land.

The harvest and post-harvest activities largely followed conventional practice across the rainfed lowlands, relying on manual techniques using family, exchange, and, in some cases, hired labour. In the 1990s there was not a single mechanical harvester or thresher used in paddy fields in Cambodia (Rickman et al. 1997). Though the numbers of reapers, threshers, and combine harvesters have grown dramatically since then, especially for commercial DS rice, every case-study village harvested the WS crop with sickles. In Ta Daeng Thmei all farmers also threshed manually, but in the other two villages a minority hired mechanical threshers. The harvest was brought back to the homestead by a cattle-drawn cart or carried on shoulder poles, with a few using two-wheeled tractors. The paddy was commonly dried on palm-leaf mats for two to three days after threshing and stored in sacks (if intended for sale) or in the household's rice barn.

Though cultural practices were common, there were differences in the level of material and labour inputs, as summarised in Table 13.3. All villages used cattle manure, averaging about 6 t/ha in Trapeang Run and Snao, but only 2 t/ha in Ta Daeng Thmei. The lower rate in Ta Daeng Thmei probably reflected the larger cultivated area and the high application of manure for DS peanut cultivation (7 t/ha). Farmers in Snao used the highest rates of seeds and mineral fertilisers, whereas these rates were not very different between the other two villages. This probably reflected the smaller cultivated area in Snao, hence both the ability and the need to intensify the use of inputs, as well as a higher cash flow (see below). Snao also had a higher average use of fuel for supplementary irrigation. With home consumption as the main objective of WS rice production, farmers

Table 13.3 Average material and labour inputs for WS rice cultivation in the case-study villages

Input	Units	Trapeang Run (n = 79)	Snao (n = 62)	Ta Daeng Thmei (n = 59)
Seed	kg/ha	81	101	71
Fertiliser	kg/ha	124	166	125
Fuel	l/ha	13	59	29
Cattle manure	t/ha	6.2	6.0	2.3
Labour	days/ha	132	97	83

appeared to utilise all available household resources to the full, but to minimise the cash outlay (e.g., in comparison with EWS and DS rice and other cash crops) because they anticipated little or no cash return from this crop (though the subsistence value of the crop, hence the saving in expenditure, was around USD 350 per year).

Although material inputs were used more intensively in Snao, it was Trapeang Run that had the highest labour use (132 days/ha), 35–60% more than the other villages. However, Ly et al. (2012) also found labour inputs for WS rice cultivation in Takeo and Kampong Thom Provinces ranging from 78 to 127 days/ha; all farmers in that study used transplanting for their WS rice crops, but land preparation performed by two-wheeled tractors was found to save up to 6 days/ha. The additional labour input in Trapeang Run was spread over the activities of land preparation, pulling, transplanting, weeding, harvesting, threshing, and transport. The limited supply of irrigation water may have added to the time needed for ploughing and transplanting, because of drier, harder soil, and may have also added to the weed burden. It is also possible that the higher labour input reflected a somewhat older farm workforce with lower daily productivity, given the demographic characteristics described above.

The unit costs and returns for WS rice production are summarised in Table 13.4. Snao, with the smallest cultivated area and the highest seeding and fertiliser rates, produced the highest mean yield (2.8 t/ha), around

Table 13.4 Average unit costs and returns for WS rice production in the case-study villages

Item	Trapeang Run (n = 79)	Snao (n = 62)	Ta Daeng Thmei (n = 59)
Yield (t/ha)	2.2	2.8	2.4
Net output (t/ha)	2.1	2.7	2.3
Net output per capita (kg)	392	552	456
Rice-deficit households (%)	41	18	15
Households selling paddy (%)	47	24	90
Mean quantity sold (kg)	313	200	1100
Farm-gate price (USD/kg)	0.28	0.28	0.28
Gross income (USD/ha)	592	757	639
Input expenses (USD/ha)	90	176	70
Net return to household (USD/ha)	502	581	569
Total labour (days/ha)	132	97	83
Net return to labour (USD/day)	4.0	6.8	7.7

20–30% higher than the other two villages; this difference was significant at the 10% level. Snao also had the highest output per capita, while Trapeang Run had the lowest at 390 kg, though this output was above the assumed per capita consumption requirement of 250 kg. Trapeang Run also had the highest incidence of rice-deficit households (41%), despite cultivating a 50% larger area than Snao, reflecting the fact that the lower yield affected the household rice supply. Moreover, households in Trapeang Run did not have the same degree of back-up from EWS rice as in the other two villages. On the other hand, the potential of the traditional WS rice crop as a source of cash income was shown in the case of Ta Daeng Thmei, with its larger area more than compensating for a lower yield. Hence 90% of Ta Daeng Thmei households sold WS paddy, with a mean of 1.1 t being sold, more than a third of mean production.

Applying a farm-gate price of USD 0.28/kg across the three villages, the differences in gross income reflected the differences in yield. However, as noted above, input expenses (especially fertiliser) were highest in Snao (USD 176/ha), significantly higher than the other two villages (at the 1% level). This reduced the advantage of Snao in terms of the mean net return to household resources (USD 581/ha), although this was still the highest return of the three cases. The lower yield and gross income, and higher labour input of Trapeang Run, gave it a significantly lower net return to labour (USD 4/day), well below the return of USD 7–8/day in the other two villages and not greatly above the opportunity cost of labour (USD 3/day).

Though traditional farming practices predominated in all three villages, certain key factors gave farmers in Snao and Ta Daeng Thmei an edge over farmers in Trapeang Run, who more closely represented the majority of WS rice farmers in the rainfed lowlands:

- Access to adequate supplementary irrigation in Snao and Ta Daeng Thmei was important to save the crop from drought periods during the WS, whereas the small ponds in Trapeang Run were only sufficient to protect the crop at the nursery stage.
- Snao farmers used only three traditional varieties, suggesting that they had selected a small number of higher-performing varieties and avoided using low-yielding varieties. Trapeang Run farmers used 15 traditional varieties, most of them yielding less than 2 t/ha.
- Higher rates of input use, including seeds, fertilisers, and fuel (for irrigation), along with better varieties, helped give Snao farmers a significantly higher yield than the other two villages.

- The small landholdings in Snao pushed farmers to intensify and diversify their cropping system, with farmyard manure, fertilisers, and on-farm irrigation being used to support up to four crops per year, thus improving the soil fertility in the WS rice fields. In contrast, in Trapeang Run, with only a single rice crop, the paddy land was baked hard by the strong sun for half the year, degrading soil properties.
- It may have also been a factor that an older farm workforce and involvement in local non-farm activities in Trapeang Run helped to drag out the duration of transplanting, fertiliser application, weeding, and harvest activities, reducing the timeliness of these operations and thus decreasing yield.

The integration of traditional and improved practices for WS rice cultivation in Snao could indicate a possible future pathway for resource-poor lowland households, such as those in Trapeang Run. Even with small paddy holdings, Snao farmers were mostly self-sufficient in rice and could earn some cash income from the WS crop. With somewhat larger holdings, though still only 1.3 ha on average, farmers in Ta Daeng Thmei could produce substantial surplus paddy to sell. The case studies show that there is potential to improve the productivity of WS rice within the context of a more intensive and diversified farming system with access at least to on-farm irrigation.

COMPARATIVE ANALYSIS OF EARLY-WET-SEASON RICE PRODUCTION

Between 55% and 65% of households interviewed in the three villages planted an EWS rice crop, even though the WS rice crop was generally sufficient for their domestic needs (Table 13.5). The EWS crop provided an additional source of cash income for those households that were already self-sufficient in rice and a supplement to the domestic supply for rice-deficit households. Even without irrigation, the incidence of EWS rice cultivation was highest in Trapeang Run, but the small cultivated area (0.15 ha) was clearly restricted by the lack of irrigation. For the two villages with irrigation, the EWS rice area appeared to be in inverse relationship to the WS rice area. Snao, with a smaller WS rice area (0.61 ha) had a larger EWS rice area (0.37 ha), while Ta Daeng Thmei, which had double the WS area (1.3 ha), had a smaller EWS area (0.21 ha).

Table 13.5   Characteristics of EWS rice cultivation in the case-study villages

Practices	Trapeang Run (n = 79)	Snao (n = 62)	Ta Daeng Thmei (n = 59)
% of households	62	65	56
Mean area (ha)	0.15	0.37	0.21
Rice variety	Modern (IR)	Modern (IR)	Modern (IR)
Land preparation	Draught animal	Draught animal	Draught animal
Crop establishment	Transplanting	Direct seeding	Transplanting
Source of irrigation	Rainfed and ponds	Rainfed and groundwater	Rainfed and reservoir
Weed control	Manual	Manual and herbicides	Manual
Harvesting	Manual	Mechanised and manual	Manual
Threshing	Manual	Mechanised and manual	Manual
Transport of grain	Oxcart and shoulder pole	Oxcart and trailer	Oxcart and shoulder pole
Drying	Sun	Sun	Sun
Storage	Bags	Bags	Bags

Three photoperiod-insensitive rice varieties were reported in Trapeang Run—the Cambodian-released varieties of IR66 and Senpidao, and the variety introduced by Vietnamese traders, IR504. However, most of the production in this village was for household consumption. In Snao and Ta Daeng Thmei, IR504 was the most widely cultivated, with a smaller number of farmers planting IR66 in Snao, and Senpidao in Ta Daeng Thmei. The cultivation of IR504 indicates that the harvest was all sold to the Vietnamese rice traders.

The EWS crop relied heavily on early rainfall in Trapeang Run, despite the availability of small household ponds, but the crop was secured by on-farm irrigation in Snao and reservoir water in Ta Daeng Thmei. Certain cultural practices in Snao were noticeably different from the other two villages. Direct seeding, the application of herbicides, and the use of machinery for harvesting and threshing were carried out only in this village. The paddy grain was stored in plastic bags rather than in barns where the WS crop was mostly stored, which usually indicated an intention to sell the EWS produce. Following the operation of the combine harvesters or reapers in Snao, the paddy was commonly sold directly to the rice traders without being transported home.

The material and labour inputs for EWS rice cultivation are compared in Table 13.6. Snao stands out as using higher rates of all material inputs (seeds, manure, fertilisers, fuels, and herbicides). Because farmers in Snao used direct seeding, they used more than three times the seeding rate of the other two villages (380 kg/ha). The practice of direct seeding with a high seed rate, as observed in Snao, can increase crop yield through a high density of plants and hence panicles per unit area, compared with the minimal tillering of short-duration varieties using the transplanting method. Many farmers in the Mekong Delta in Vietnam broadcast at up to 300 kg/ha to ensure crop establishment and minimise weed infestation, with yields of 4–6 t/ha (Nguyen and Vo-Tong 2002).

Snao farmers also used nearly twice the rate of mineral fertilisers and applied much more cattle manure than in the other two villages. Every farmer cultivating EWS rice in Snao required fuel for pump-irrigation, averaging five times the mean fuel input in Trapeang Run, where only 43% of EWS rice growers used fuel. Farmers in Ta Daeng Thmei did not require fuel because they had access to gravity-fed irrigation; if not, they did not cultivate those plots in the EWS to avoid pumping costs. Snao farmers also incurred USD 100/ha for spraying herbicides and pesticides to control weeds and/or insects but the other two villages reported no cash outlays on agrochemicals.

The use of direct seeding, chemical weed control, and mechanised harvesting and post-harvest operations in Snao meant that the total labour requirement was very low (32 labour-days/ha), almost one-fifth that of Trapeang Run and one-third that of Ta Daeng Thmei. Trapeang Run had the highest labour input across all the activities—seedbed, pulling, transplanting, weeding, and harvesting—45% more than in Ta Daeng Thmei. As discussed in relation to the WS rice crop, one reason for this difference could be the lack of irrigation in Trapeang Run, which meant

Table 13.6 Material and labour inputs for EWS rice cultivation in the case-study villages

	Trapeang Run (n = 49)	Snao (n = 40)	Ta Daeng Thmei (n = 33)
Area (ha)	0.15	0.37	0.21
Seed (kg/ha)	114	377	106
Fertiliser (kg/ha)	151	265	151
Fuel (l/ha)	34	171	0
Herbicides (USD/ha)	0	103	0
Cattle manure (t/ha)	6.9	8.7	3.4
Labour-days/ha	153	32	105

there was a firm soil surface, increasing the labour-days needed for seedbed management, pulling seedlings, and transplanting. This also provided favourable conditions for weed infestation, increasing the labour input required for weeding. In addition, the engagement of younger family members in daily non-farm activities, and reliance on older family members for farm work, could have increased the number of work-days for a given task.

Both Trapeang Run and Ta Daeng Thmei used slightly more labour per hectare on the EWS crop than for their respective WS crops. The EWS rice crop required four to five more labour-days than WS rice for irrigating in the two villages. The firm soil surface in the EWS also doubled the labour-days required to pull young seedlings in Trapeang Run (21 labour-days, compared with 10 labour-days for WS rice). Ta Daeng Thmei also needed an extra three labour-days for pulling seedlings. However, the small cultivated area made these per hectare differences less significant.

An economic analysis of EWS rice production in the three villages is presented in Table 13.7. Though the yields for Trapeang Run and Snao relate to the 2011 harvest, and for Ta Daeng Thmei to the 2010 harvest, the provincial yields varied little between these years (MAFF 2011–2013), consistent with the close to average rainfall in both years. Snao had a significantly higher yield (4 t/ha) than the other two villages, despite cultivating the same IR rice varieties (mainly IR504), presumably reflecting the high seed rate and higher rates of nutrient application. Also, the intensive utilisation of the paddy fields throughout the year in Snao meant there was a likely carryover effect of mineral and organic nutrients applied in each season. Poor inherent soil properties had also been improved, with

Table 13.7 Average unit costs and returns for EWS rice production in the case-study villages

Item	Trapeang Run (n = 49)	Snao (n = 40)	Ta Daeng Thmei (n = 33)
Yield (t/ha)	2.6	4.0	2.2
Net output (t/ha)	2.5	3.7	2.1
Farm-gate price (USD/kg)	0.24	0.23	0.24
Gross income (USD/ha)	602	843	505
Input expenses (USD/ha)	125	501	79
Net return to household (USD/ha)	490	342	425
Total labour (days/ha)	153	32	105
Net return to labour (USD/day)	3.50	11.70	4.50

manure and crop biomass frequently being incorporated in the course of successive cultivations, and the soil was protected by almost continuous crop cover. The EWS yield in Snao was also significantly higher than the WS yield in the same village—a result of the higher yield potential of the modern varieties.

The gross income per hectare in the three villages followed the same pattern as the yields. The higher expenses in Snao (USD 500/ha) reduced the net return to household resources to USD 340/ha, significantly lower than the other two villages. However, the use of labour-saving innovations (direct seeding, herbicides, and mechanised harvesting) significantly reduced the labour input, enabling farmers in Snao to achieve the highest net return to labour (USD 12/day), about three times the return in the other two villages. This return was also double the labour return for the WS crop in Snao.

EWS rice production had been adopted in 16 of 24 provinces in Cambodia by 2012, accounting for 8% of the total harvested rice area, and the equivalent of 50% of the area used for DS rice (MAFF 2011–2013). The EWS rice area (242,113 ha) had more than doubled over the previous three years. Takeo had the second largest area of EWS rice (47,764 ha) but Kampong Speu had only 1770 ha. It is likely that the area and output of EWS rice will continue to expand, both to supplement subsistence production and generate cash income. The case-study villages illustrate this trend. The main purpose of EWS rice cultivation in Snao and Ta Daeng Thmei was to generate cash income and, in Trapeang Run, to supplement domestic rice supply.

In particular, though most farmers in each village cultivated EWS rice, Snao farmers cultivated the largest area and the highest proportion (about two-thirds) of their paddy holdings to EWS. The motivation was the small area available for WS rice production and the availability of on-farm irrigation. EWS cultivation in Trapeang Run was restricted by the lack of irrigation and only some plots in Ta Daeng Thmei were favourable for gravity-fed irrigation. Moreover, with a large surplus of WS rice, there was less incentive for farmers in Ta Daeng Thmei to spend money on fuel to increase the EWS rice area.

A number of specific approaches had been adopted in Snao to boost the EWS rice yield and net returns to family labour. The key cultural practices comprised mechanised land preparation, harvesting, and post-harvest operations, direct seeding, and applying herbicides, significantly reducing the total labour input. The crop also received high levels of material inputs including seed, manure, mineral fertilisers, and fuel to improve the crop yield. The yield was certainly improved by the reliable supply of on-farm irrigation. These practices suggest a way forward for less-productive rainfed villages such as Trapeang Run.

COMPARATIVE ANALYSIS OF NON-RICE CROPS

Apart from cultivating rice in the WS and EWS, non-rice crops were also cultivated in the DS within all three villages, mainly to produce cash income but also for household consumption. Table 13.8 summarises the major crops and farming practices in each village.[2] The various non-rice crops in Trapeang Run comprised watermelons, cucumbers, pumpkins, mung beans, and convolvulus. In Snao, radish was the dominant crop, with some cucumber cultivation, and in Ta Daeng Thmei peanuts were the major DS crop. The radish crop was cultivated on raised beds and peanuts on slightly raised beds, but most other crops were planted on flatbeds. Because radish cultivation involved intensive cropping, a power tiller was necessary to prepare the land but draught animal power with a conventional mouldboard plough was used to raise the beds (Fig. 13.4). Trapeang Run depended on small household ponds to irrigate the DS

Table 13.8 Characteristics of DS non-rice crop cultivation in the case-study villages

Practices	Trapeang Run	Snao	Ta Daeng Thmei
Crops	Various	Radish, cucumbers	Peanuts
Land preparation	Draught animals	Two-wheel tractor/draught animals	Draught animals
Cultivation method	Flatbed	Raised bed	Low raised bed
Irrigation source	Pond	Groundwater/pond	Reservoir
Water requirement	Daily/ occasional	Daily	Three to four times per season
Pest control	Chemicals	Chemicals	n.a.
Weed control	Manual	Manual	Manual
Harvesting	Manual	Manual	Manual
Threshing	n.a.	n.a.	Manual
Transport	Bicycle/oxcart	Transported by buyer	Bicycle/shoulder pole
Drying	n.a.	n.a.	Sun
Storage	Sold at harvest	Sold before harvest	Bags

Fig. 13.4 Farmer in Snao preparing paddy field for radish cultivation in the dry season. (Source: Rob Cramb)

crops but, as already noted, Snao had access to a reliable groundwater supply and Ta Daeng Thmei to surface irrigation.

Radish cultivation required considerably more material inputs and labour-days than the crops in the other two villages (Table 13.9). The use of mineral fertilisers, cattle manure, fuel, and pesticides was much greater for radish cultivation than for peanuts or the other non-rice crops. The cucumber crop appeared to require little cattle manure because the application was made precisely in the planting holes rather than being spread across the entire planted area. The crops requiring daily watering were radish, cucumber, and convolvulus, with Snao farmers pumping groundwater for radish and cucumber for 1–2 hours/day and Trapeang Run farmers mostly fetching water from ponds to the cropped plots by watering can. Gravity-fed irrigation was applied three to four times for the peanut crop in Ta Daeng Thmei. Watermelon, pumpkin, and mung bean cultivated in Trapeang Run were watered only at planting time, with possibly one to two more supplementary waterings. The labour input for planting radish and cucumber was comparable to the input for other non-

Table 13.9 Material and labour inputs per ha for DS crop cultivation in the case-study villages

	Trapeang Run	*Snao*	*Ta Daeng Thmei*
Households (%)	44	82	80
Crop cycles	1	2	1
Area (ha)	0.13	0.36	0.19
Seed (kg/ha)	n.a.	6	200
Fertiliser (kg/ha)	35	385	100
Fuel (l/ha)	15	367	n.a.
Pesticides USD/ha	31	275	0
Cattle manure (t/ha)	3	12	7
Labour-days/ha	215	241	95

Table 13.10 Average unit costs and returns for DS non-rice crop production in the case-study villages

Measure	*Trapeang Run (various crops)*	*Snao (radish)*	*Ta Daeng Thmei (peanut)*
Yield (kg/ha)	a	b	1214
Seed (kg/ha)	a	6	204
Output (kg/ha)	a	b	1010
Farm-gate price (USD/kg)	a	b	1.0
Gross income (USD/ha)	454	2760	1010
Input expenses (USD/ha)	92	1018	57
Net returns to household (USD/ha)	362	1742	953
Total labour (days/ha)	215	241	95
Net returns to labour (USD/day)	1.70	7.30	11.00

Notes: a. There were many crops grown on a small scale and intermixed on the same plot, hence it was not possible to determine yield, seed, output, and price; b. The radish crop was bought before harvest by the trader who harvested the crop, hence only gross income is known, not the physical yield and output

rice crops in Trapeang Run (over 200 labour-days/ha), but more than twice that for peanut cultivation.

An economic analysis of DS non-rice crops in the three villages is summarised in Table 13.10. The radish cultivation in Snao produced the largest gross income (USD 2760/ha), six times that of the various crops in Trapeang Run and three times that of the peanut crop in Ta Daeng Thmei. Cucumber, cultivated by some non-radish farmers in Snao, provided around half the gross income of radish. However, radish production had much higher input expenses. As well, planting, watering, and weeding for the radish crops all required a high labour input with a high concentration, necessitating the use of hired or exchange labour. The lower labour concentration for cucumber, peanut, and other crops meant they could be managed by the farm family; for example, the harvest of cucumber was carried out daily by one or two family workers over a period of about 20 days.

Despite the high expenses, radish cultivation still provided the highest net return to household resources (USD 1740/ha), five times that of Trapeang Run crops and double the returns of peanut and cucumber cultivation. However, the high labour input reduced the net return to labour to about USD 7/day for radish, compared with USD 11/day for peanuts. Cucumbers (USD 4/day) and the non-rice crops cultivated in Trapeang Run (USD 2/day) gave significantly lower returns to labour, in the latter case less than the presumed opportunity cost of labour (USD 3/ day). Most of the households in Trapeang Run produced very small outputs for their own consumption; only a quarter of the DS crop growers were able to generate some cash income from their crops.

Over a decade ago, Pingali (2004: 43) made the assessment that "dry-season cropping activities in the rainfed [rice-growing] areas [of South and Southeast Asia] are limited because of technical problems related to timely and effective crop establishment, limited moisture (or excessive moisture in some cases), and generally modest or high yield instability". However, the three case-study villages show that WS paddy land has potential for the cultivation of non-rice crops in the DS, both to improve household cash income and supply domestic consumption. The crops were able to be grown under a range of irrigation conditions, from small ponds to a large-scale reservoir. The crops cultivated also had different water requirements, ranging from daily watering to two to three irrigations per crop. However, the key to obtaining viable returns was a reliable irrigation source as in Snao and Ta Daeng Thmei. The limited water supply in Trapeang Run provided negligible returns and risked wasting production inputs. The improvement of on-farm irrigation would be necessary for Trapeang Run and other rainfed lowland villages to produce a significant household cash income from the cultivation of non-rice crops in the DS.[3]

Given an adequate supply of water, villages such as Trapeang Run could be expected to replicate the success of radish growers in Snao and peanut growers in Ta Daeng Thmei. To viably adopt the Snao radish cropping system, farm-households would also need to have suitable soils, an available market, sufficient working capital, and an adequate supply of family labour to undertake the intensive operations required. The lower requirements for water, cash outlays, and labour for the peanut system in Ta Daeng Thmei make this a more feasible DS cropping option for resource-poor farmers and those with other non-farm employment options in villages such as Trapeang Run.

COMPARATIVE ANALYSIS OF CROPPING SYSTEMS IN THE THREE VILLAGES

Representative farm budgets were constructed to reflect the whole-year cropping system of typical households in the three villages (Table 13.11). Trapeang Run, with only small ponds to provide supplementary irrigation, was restricted to an annual cultivated area of 1.2 ha per household, not much more than the mean farm size of 0.9 ha. Snao, despite a small farm size of 0.6 ha, could draw on groundwater to achieve an annual cultivated area of 1.4 ha from the same land (DS rice-land was excluded from the representative budget). Ta Daeng Thmei, being fully irrigated, could crop a total of 1.7 ha for a farm size of 1.3 ha.

Given the higher cropping intensity of the representative farms in Snao and Ta Daeng Thmei, these farms achieved higher paddy output (3.2 and 3.5 t, respectively) and greater paddy surpluses (1.8 and 2.3 t, respectively) than the Trapeang Run representative farm, which was much more dependent on the WS rice crop. The higher output from Snao also reflected higher yields in both the WS and the EWS, probably due to the higher year-round input of organic and inorganic nutrients. Each of the three budgets indicates household self-sufficiency in paddy, though the lower surplus in the Trapeang Run case (1 t) reflects a greater incidence of rice-insufficiency within that village population.

The Snao farm had the highest annual expenditure, mainly for the DS radish and EWS rice crops, although the WS rice crop also incurred higher expenditure than in the other two villages. Fertiliser, fuel, and pesticide were all large items of expenditure in this case. In the other two villages the major expenses were for the fertiliser input for the WS rice crop, with relatively less expenditure on the EWS rice and DS non-rice crops. As noted above, the application of farmyard manure was two to three times higher in the Snao farm (11 t/year), nearly 70% of which was applied to

Table 13.11 Annual inputs, outputs, and net cash flow of representative cropping systems in the case-study villages

	Unit	Trapeang Run	Snao	Ta Daeng Thmei
Annual cultivated area	ha	1.21	1.34	1.70
• WS rice	ha	0.93	0.61	1.30
• DS non-rice crops	ha	0.13	0.36	0.19
• EWS rice	ha	0.15	0.37	0.21
Cropping intensity	ha	1.3	2.2	1.3
Paddy output	kg	2423	3230	3512
Paddy surplus[a]	kg	1081	1777	2342
Total gross income	USD	726	1847	1199
• Paddy	USD	667	829	965
• Non-rice crops	USD	59	1018	234
Total cash income	USD	351	1423	779
• Paddy	USD	292	429	584
• Non-rice crops	USD	59	1018	195
Total labour input	days	173	165	148
• WS rice	%	71	39	73
• DS Non-rice crops	%	16	52	12
• EWS rice	%	13	9	15
Labour-intensity	days/ha	186	270	114
Farmyard manure	kg	7129	11,279	5033
Total cash costs	USD	124	722	118
• Fertiliser	USD	88	272	118
• Fuel	USD	28	231	–
• Pesticide	USD	4	153	–
• Seed	USD	4	26	–
• Machinery	USD	–	40	–
Net cash flow	USD	228	701	660

[a]Surplus computed based on consumption of 1250 kg of paddy per household (assuming five household members)

the DS radish and EWS rice crops, with many radish and EWS growers buying extra farmyard manure from other nearby villages. In the other two villages, most farmyard manure was applied to the WS rice crop—81% in Trapeang Run and 60% in Ta Daeng Thmei.

The cropping systems required similar annual labour inputs of 150–175 days/year, that is, less than one full-time worker. It was estimated that cattle activities required a further 150 days/year in each village, and non-farm activities accounted for a significant proportion of household labour, especially in Trapeang Run. Though the total labour input for cropping was similar, the labour-intensity was highest for the Snao farm (264 days/ha), reflecting the small farm size and the high cropping intensity. WS rice absorbed a little over 70% of the total labour input in the Trapeang Run and Ta Daeng Thmei farms, but less than 40% in the Snao farm, where DS radish cultivation accounted for the largest share (54%).

The monthly labour profile was also similar between the Trapeang Run and Ta Daeng Thmei farms, with two comparable peak periods in July–September, when the EWS rice harvest coincided with land preparation and transplanting for the WS rice crop, and December–January, when the harvesting of WS rice and the planting of DS peanut and other non-rice crops were carried out. In the Snao farm, the labour concentration was also high in the July–September period but peaked from December to April due to the WS rice harvest and the intensive DS radish activity. Collecting native grasses for cattle in the WS increased the labour requirement in the July–September period in all three villages.

Besides the WS and EWS rice crops, the DS cultivation of radish, peanuts, and other non-rice crops contributed to farm income, especially for the representative farms in Snao and Ta Daeng Thmei. The Snao farm generated the highest gross income (USD 1820/year) and cash income (USD 1420/year), two to four times that of the other two villages. After deducting the high level of cash expenditure (USD 720/year), the Snao farm still had the highest net cash flow (USD 700), somewhat higher than Ta Daeng Thmei (USD 660) but three times the net cash flow for Trapeang Run (USD 230). The DS radish crop contributed about 90% of the net cash flow in the Snao farm, whereas the peanut crop contributed only 25% of the net cash flow in the Ta Daeng Thmei case, the majority of the cash flow coming from the sale of surplus rice from the WS and EWS. In Trapeang Run, the sale of surplus rice from the two seasons was the main source of farm cash income, the non-rice crops giving a negligible net cash return.

The representative farms in Trapeang Run and Ta Daeng Thmei experienced no land use constraint, because the cultivation of DS non-rice crops and EWS rice occupied only a fraction of the total paddy land. Even with the late harvesting of the EWS rice crop, there was little impact on the preparation of the WS rice nursery, because the area of EWS rice was only 16% of the total paddy land in each village. There was also a short break in December between the harvest of WS rice and the planting of peanuts (Ta Daeng Thmei) or other non-rice crops (Trapeang Run), due to the wet field conditions following the rainy season.

However, the small total landholding in the Snao farm and the relatively large cultivated areas of DS radish and EWS rice meant that the farmer needed to manage the restricted land resource appropriately—through timely direct seeding of EWS rice and the careful planning of WS rice activities, such as nursery plot allocation, gradual land preparation of the transplanted field, and use of varieties with a diversity of maturation periods. The early broadcasting of the EWS rice was necessary to provide a short window between the harvest of EWS rice and land preparation for WS rice. The nursery plot designated for the WS rice was not used for the EWS rice crop. The land preparation and transplanting of WS rice were gradually carried out from available plots. An early-maturing variety of WS rice was used on the land targeted for the first DS radish crop, starting from mid-December, which also minimised irrigation costs.

Conclusion

This comparison of representative cropping systems shows that, compared with the largely rainfed condition of Trapeang Run, typical of most of the lowland ecosystem, on-farm and (where feasible) canal irrigation can greatly increase the intensity, diversity, and profitability of land use. This can occur without being seriously constrained by available family labour, though in Snao there had been a move to adopt some labour-saving innovations in the DS and EWS to accommodate the tight turnarounds between successive crops on the limited paddy land. However, even in Snao, the potential for irrigated cropping had not been fully realised, due to the scattering of plots and the restricted investment in tube wells. The lands accessible to gravity-fed irrigation in Ta Daeng Thmei could also be extended, increasing further the potential cropping intensity.

Nevertheless, even these partially irrigated systems not only increased land and labour utilisation, making greater use of the limited set of household resources, but improved the physical and chemical properties of the soil, reduced the risk of a household rice-deficit, increased the production of a marketable surplus of rice, and increased the level and diversity of crop income. The resultant cash flow provided the necessary working capital to keep the cropping system turning over, with minimal need for credit, while providing income for household needs. It is significant that, on average, two-thirds of cash income in Trapeang Run came from non-farm employment, compared with only 12% in Snao and 21% in Ta Daeng Thmei.

Thus the comparison suggests a potentially feasible strategy for lowland villages like Trapeang Run to increase food security and farm and household incomes. While outmigration from the densely populated, rainfed lowlands of southern Cambodia will undoubtedly continue, the case studies show that the development of more intensive, diverse, and market-oriented farming systems, based on on-farm irrigation, can provide a promising alternative pathway for many rural households.

NOTES

1. This reservoir was initially built during the Khmer Rouge era.
2. As mentioned above, half the farmers in Snao had access to floodplain land suited to DS rice cultivation but flooded and uncultivable in the WS. This option is not available to farmers in the lowland agroecosystem, which is the focus of the comparison in this chapter.
3. This need not necessarily be groundwater. An integrated farming project in central Thailand used 30% of the total farm area for pond excavation and generated an annual profit four times that of a single rice crop, thereby more than compensating for the loss of land (Setboonsarng and Gilman 2009).

REFERENCES

Chea, Sareth, 2015. *Evaluating Cropping System Options for Farmers in the Lowland Rice-based Systems of Cambodia.* PhD Thesis, University of Queensland.

Javier, E. L., 1997. Rice ecosystems and varieties. In H. J. Nesbitt, ed., *Rice Production in Cambodia*, pp. 39–81. Manila: International Rice Research Institute.

Ly, P., Jensen, L. S., Bruun, T. B., Rutz, D., and de Neergaard, A., 2012. The system of rice intensification: adapted practices, reported outcomes and their relevance in Cambodia. *Agricultural Systems* 113: 16–27.

MAFF, 2011–2013. *Bulletin of Agricultural Statistics and Studies.* Department of Planning and Statistics, Ministry of Agriculture, Forestry and Fisheries, Phnom Penh, Cambodia.

Nguyen, D. C., and Vo-Tong, X., 2002. Environmental conditions as determinants of direct seeding techniques in different ecosystems in the Mekong Delta of Vietnam. In S. Pandey, M. Mortimer, L. Wade, T. P. Tuong, K. Lopez, and B. Hardy, eds., *Direct Seeding: Research Issues and Opportunities.* Proceedings of the International Workshop on Direct Seeding in Asian Rice Systems, Bangkok, 25–28 January 2000. Manila: International Rice Research Institute.

Pingali, P., 2004. *Agricultural Diversification: Opportunities and Constraints.* Paper presented at FAO Rice Conference, Rome, 12–13 February 2004.

Rickman, J. F., Meas, P., and Om, S., 1997. Farm mechanization. In H. J. Nesbitt, ed., *Rice Production in Cambodia*, pp. 93–98. Manila: International Rice Research Institute.

Setboonsarng, S., and Gilman, J., 2009. *Alternative Agriculture in Thailand and Japan*. Bangkok: School of Environment, Resources and Development, Asian Institute of Technology. Available at http://www.solutions-site.org/node/47 (accessed January 2014).

CHAPTER 14

The Supply of Fertiliser for Rice Farming in Takeo

Theng Vuthy

INTRODUCTION

One reason historically for the low rice yields in Cambodia compared with Vietnam and Thailand has been the low use of fertilisers (Theng and Koy 2011), even though many demonstration trials have shown a high yield response to fertiliser application. One key constraint to increased use appears to be limited access to adequate stocks of affordable, good-quality fertilisers. Much of the fertilisers used by farmers are imported from Vietnam and Thailand, but there are important issues of quality, incorrect and indecipherable labelling, unreliable supply, variable prices, and insufficient information about fertilisers and other input use. A study by Schamel and Hongen (2003) shows that farmers chose to abstain from fertiliser markets altogether or apply fertilisers at rates below recommended levels because they had been sold bad-quality products in the past, which deterred buyers who were not willing to pay full market price for the quality of fertilisers available. Identifying the constraints that inhibit the use farm inputs will help to highlight possible policy interventions to improve farmers' access to and informed use of these inputs.

T. Vuthy (✉)
Office of Food Security and Environment, USAID, Phnom Penh, Cambodia
e-mail: vtheng@usaid.gov

© The Author(s) 2020   291
R. Cramb (ed.), *White Gold: The Commercialisation of Rice Farming in the Lower Mekong Basin*,
https://doi.org/10.1007/978-981-15-0998-8_14

A study of the fertiliser value chain in Takeo Province was conducted. The hypothesis of the study was that limited access to good-quality, affordable fertilisers is a major constraint to improving rice yields in the province. Fertilisers can help increase rice production, but issues surrounding quality discourage rice farmers from investing in farm inputs. Policy changes to ease this issue could increase productivity and farm income, contributing to improved wellbeing and reduced vulnerability of farm households. The objectives of the study were to (1) analyse the value chain for rice fertilisers; (2) identify the channels for the low-quality fertilisers being distributed; (3) estimate the yield loss associated with low-quality fertiliser application; (3) review government policy to control fertiliser trade; and (4) identify ways to improve the fertiliser market.

A mixed methods approach was applied to analysing the fertiliser value chain (Kaplinsky 2000; Kaplinsky and Morris 2001). Qualitative and quantitative information was collected from different actors in the value chain via group interviews and interviews with key informants. Four group interviews were conducted with farmers in three districts—Tram Kak, Prey Kabbas, and Kaoh Andaet. These farmers represented different rice ecosystems—wet-season (WS) rice, WS rice with supplementary irrigation, and fully irrigated dry-season (DS) rice. Thirteen key informants were interviewed, including fertiliser importers, distributors, and retailers. Other stakeholders such as provincial extension workers, agronomists, and agricultural legislators were also interviewed. Official statistical data were also obtained and analysed. The major themes for the group interviews were fertiliser availability, product preferences and prices, fertiliser quality issues, credit access for farm inputs, government policy on fertiliser use, incentives and risks of fertiliser application, and yield lost due to poor-quality fertilisers. The key questions for the key-informant interviews were fertiliser suppliers and marketing strategies, transportation and logistics issues, fertiliser quality issues, government policy on fertiliser trade and quality control, and challenges of fertiliser trade and competitors.

The Fertiliser Market in Takeo

Growth in Farmer Demand

In the past decade, farmers in Takeo Province have shifted rapidly from subsistence production to market-oriented farming, which has entailed a substantial increase in rice production. This rapid transition is due to quick uptake of high-yielding varieties, increased use of fertilisers and pesticides, increased mechanisation, and improved irrigation. Growth in cross-border trade with Vietnam has been an additional major factor. Dry-season rice is the province's main export, while the main imports are seeds, fertilisers, pesticides, and construction materials from Vietnam.

The rapid uptake of high-yielding rice varieties has entailed greater use of fertilisers and pesticides. Figure 14.1 shows that, with the exception of Doun Kaev District, more than 80% of rice farmers in Takeo Province used inorganic fertilisers. This implies that there was no supply constraint in the market place, a fact confirmed by farmers in all study villages as well as key informants.

There is no problem to buy fertilisers in our villages. If you have money you can buy any amount or any kind you wish to buy. You can also find different product brands in a shop near our village here. In addition, you can also buy on credit and pay back at the harvest. (Group interviews with farmers in Samrong, Prey Kabbas, and Kaoh Andaet districts)

We have few fertiliser products in my store at this time, because it is off-season and farmers do not need [fertiliser] at the present. During planting time, it is not difficult; we can order any products and amount from different suppliers. We just call to them and they will bring their fertilisers to my shop here within

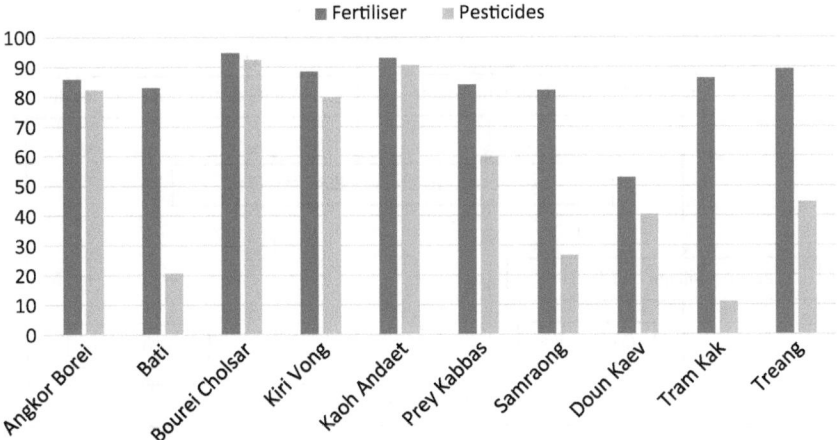

Fig. 14.1 Percentage of households using chemical fertilisers and pesticides in Takeo Province by district, 2010. (Source: Commune Database, 2010)

*a day or two. They are in Takeo or from Phnom Penh. (Key-informant
interviews with retailers in Samrong and Prey Kabbas districts)*

Fertiliser Supply Chain

The fertiliser market structure is evolving rapidly to meet farmers' demands
and service the growing rice sector in Takeo. The market structure is well
organised and led by the private sector operating a competitive marketing
strategy, with prices set by market forces (Fig. 14.2).

There were six major fertiliser supply companies distributing agro-
products in Takeo Province from their provincial wholesale outlets to one-
stop retail shops in local village markets. Heng Pich Chhay (HPC)
Company had business headquarters and warehouses in Takeo, while the
other five suppliers had their head offices in Phnom Penh or elsewhere but
had major distribution points (though no branch office) in Takeo. HPC
Company imported different kinds of fertilisers produced in Japan, the
Philippines, the USA, China, and Vietnam through Vietnam traders who
entered Cambodia through the Phnom Den checkpoint. This company
supplied fertilisers not only in Takeo but in almost all provinces in

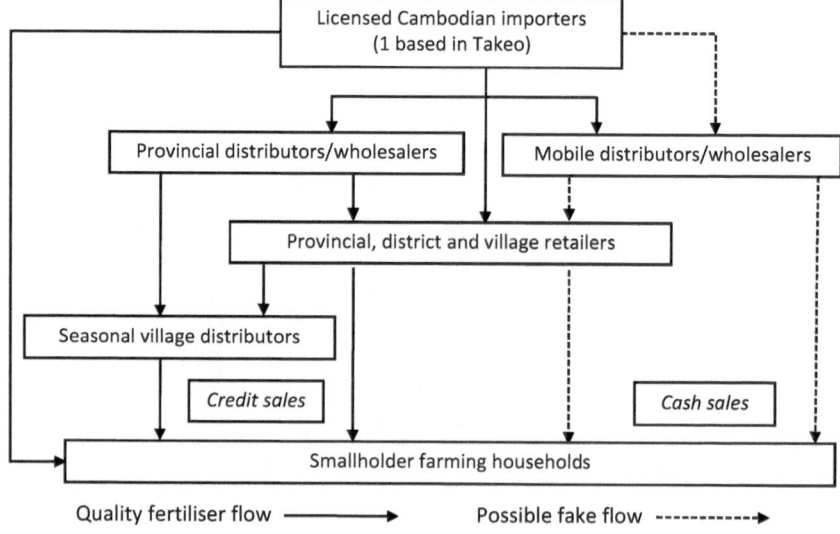

Fig. 14.2 Fertiliser distribution channels in Takeo

Cambodia. The two largest suppliers were the HPC Company and the Yetak Group; their products were widely available in most wholesale and retail outlets, even in small village shops. Other suppliers were Chhun Sok Ann, Cheam Tech, Sayimex, and Lim Bun Heng. The Lim Bun Heng Company only imported and distributed specific fertilisers from Thailand, such as urea, 15-15-15, and 16-20-0. Other importers had different suppliers, from China, Japan, the USA, Vietnam, and the Philippines, but these products mostly came to Cambodia through Vietnam-based traders.

Many kinds of fertilisers, distributed by different importers and distributors, were available in the market. The single-nutrient products were urea and muriate of potash (KCl). Compound nitrogen-based fertilisers included di-ammonium phosphate (DAP) (18-46-0) and ammonium sulphate (16-20-0). Compound nitrogen, phosphorus, and potassium (NPK) products were available on the market in ratios of 15-15-15, 16-16-8-(13S), and 20-20-15. All fertilisers were sold in 50 kg bags, though farmers could buy products by the kilogramme.

There was little vertical integration reported during interviews, except for the larger importers such as the Yetak Group and HPC Company. They tried to promote their brands, with a buffalo logo for the Yetak Group and a triangle logo for the HPC Company. These two companies had their representative lead dealers in almost every province and employed agronomists to conduct field demonstrations as part of their marketing strategy, as well as producing leaflets about fertiliser use and application rates in the Khmer language to distribute to farmers. Most of the fertilisers sold in the market were labelled in Khmer, with the exceptions of 16-16-8-13 produced in the Philippines and urea from China and Vietnam, though these products were marked with small stickers in Khmer.

Neither traders nor the Provincial Department of Agriculture (PDA) had any records of the quantity of fertilisers imported or distributed in the province. It has been reported that there was large-scale smuggling of fertilisers from Vietnam into Cambodia, which were then sold on the market (Asian Development Bank 2002: 27). Smuggled goods were readily identified because the bags were not labelled in Khmer or marked with Khmer stickers. It was legal for farmers to come to Vietnam and buy up to 50 bags of fertilisers for use on their farms near the Cambodia-Vietnam border. However, some farmers came to Vietnam many times to buy fertilisers to sell to dealers in Cambodia for profit. This kind of illegal trade was reportedly common in Takeo in the areas close to the Vietnam border. Police at the border knew of this activity but made no arrests, in exchange for some benefits. In the case of large movements of fertilisers, as would be carried out by the six major fertiliser companies, the bulk of unofficial imports from neighbouring countries (Thailand and Vietnam) would need to be conducted by traders aligned with those companies in their particular zone of operations in order for those traders to have "permission" to operate (Asian Development Bank 2002: 27). The incidence of smuggling may be taken to imply that fertiliser prices in Cambodia are kept artificially high through regulating the quantity traded.

Licensed Cambodian importers stored fertilisers in warehouses near the border (Thailand and Vietnam) and/or in Phnom Penh. The HPC Company had its business headquarters and warehouse in Kiri Vong District near the Vietnam border and had many trucks to transport imported fertilisers both within Takeo and to other provinces. The other five companies did not have fertiliser stockists in Takeo town, but they had appointed lead representatives/dealers to serve as distribution points throughout the province. The amount of fertilisers held by the provincial lead dealers varied according to the planting season. Larger distributors had warehouse facilities that could store from 100 to 1000 tons during peak season. Transport costs varied according to the distance from the main warehouse to the distribution points. Haulage cost about USD 0.25 per bag per 100 km, and loading fertilisers on and off trucks cost about USD 0.05 per bag. Most of the larger distributors had trucks to deliver to district and village retailers. District and village shops were smaller, with limited storage, and usually fertilisers were ordered during the planting season (May to September for wet-season rice and November to February for dry-season or recession rice) to save space for other merchandise.

Village retailers were typically one-stop shops selling a wide range of farm inputs including animal feed, pesticides, seeds, and fuel in addition to fertilisers. In 2011–2012, about 634 traders in the province—mostly shop owners selling farm inputs—were called by the PDA to attend training on trade and safety in farm inputs. Village retailers typically bought fertilisers from the representatives of the main provincial dealers; however, some also used different suppliers depending on prices and services offered and/or to meet specific demands of their customers.

Retailers' transactions with farmers were done in cash or on credit. Field interviews revealed that about half of retail sales were made on credit, with an added mark-up of KHR 15,000–20,000 per bag per planting season (three to six months).[1]

If we sell on a cash basis, we could make a profit of only about 1,500 to 2,000 riels per bag. Prices are very competitive among retailers in the market; if we do not sell with this profit, other shops will sell ... If we sell on credit we can mark-up about 15,000–20,000 riels per bag and receive repayment from farmers within 3–6 months, but we sell on credit to those whom we know well and who pay back on time after harvesting their crops. Payment can be made either in paddy or cash. Every year, about half of sales are made on credit. If depending on cash sales we can earn very little from this business. (Interviews with retailer shops in Tram Kak District)

Some provincial distributors and district retailers resold their fertilisers to seasonal village traders who sold and delivered fertilisers directly to farmers. All traders who sold agro-chemical products needed to be annually registered at the PDA's regulatory office; otherwise, their business activity was illegal. However, the seasonal village traders were not required to be registered and could sell fertilisers in many locations in Takeo. Many were better-off farmers in the villages with good connections with the main dealers, and hence they could make a profit from this business. Most of the seasonal traders resold fertilisers on credit to farmers and received repayment during the subsequent harvest. These credit sales involved a mark-up of as much as USD 5 per bag per planting season (six months for the wet season and three months for the dry season). HPC Company also sold on credit directly to farmers; about 500 tons were sold to farmers during the 2011 planting season. This involved a premium of about USD 1.50 per bag for three months—a much lower rate than demanded by the village traders.

Mobile distributors formed another distribution channel. They had no specific business office nor was it clear exactly where they came from, but they could be contacted by phone and delivered fertilisers as and when retailers needed their services. They were well connected and had long-standing business relationships with some importers. They purchased fertilisers from importers and loaded them onto trucks for delivery and re-sale to provincial, district, and village retail shops, and directly to farmers.

Marketing Margins

An analysis of fertiliser margins in Takeo was undertaken based on estimates provided by informants of the purchase prices, selling prices, handling costs, transport costs, unofficial road haulage fees, and mark-ups by different actors along the supply chain, together with the annual and monthly retail prices of different fertiliser products from secondary data sources.

The prices of all common fertilisers available in the Takeo market increased steadily from 2002 and spiked in 2008, in line with the worldwide spike in food and fuel prices in that year (Fig. 14.3). During 2008, prices of fertilisers increased to about USD 40 per 50 kg bag, while DAP rose to about USD 60 per bag. At these prices, all kinds of fertilisers were unaffordable for most smallholder farmers and the financial returns to fertiliser use were negative. The fertiliser prices then returned to normal trend in 2009. During the field visit in February 2012, the village price of urea was about USD 28 and that of DAP was USD 36 per 50 kg bag; these prices matched price trends recorded by the Ministry of Agriculture, Forestry and Fisheries (MAFF).

An estimate of marketing margins for DAP imported from the USA is presented in Table 14.1. This shows that the overall margin from the importer to the village retailer was about 15%. The mark-up for import companies (of which there were six) was about 5%, which does not appear excessive, whereas for traders further along the supply chain it was only

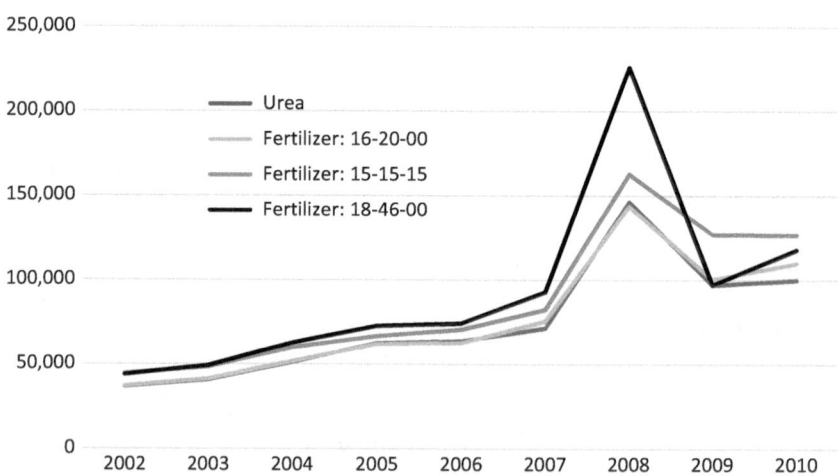

Fig. 14.3 Yearly average nominal retail prices of major fertilisers in Takeo, 2002–2010 (KHR/bag). (Source: Agricultural Marketing Office 2002–2010 (USD 1 = KHR 4000))

Table 14.1 Analysis of marketing margins for imported di-ammonium phosphate

	USD/50 kg	% of imported price
Cost to importer at Vietnam border	31.5	100
Transport to Cambodia (<100 km @ $0.25)	0.3	0.8
Transfer into border warehouse	0.1	0.2
Cost into border warehouse	31.8	101.0
Label changes and importer's mark-up	1.7	5.3
Importer's selling price	33.5	106.3
Transport to province (100 km @ $0.25)	0.3	0.7
Distributor mark-up and handling	0.8	2.2
Into store of provincial distributor	34.5	109.3
Provincial distributor mark-up	0.5	1.4
Distributor selling price	35.0	110.7
Transport to village dealer and handling	1.0	2.9
Into store of village dealer	36.0	113.6
Dealer mark-up for cash sale	0.5	1.4
Retail cash price in village	36.5	115.0
Value added—importer to retailer	5.00	15

Source: Author's calculations based on data from field interviews in February 2012

1.5–2%. When operating costs were taken into account, the margins for fertiliser traders at the provincial, district, and village levels were very low. The highest margin other than the importers' mark-up was the transport cost from provincial distribution points to village shops (3% of the imported price), which was largely due to unofficial fees paid to roadside police during transportation. In general, the analysis indicates that the fertiliser supply chain in Takeo was very competitive, particularly for a commonly used product such as DAP. These findings are consistent with those of the International Fertiliser Development Centre (IFDC 2010).

FERTILISER QUALITY ISSUES

Fertiliser quality problems arose in Cambodia as a result of the huge price spike in 2008, creating an opportunity for malfeasance in the fertiliser sector in response to the demand from farmers for "cheaper" fertilisers (IFDC 2010). IFDC (2010) conducted nutrient analysis of sampled fertilisers from ten provinces and found that almost all compound NPK and NP (16-20-0 and DAP) fertilisers sold on the market were well below acceptable quality index values (Table 14.2). However, the nutrient

Table 14.2 Nutrient analysis of selected fertiliser samples, mid-2010

Sample	Product (N-P-K-S)	% N	% P_2O_5	% K_2O	% S	Total nutrients (%)
5	20-20-15	21.90	10.50–11.40	9.00		77.00
12	20-20-15+TE	16.40–16.80	16.70–17.40	14.20		88.00
21	20-20-15+TE	17.10	18.20–18.00	13.60		68.22
24	20-20-15	17.70–18.50	19.50	11.60		90.18
28	20-20-15	20.70	19.60	9.77		91.04
34	20-20-15+TE	17.20	17.50–17.30	12.80		86.36
38	20-20-15	17.80–18.10	21.20	10.40		90.36
42	16-16-8-13	13.90	5.51–5.47	0.15	12.10	48.90
43	16-8-8-13	17.20	8.66	2.85	15.60	89.72
46	20-20-15-13	24.60	12.80–12.20	11.90	2.12	89.64
60	16-8-8-13	16.30	6.80–6.92	3.47	15.00	83.41
73	20-20-15+TE	17.80–19.00	20.30	9.41		88.56
88	20-20-15+TE	16.10–16.20	20.30	11.70		87.64
92	20-20-15+TE	15.80–16.10	21.10	9.79		85.44
97	25-20-10+TE	23.00	19.60	12.30		99.82
102	20-20-15+TE	21.60	16.20–16.60	9.43		86.60

Source: IFDC (2010)

Note: TE = trace elements; benchmark index value is 98%

content of most of the single-nutrient fertilisers analysed (urea) and some DAP was within an acceptable range (IFDC 2010: 25–35).

In response, the Department of Agricultural Legislation (DAL) has made concerted efforts to minimise the incidence of "fake" products in the market place, including increased certification of dealers, providing training to dealers on how to assess fertiliser quality, and instructing dealers on the signs of adulteration, oil coating contamination, and/or re-bagging. However, the method of fertiliser quality control employed so far is based on visual inspection only owing to the lack of analytical capacity in DAL headquarters in Phnom Penh. Visual inspection can only detect very obviously adulterated products but has limited ability to detect adulterated fertilisers or fertilisers with lower than specified nutrient analysis. Effective control requires capacity to analyse products chemically.

> *You can see these samples we took from some dealers and retailers in Takeo, we suspected that they are fake products, we can inspect by visual inspection only, and we did not know exactly whether these samples have low nutrient analysis. If we want to analyse these samples at the headquarters laboratory in Phnom*

Penh we need money to pay for the services, but we have no budget to do so. These are the problems and the capacity limitation of our staff to control the fertiliser quality problems here. (Interviews with provincial agricultural regulatory officers)

Fake products were widely reported by customers, importers, dealers, and senior PDA officials during the field visits and interviews, confirming the findings of IFDC (2010). The most common practice was re-bagging less expensive fertilisers such as DAP and urea in sacks labelled with a high-quality brand, for instance, urea from Thailand and DAP produced in the USA, which are well-known high-quality products. Importers interviewed reported that some retailers had been brought to the authorities to get them to confess and promise not to buy and sell fake products using their brand name. Senior PDA officers, dealers, and retailers reported that, although there was a significant drop in the incidence of fake products, the problem persists, affecting about 5–10% of fertilisers in the market (compared with about 30% during the price spike in 2007–2008).

Last production year, about 200–300 tons of fake DAP products were sold in this area. Some mobile dealers drove their trucks loaded with DAP products with trademark of HPC brand name and sold to either farmers or retailers with cheaper prices than usual. When we inspect fertiliser inside the bags, they are not the products of HPC brand. We cannot find those who carried out this malfeasance, but we arrest retailers who on-sell to farmers. Farmers complained about no crop response though they applied more fertiliser than usual ... Now farmers realised that cheap fertilisers are not good fertilisers. They want good quality fertiliser though it has a bit higher price. (Interviews with fertiliser importer and dealers in Prey Kabbas)

Some retailers sell Thai urea, a blue bag urea, a most popular well-known urea for most farmers; but in fact the product inside is not the Thai one. It may be a granular urea produced either in Vietnam or China, but it is re-bagged with the Thai brand and sells as the Thai product. It is difficult for farmers to differentiate the Thai product from the urea produced either in Vietnam or China because it's granular and the shape is almost the same. Farmers can know it is fake only by the crop response, but fertiliser dealers know which is Thai and which is not, and mobile distributors cannot cheat us. Thus, it is easy for malfeasance to occur for urea. This is the most common fake product in the market. (Interviews with retailers in districts visited and PDA officials)

The problem of fertiliser is still affecting the farmers, who are the fertiliser users, but the problems have reduced greatly compared to the peak price level in 2008. Presently, it is affecting about 5-10% of sales in the market. (Key-informant interviews with importers and senior PDA officials)

Dilution and adulteration of fertilisers were also reported by interviewed farmers. Farmers said that their crops were not responding as well to fertiliser compared to the previous year's crops and they blamed the low crop response on low-quality fertilisers. Technical experts, however, argued that such claims are almost impossible to put down to poor-quality fertilisers alone because other factors, such as different seasonal conditions, seed quality, and cropping practices, also affect yield. Nevertheless, the combined evidence from interviews with farmers, fertiliser dealers, and importers, and the fertiliser nutrient analysis conducted by IFDC (2010) strongly suggests that the low quality of fertilisers sold on the market is a critical problem affecting crop yield and resulting in financial loss for farmers in the study area.

The selling of short-weight bags and coating low-grade NPK fertilisers with oil to change the product's appearance were also reported by farmers and fertiliser dealers during field visits. However, these instances occurred during 2008; such problems were no longer considered commonplace. Farmers stressed that the most common issues they faced were re-bagging and adulteration.

Possible distribution channels for fake products are shown in Fig. 14.2. Senior agricultural regulators and importers were confident that most of the main dealerships and retailers did not distribute fake products to customers. However, they blamed the distribution of cheap, problem fertilisers to small retailers and farmers on intermediaries and mobile distributors. According to the regulations for agro-chemical distribution, agro-chemical dealers and retailers have to register with the Provincial Department of Agriculture to get certification for distributing agro-chemicals; otherwise their business activity is illegal (RGC 1998). Therefore, because the intermediaries and mobile distributors, including seasonal village traders, are unidentified and unregulated, the concerted efforts by MAFF and PDA to crackdown on fertiliser problems have so far had little effect.

It was reported that mobile distributors in particular had a clear opportunity to adulterate fertilisers—by mixing low- and high-quality products and selling them as high-quality fertilisers, re-bagging low-quality fertilisers in bags labelled with a high-quality brand, and even selling short-weight bags. During the field study, it was also reported that someone would pay farmers for their empty high-quality brand bags, that is, those with YITAK and/or HPC brands, and use them for such forms of malpractice.

> In Takeo there are a lot of mobile distributors and we cannot control their business activity. They have no specific office and we cannot find exactly where they come from. They have a long and good relationship with some importers, and thus they can buy fertiliser and load onto their trucks to resell to any shops and even farmers. They can adulterate the fertiliser before they deliver and resell it to retailers. They have both good- and low-quality fertilisers to deliver and sell to retailers; usually the fake ones are kept inside the trucks and the authorities have difficulty to find them. (Interview with provincial agricultural regulatory officer)

> Almost every day someone comes to ask us whether we have used fertiliser bags to sell to them. They ask only for used bags that have good brands and are in good condition. If we have them we can get 4,000 riels per bag. (Group interviews with farmers in Tram Kak, Prey Kabbas, and Kaoh Andaet)

While there is no evidence that seasonal village traders sold problem fertilisers to customers, there was a very high possibility that this was the case since they too were unregulated and could easily make connections with mobile distributors to resell fake products at any time.

> There are a lot of mobile dealers who come to ask us [to buy their products] almost every day. They have many kinds of fertiliser on their trucks and different product brands with different prices. They have cheaper prices but they are not as good as the higher-priced ones. We are retailers, we know which is good quality and which is low quality (fake products). The fake product is for those who want lower prices. Sometimes when we run out of stock and need fertilisers to sell, we can order from these mobile dealers. We do not know their office but we normally contact them by calling. (Interview with retailer in Prey Kabbas)

Farmers indicated they would monitor the quality of fertilisers supplied by local retailers (by observing crop response) and adjust their future purchases accordingly.

> If you buy cheap fertilisers, you have a high chance to get fake fertilisers. Cheap fertiliser is not as good as the higher-priced one. If we note that we bought fertilisers of poor quality from a retailer, we may buy from another retailer next

time. (Group interviews with farmers in Tram Kak, Prey Kabbas, and Kaoh Andaet)

In the absence of field trials to measure the actual yield loss due to the use of fake products, an estimate was made based on farmers' perceptions and recall. Farmers reported that they suffered a yield loss of 40–60% if they applied poor-quality fertilisers and did not follow up with a second round of good-quality fertilisers. If, however, on seeing that their crop did not respond to the first application, they applied a second round of good-quality fertilisers, the yield loss was restricted to about 20%. As noted above, importers and senior PDA officials reported that currently about 10% of sales are of fake products. Hence, it can be assumed that 10% of rice farmers in Takeo used fake products in 2011 and incurred between 20 and 50% yield loss.

In Takeo, there were about 179,800 rice farming households (CDB 2010) producing on average about 6.2 tons of paddy in 2011 (MAFF 2012). A yield loss of 20% would correspond to a loss of about 1.2 tons per household, worth USD 285.[2] If 10% of farmers used poor-quality fertilisers and suffered a yield loss of 20%, the total annual crop loss for Takeo as a whole would be about USD 5.2 million. This loss would increase to about USD 13 million if farmers did not have the funds to then buy good-quality fertilisers after seeing the poor crop response (Table 14.3). If the same assumptions are extended to Cambodia as a whole, the losses would be of the order of USD 40 million and USD 106 million, respectively.

Table 14.3 Estimated value of production losses due to use of fake fertilisers in Takeo in 2011

Season	Paddy output[a] (t)	20% yield loss (t)	50% yield loss (t)	Paddy price (KHR/kg)[b]	Gross value of lost output (USD)	
					20% yield loss	50% yield loss
Wet	64,935	12,987	32,467	1000	3,246,725	8,116,813
Dry	45,569	9114	22,784	850	1,936,666	4,841,664
Total	110,504	22,101	55,251		5,183,391	12,958,476

Source: Author's estimate

[a]Output of the 10% of households buying fake fertiliser

[b]USD 1 = KHR 4000

FERTILISER POLICY

MAFF is the government authority responsible for controlling fertiliser trade in Cambodia. All agro-chemical importers have to be registered with the Ministry of Commerce and then have to apply to MAFF to become agro-business companies. To import agro-chemicals, including fertilisers and pesticides, these companies need a licence, which is renewable annually. To be granted a licence, importers must provide details of the products and quantities to be imported, along with laboratory test results of the imported products to confirm their quality. Each application, whether for single or multiple products, is restricted to a maximum of 30,000 tons. An official fee of USD 75 is charged for each imported product registered.

IFDC (2010) has argued that the import licensing procedures are complex, out of touch with market demand, and restrict market competition. The procedure creates rent-seeking opportunities and many unofficial fees are paid through a facilitator to ensure the granting of the licence. Furthermore, the restriction on import tonnage per importer is contrary to market principles, creating a considerable commercial drawback in that it hinders the full realisation of economies of scale by importers. According to the IFDC, in a market economy, the private sector should be free to determine supply based on market and commercial risk assessments. The government's role should be to concentrate on monitoring quality, based on "truth-in-labelling" legislation. The licensing and tonnage quota system also prevents larger importers from cost-effective importing from the international market and hence forces importation through either Vietnam or Thailand, which adds to the transaction costs for customers. In addition, the licensing and tonnage quota system encourages illegal imports and prevents small firms from formally entering the market.

In response to the rash of fertiliser problems since 2008 and to crack down on fake products, MAFF and DAL in the PDA have put in place urgent policy measures as follows:

- increasing certification for fertiliser dealers and retailers and providing training for wholesalers and retailers about the signs of fake fertilisers, adulteration, and re-bagging;
- providing training for fertiliser users to increase awareness about fake fertilisers;
- increasing competition among the major fertiliser importers for market share in a small total market; and

- adopting a new law in late 2011 to regulate the registration, trade, and use of agro-chemical products in Cambodia. The new law is comprehensive and needs to be applied in conjunction with specific regulations (sub-decrees) that can be amended by MAFF without parliamentary approval.

The efforts by MAFF and PDA to provide training about fake products, combined with farmers' direct experience with fake products, has helped farmers to realise that cheap fertilisers do not necessarily provide the nutrients needed for a good crop response. However, the broader issues raised by IFDC (2010) have not yet been addressed.

CONCLUSION

Based on this study, farmers' access to affordable, good-quality fertilisers could be improved, not only in Takeo Province but also in Cambodia as a whole, by addressing the following concerns. MAFF could amend the import licensing procedures and regulations for fertilisers, simplifying and speeding up the licensing process and thus removing the scope for rent-seeking behaviour and the need for facilitators to expedite the issuing of licences. Easing licensing procedures and regulations would also reduce the incentive for illegal imports. It would also be more appropriate for import licences to be approved by MAFF based on the suitability of a product's use in Cambodia; then importers could be allowed to import any quantity of a registered fertiliser product based on market demand and their own commercial risk assessment. This would also reduce the scope for illegal (and therefore unregulated) imports. All imported fertiliser products should be labelled to identify the manufacturer. This would enable the sources of sub-standard fertilisers to be traced, particularly from Vietnam and China, where it is claimed that sub-standard blends and granular products are produced. Besides the certification of dealers and retailers, it is timely for MAFF and PDA to take further steps to also certify third party traders (intermediaries, mobile distributors, and seasonal village retailers) who purchase and deliver fertiliser to villages for re-sale to farmers. Finally, fertiliser dealers, retailers, and other fertiliser traders should buy fertilisers only from certified importers or distributors and transport to villages for direct re-sale to farmers. PDA inspectors should monitor and spot-check fertiliser operators to help reduce fertiliser quality problems for smallholders.

NOTES

1. 1 USD = KHR 4000.
2. In addition, there is the cost of replacing the fake fertiliser in order to limit the yield loss to 20%. If farmers bought one bag of urea and one bag of DAP, the additional cost would be USD 65, giving a total financial loss of about USD 350 per household.

REFERENCES

Agricultural Marketing Office, 2002–2010. *Agricultural Marketing Information Bulletin 2002–2010.* Phnom Penh: Department of Planning and Statistics, Ministry of Agriculture, Forestry and Fisheries.

Asian Development Bank, 2002. *Report on Marketing in the Agricultural Sector of Cambodia.* Agriculture Sector Development Program (ADB—TA No. 3695—CAM). Phnom Penh: Ministry of Agriculture, Forestry and Fisheries.

Commune Database, 2010. Commune Database Online, National Institute of Statistics, Ministry of Planning, Phnom Penh. Available at http://db.ncdd.gov.kh/cdbonline/home/index.castle (accessed 15 February 2012).

International Fertiliser Development Centre, 2010. *Rapid Appraisal of Fertiliser Quality in Cambodia.* IFDC report prepared for the World Bank.

Kaplinsky, R., 2000. Globalisation and unequalization: what can be learned from value chain analysis? *Journal of Development Studies* 37: 117–146.

Kaplinsky, R., and Morris, M., 2001. *A Handbook for Value Chain Research.* Brighton: Institute of Development Studies, University of Sussex.

MAFF, 2012. *Ministry of Agriculture Forestry and Fisheries Report 2011/12.* Phnom Penh: Ministry of Agriculture Forestry and Fisheries.

RGC, 1998. Sub-Decree on the Standard and Management of Agricultural Materials, No. 69 ANKr-BK, adopted 1998. Phnom Penh: Royal Government of Cambodia.

Schamel, G., and Hongen, F., 2003. *Adverse Selection in Developing Country Factor Markets: The Case of Fertiliser in Cambodia.* Paper presented at American Agricultural Economics Association (AAEA) Annual Meeting, Montreal, Canada, 27–30 July.

Theng, V., and Koy, R., 2011. *Review of Agricultural Policy and Policy Research.* Policy Discussion Paper. Phnom Penh: Cambodian Development Resources Institute.

CHAPTER 15

The Use of Credit by Rice Farmers in Takeo

Kem Sothorn

INTRODUCTION

Farmers' lack of access to both working and investment capital is considered one of the major factors hindering the transition from low-input agricultural systems to more productive commercial ones (ACI 2005). Rice productivity in Cambodia is significantly constrained by low application of agricultural inputs, notably fertiliser, mechanisation, and irrigation (ACI 2005; World Bank 2007). Improving access to rural credit would be a significant step forward for smallholder agricultural development in the country. While there has been a boom in microfinance in Cambodia (CMA 2011), the impact on smallholder farmers remains limited, mainly due to the risks posed by insecure land tenure and uncertain returns to on-farm investment. Understanding the pattern of credit access and the way it affects rice farmers' borrowing and investment decisions could usefully inform policy options to improve the viability of rural credit delivery.

The hypothesis of this study was that ready access to credit raises rice farmers' productivity and farm income, thus improving the well-being and reducing the vulnerability of rural households. The study sought to (a) understand the patterns and characteristics of credit access of different farmers, (b) investigate the impact of credit on farmers' production and

K. Sothorn (✉)
Parliamentary Institute of Cambodia, Phnom Penh, Cambodia

livelihood systems, (c) identify challenges and opportunities for successful credit utilisation, and (d) provide options for improving credit access and promoting successful farm credit utilisation.

Qualitative research was undertaken in Takeo Province. The villages chosen for the study represented three rice production systems: rainfed wet-season (WS) rice, WS rice with supplementary irrigation, and irrigated dry-season (DS) rice. Five group interviews were undertaken with three types of rice farmers: subsistence farmers, semi-commercial farmers, and commercial farmers. The major themes for the group interviews were: the pattern of credit access, the impact of loans on productivity and livelihoods, and the challenges faced in accessing loans. Key informant interviews were undertaken with representatives of five major microfinance institutions (MFIs) in Takeo—CREDIT, AMRET, Sathapana, Thaneakea Phum, and Vision Fund. The interviews focused on constraints and opportunities for expanding credit, the institutional arrangements and policy environment for MFI operations, and trends in the development of the rural credit sector. Secondary data and statistics were collected from the Cambodia Microfinance Association (CMA), provincial development statistics, and previous research reports.

In this chapter, an overview of the credit sector in Cambodia is presented, with particular focus on the evolution of policy and outreach; the findings of the survey are discussed, including an analysis of the challenges and opportunities for credit use by subsistence, semi-commercial, and commercial rice farmers; and policy options and research gaps are identified.

OVERVIEW OF CREDIT DEVELOPMENT IN CAMBODIA

Informal credit systems have long been an important part of rural livelihoods (Phlong 2009). Informal credit draws on a culture of reciprocity and risk-sharing within kinship groups and the residential community or village and is still widely practised. Some forms of informal credit such as village banks, savings-based microfinance, and self-help groups are vibrant forms of economic exchange, often initiated by NGO-sponsored community development programmes. These sources of finance are limited in coverage and provide relatively little capital. In most cases they cannot meet the demand for investment. Many of these community-based organisations (CBOs) simply dissolved after the project was completed. The limited capacity of informal credit systems has led to efforts to improve access to capital from formal credit institutions.

Most formal credit in Cambodia has emerged from non-profit microfinance projects initiated to fill the void left by the virtually non-existent rural banking sector. According to CMA (2011), the sector has gone through three major stages of development (Table 15.1). A number of reforms have been made to improve the institutional environment for the development of microfinance, resulting in a steady increase in the number of licenced MFIs and registered micro-credit NGOs.

The Asian Development Bank estimated that the demand for rural finance in Cambodia was around USD 120–130 million per annum in 2000 (ADB 2001). By 2011 there were 29 MFIs and 1 commercial bank providing financial services in 24 provinces, covering 59,458 villages and 1.1 million borrowers with outstanding loans of USD 572.7 million, more than four times the ADB's estimate (Fig. 15.1). By the first quarter of 2017, CMA reported 61 MFIs with 1.9 million borrowers and outstanding loans of USD 3328 million (CMA 2017), a further sixfold increase.

Even though the number of villages covered by MFIs in 2011 was five times higher than the total number of villages in the country (due to multiple MFIs working in a given village), this does not mean that all rural communities had access to financial services. Rather, it reflects that MFIs tended to concentrate in densely populated areas and in economically active villages (Figs. 15.2 and 15.3). Hence, the number of MFIs in a

Table 15.1 Evolution of formal credit sector in Cambodia since 1995

Period	Features
1995–2000	*Government Support; Institutionalisation* • Credit Committee for Rural Development (CCRD) established in 1995 • National Bank of Cambodia (NBC) set up Supervision Office of the Decentralisation of Banking Systems Bureau in 1997
2000–2005	*Commercialisation* • The government adopted a two-tier system under the Law on Banking and Financial Institutions in 1999 • Reform of banking system in 2000 • Number of microfinance institutions (MFIs) increases from 3 to 15 • Eight decrees issued in 2005 to enforce registration and licencing of MFIs
2005–	*An Integral Economic Player* • Credit information system introduced in 2006 to collect and share negative information from commercial banks • Cambodia Microfinance Association (CMA) established in 2007 • Decree on Licencing Microfinance Deposit Taking Institutions issued by NBC in 2007

Source: Author's compilation from CMA (2011)

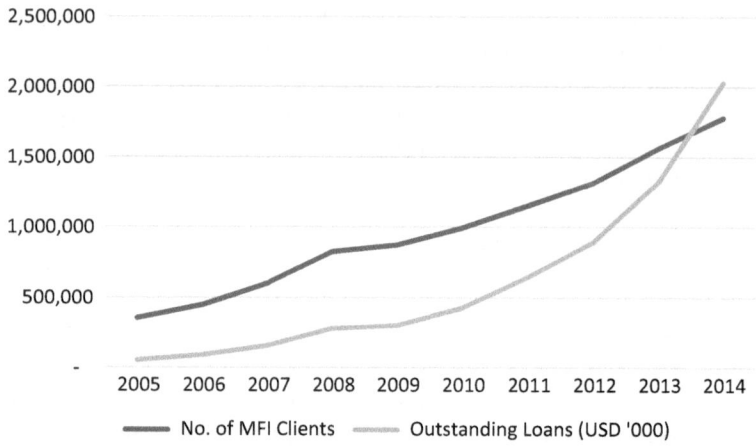

Fig. 15.1   Growth of MFI clients and loans in Cambodia, 2005–2014. (Source: CMA 2015)

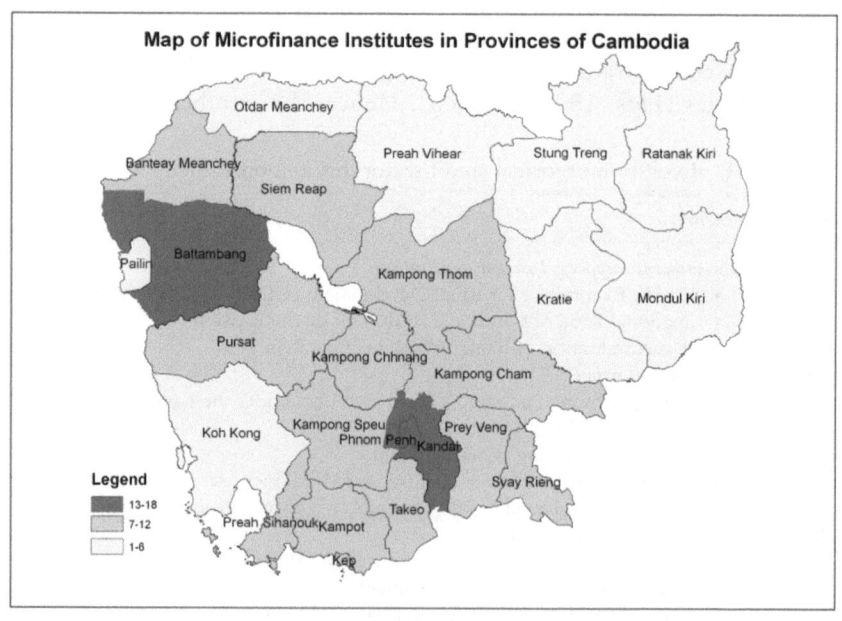

Fig. 15.2   Number of MFIs by province, 2011. (Source: Constructed from CMA data)

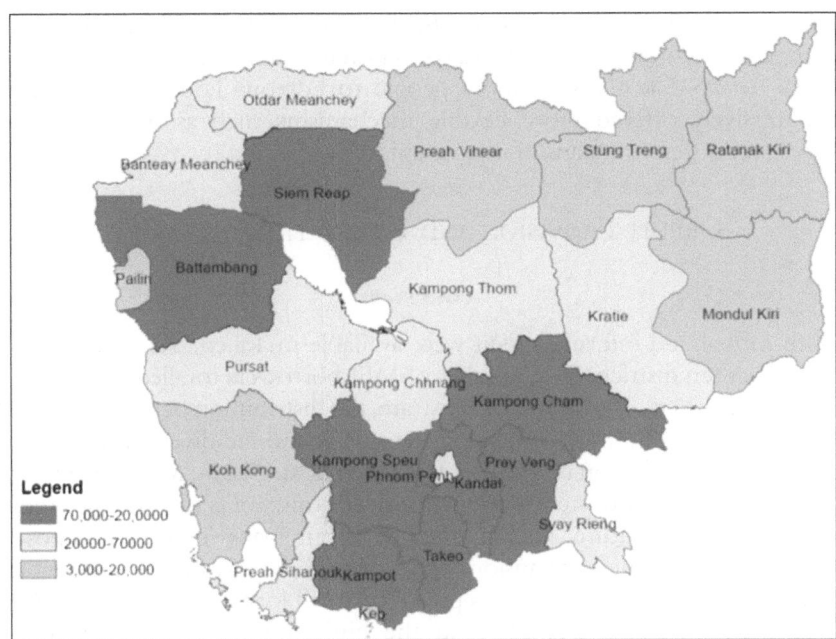

Fig. 15.3 Number of MFI borrowers by province, 2011. (Source: Constructed from CMA data)

province was associated with agricultural and economic growth in that province. For example, the highest numbers of borrowers in 2011 were in the provinces with the fastest rates of agricultural growth, such as Battambang, Kampong Cham, and Takeo (Ovesen et al. 2012). A province like Siem Reap also had a high number of both borrowers and MFIs because of economic diversification through tourism. Hence large numbers of poor and very poor households in less developed provinces were still not able to access financial services, preventing them from improving their livelihoods and gaining greater benefits from economic development.

Economic growth and macroeconomic stability have been important stimulants for the development of the rural credit sector. First, the change towards a liberal economic environment enabled the development of MFI institutions. Second, the development of rural infrastructure such as roads and irrigation improved access to farm inputs, supportive rural policy, and expanding markets have stimulated agricultural and rural development and hence the demand for credit. Third, increased competition among MFIs led to a reduction in monthly interest rates from 5–6% to 2.5–5%, allowing more people to borrow. Fourth, MFIs have progressively offered more flexible mechanisms such as different loan types and different means of repayment.

CREDIT PROVISION AND UTILISATION IN TAKEO

Overview

Both formal and informal credit were available in Takeo. Ten MFIs operated in all ten districts. The number of MFI borrowers totalled 116,695 in 2011, but as Figs. 15.4 and 15.5 illustrate, the distribution between districts varied widely, reflecting population density, economic diversification, and agricultural production (CMA 2011; Ovesen et al. 2012). Most farming in the province was subsistence-oriented. Most households did not use credit to invest in agriculture; instead, they took out loans for other purposes such as small business expansion, wage migration, and buying household assets. Around 70% of borrowers preferred small loans of USD 250–1500. Bati and Tram Kak Districts had the highest number of borrowers.

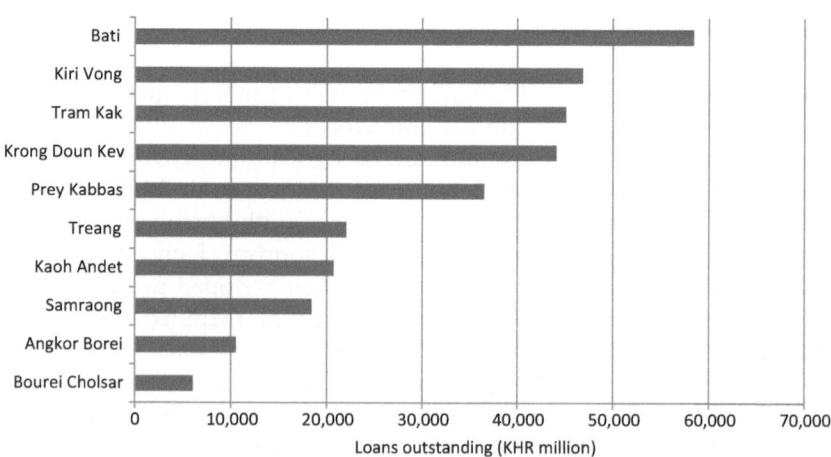

Fig. 15.4 Total amount of MFI outstanding loans in Takeo in 2011 by district (KHR million). (Note: USD 1 = KHR 4000; Source: constructed from CMA data)

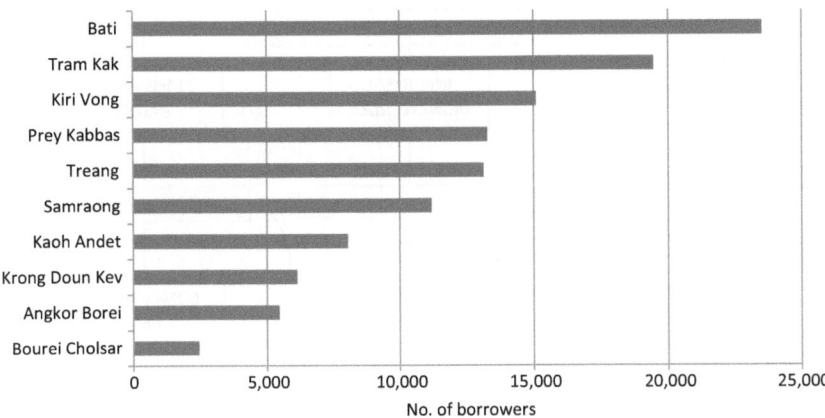

Fig. 15.5 Number of MFI borrowers in Takeo in 2011 by district. (Note: USD 1 = KHR 4000; Source: constructed from CMA data)

However, the greatest demand for agricultural loans was from commercial rice producers in Angkor Borei, Bourei Cholsar, Kiri Vong, and Kaoh Andet districts, which were reported to have the highest number of farmers accessing credit for irrigated DS rice production.

Credit Access by Type of Rice Farmer

Access to credit varied between different types of farmers—the subsistence-oriented farmer, the semi-commercial farmer, and the commercial farmer. The pattern of access to loans among these three types is illustrated in Fig. 15.6. It can be seen that the three main sources of credit were merchants/traders, moneylenders, and MFIs. The subsistence and semi-commercial farmers used all three sources, though to differing degrees, as well as traditional rotating savings and credit associations (ROSCAs). The commercial farmers used only MFIs. It is interesting to note that the MFIs also provided credit to merchants, traders, and moneylenders, reflecting the overall importance of MFIs in the supply of credit in the province.

(a) *Subsistence farmers.* The main source of finance for small, subsistence-oriented farmers was the local moneylender. Loans were usually small, ranging from USD 250 to USD 1000, and the interest rate was around 10% per month, which was more than three times higher than the rate charged by MFIs. Interestingly, these loans were mostly not used for

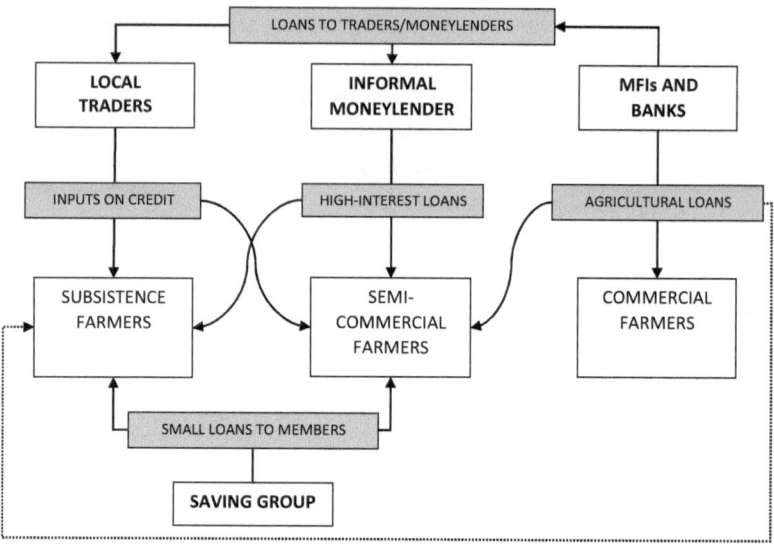

Fig. 15.6 Pattern of access to credit by type of rice farmer. (Source: Field interviews)

agriculture as such but for expanding and diversifying income sources through raising animals, migrating to work in industrial plantations in Kratie or Ratanakiri to the north, or opening a small business. They were also used for consumption expenditure and house construction. Subsistence farmers preferred to use labour-intensive production methods which required less use of purchased or hired inputs, thus minimising their demand for formal loans. However, agricultural inputs (fertiliser, pesticides) were bought on credit from local input suppliers, to be repaid after harvest.

These informal sources of credit required no collateral, were flexible, and could easily be accessed by farmers, explaining why they have long been favoured by small farmers, despite higher interest rates than for MFI loans. Strong social capital within the community underpinned access to credit for subsistence farmers. Moneylenders were usually located in the same village and were well-informed about the borrower's situation. Subsistence farmers in a group discussion in Chumpu Prik Village, Prey Kabas District, expressed their preference as follows: "We prefer to borrow money from a moneylender in the village because it is fast and flexible. We can take a loan whenever we want and repay the loan after harvesting or when we get money from other jobs. It is not like getting money from *angkar* (MFIs) which require us to fill in various forms; they also need collateral and we cannot repay early."[1]

There were several other reasons that formal credit was less accessible to subsistence farmers. In Takeo, 57% of farmers had less than 1 ha, hence had insufficient collateral for a medium-sized loan. Moreover, subsistence rice farming was still highly vulnerable to production failures due to unpredictable weather conditions and pest outbreaks, hence financing it was highly risky. In fact, small farmers were usually not targeted by MFIs for agricultural loans but for loans to diversify their income-earning activities (hence the dotted line in Fig. 15.6). Interviews with MFI officers indicated that small farmers' lack of skill and knowledge on how to use credit properly was the major factor limiting their access to formal credit. In difficult circumstances, poor households often used loans not only for the productive purpose specified but to cope with shocks or refinance a previous loan, risking their ability to repay the new loan. As a result, many farmers were trapped in a debt cycle.

Group lending was another model for credit provision that was designed to improve access to finance for farmers without collateral, using the group guarantee mechanism. However, high levels of risk made it difficult for small farmers to take out a group loan. Small farmers were extremely vulnerable to both idiosyncratic and covariant shocks. Idiosyncratic shocks were those specific to a given household, including illness, localised crop damage, business failure, and loss of income from employment. However, covariant shocks affected all farmers in a group. Given limited irrigation facilities, drought was the most common covariant shock affecting farmers. Given this high degree of shared vulnerability among small farmers, group loans were less likely to give them access to formal credit for farming. Even with no collateral requirement, poor farmers found it hard to form a group or choose the group's representative who would be responsible for the members' repayments. Farmers in Prey Kdouch Village, Tramkak District, commented: "It is so difficult for us to form a group to access a loan. All of us face such hardship in our living. We simply cannot form a group [to take a] loan because we do not trust our group members to be able to pay back their loan on time. It may be difficult for us if they don't."

Traditional self-help groups in the form of rotating saving and loan associations offered some kind of financial support to farmers. However, the capital in each saving group was small, ranging from KHR 2 million

(USD 500) to KHR 6 million (USD 1500). Group members could access loans up to KHR 400,000 (USD 100). Most loans were used for daily consumption or to cope with shocks but not for investment in agriculture or business. This type of loan contributed to risk-coping ability but was not likely to contribute to increased productivity.

In sum, subsistence farmers were largely excluded from formal agricultural credit. The risk of production failure, the lack of collateral, the high degree of covariant risk, and the frequent diversion of loans to non-productive uses inhibited poor farmers from applying for agricultural loans and discouraged MFIs from approving them. In this context, informal credit continued to play an important role in providing the short-term capital needs of this type of farm household.

(b) *Semi-commercial farmers.* This type of farmer produced rice for both household consumption and sale. This was possible because they cultivated both WS and DS rice, with the DS crop produced exclusively for the market. A typical enterprise budget for both seasons is presented in Table 15.2.

For the WS crop, farmers used family labour for most activities. The rice was largely unirrigated, and little machinery was used as the farms were mostly not accessible to tractors. Long-duration rice varieties were

Table 15.2 Enterprise budget for semi-commercial rice farming

Budget item	Wet-season rice			Dry-season rice		
	Value/ha (KHR × 10³)	% of costs	% of revenue	Value/ha (KHR × 10³)	% of costs	% of revenue
Land preparation	160	13	7	360	11	6
Seed	70	6	3	360	11	6
Hired labour	350	30	15	0	0	0
Chemical fertiliser	250	21	11	1290	37	23
Pesticide	0	0	0	500	15	9
Irrigation	300	25	13	640	19	11
Harvesting/ threshing	60	5	3	280	7	6
Total costs	1190	100	52	3430	100	61
Gross revenue	2300		100	5580		100
Gross margin	1110		48	2150		39

Source: Field interviews

planted, and chemical fertiliser application averaged about 100 kg/ha. Valuing the output at the market price, the gross margin was about KHR 1.1 million/ha (USD 275). Farmers preferred to access inputs such as fertiliser from local merchants on credit, claiming that this service was readily available and they could repay the credit after harvest (Fig. 15.6). The interest rate for this credit was 3–5% per month.[2] The WS rice harvest was mainly for household consumption. However, farmers sometimes sold part of the WS crop if the DS crop had been less profitable or if they needed to repay credit or loans. They often experienced a shortage of rice (and cash) for around two–three months before the WS harvest.

For the DS crop, expenditure on inputs was three times higher, especially due to higher use of fertilisers, which made up 37% of costs, but also because of expenditure on mechanised land preparation, irrigation, pesticides, and threshing (Table 15.2). However, higher yield and price meant that the gross margin was KHR 2.2 million/ha (USD 550), about double the return for the WS crop. The higher costs and returns meant that farmers sought both formal and informal credit for their production expenditure. Though total costs were about KHR 3 million/ha, loans were around KHR 1–2 million (USD 250–500). Given their small landholdings (hence limited collateral), most farmers could only obtain loans of around KHR 1 million, suggesting that the formal lenders were imposing capital rationing on small farmers. Given the excess demand for credit, some moneylenders took loans from MFIs at 3% interest and re-lent this to farmers (without collateral but with other ties) at 5–7% interest (Fig. 15.6).

As with WS rice, farmers also bought inputs on credit in the DS, incurring up to 5% per month in implicit interest to be paid along with the principal after harvest. However, DS gross margins were highly vulnerable to fluctuations in the price as farmers were mainly dependent on the Vietnam market. In 2012, for example, the rice price quickly fell from KHR 1200/kg to 700/kg (from USD 0.30 to 0.18). In these circumstances, some farmers borrowed additional money from the MFIs to repay the local input merchants, thus risking the loss of their land and other household assets to meet their production commitments. If production failure (e.g., due to a pest outbreak) coincided with falling prices, the farmers would have been plunged into a debt crisis.

In semi-commercial rice farming, farmers earned only modest income from their production due to the high cost of production, high interest on loans, and reduction in the market price of rice. In group discussions, farmers maintained that this level of income could only support a modest lifestyle and severely limited their ability to cope with shocks. Thus their livelihoods remained highly vulnerable. As group participants in Pich Sar Commune, Koh Andeth District, expressed it: "After harvesting we have to pay for chemical fertilisers, pesticides, machinery hire—we pay for almost everything in rice production. We earn around 30–40 meun riel [USD 75–100] per kong [0.13 ha].[3] We have been trying very hard each year just to make a profit." Another group of farmers in Pich Sar Commune, Koh Andeth District, commented: "Money from *angkar* (MFIs) helps only those who are already rich, because they have multiple occupations and constant income to pay back the interest. Not like us who depend only on rice farming. We cannot borrow money like them because we cannot even pay the interest. Rice farming is too risky now, everything for production is expensive but when we sell our rice the price is too cheap."

(c) *Commercial farmers.* Commercial farmers cultivated only DS rice for the market, using large areas of land that were flooded in the WS, adding to the fertility of the soil and reducing the fertiliser requirement for DS production. Interviews with rice farmers in Khmol Village, Kamnab Commune, Kirivong District, indicated that around 50% of farmers owned around 5 ha of paddy land, 30% owned 3–4 ha, and only 20% owned 1–2 ha. Farmers used IR varieties, given their high yield and market potential. The average yield was 6 t/ha, significantly more than semi-commercial DS rice. Production costs were much the same as for the semi-commercial farmers in the DS as the lower fertiliser costs were offset by higher pesticide use and mechanisation (Table 15.3). The total cost of

Table 15.3 Enterprise budget for commercial dry-season rice farming

Budget item	Value/ha (KHR × 10³)	% of costs	% of revenue
Land preparation	280	9	6
Seed	210	7	4
Hired labour	0	0	0
Chemical fertiliser	460	15	10
Pesticide	1300	42	27
Irrigation	560	18	12
Harvesting/threshing	280	9	6
Total costs	3090	100	65
Gross revenue	4800		100
Gross margin	1710		35

Source: Field interviews

production was KHR 3.1 million (USD 775) per hectare, of which 57% was for expenditure on agrochemicals. Land preparation and harvesting involved hiring mechanised services (18% of costs) and irrigation costs accounted for a further 18%. Despite higher yields, gross revenue was lower than in Table 15.2 because of a fall in price, hence the gross margin was KHR 1.7 million (USD 425) per hectare. However, the larger scale of production meant that total profits were substantial.

Commercial rice farmers took loans from MFIs to cover their entire production expenditure (Fig. 15.6), though some of the large landholders used their own capital for some or all of their costs. Having large and fertile landholdings to offer as collateral plus a profitable enterprise gave commercial farmers ready access to agricultural credit. Loans of USD 1000–1500 were common among farmers with 3–4 ha of DS paddy land, four–six times as much as the semi-commercial farmers. Given the urgency to repay their loans and avoid incurring further interest charges, farmers sold their paddy immediately after harvest, preventing them from obtaining a higher price during the wet season.[4] Commercial rice production also faced price fluctuations, pest hazards, and rising input costs, which threatened to reduce gross margins and hence the ability of farmers to repay their loans to the MFIs. Farmers in Khmol Village, Kirivong District, recounted: "Before we cultivated late-season [recession] rice, the yield was low but we were never trapped in debt like today. We sold our rice only when we needed money or when the price of rice was high ... Now we cultivate dry-season rice, the yield is high but we also pay for everything ... The costs of fertilizers, pesticides, and machinery use keep increasing year by year ... We are afraid there might be one day that we cannot repay our loan to *angkar* (MFI)." Despite these concerns, credit for rice farming seemed to be much more viable when used by commercial farmers. Access to credit worked smoothly where there was a favourable natural endowment, good access to inputs, supporting physical infrastructure, and a ready market for the output.

Impact of Credit on Rice Farmers' Livelihoods

In group discussions with the three types of rice farmers, opinions were sought regarding the positive and negative impacts of credit on household livelihoods.

For subsistence-oriented farmers, the chances of successfully using loans for agricultural production were slim. Subsistence rice-farming systems suffered from lack of collateral, lack of supporting infrastructure, high input costs, considerable risk and vulnerability, limited income diversification, and a fluctuating rice price. Many MFIs avoided providing agricultural loans to small subsistence farmers. The farmers themselves rarely sought agricultural loans but instead focused on loans for more productive purposes such as animal raising, trading, operating small businesses, and migrating to work in industrial plantations. Access to loans for these purposes helped to diversify their income sources. Farmers felt that, to be able to use agricultural loans successfully, they needed to have at least two income sources from non-farm activities, otherwise they could not repay the loans to the MFIs. Farmers who obtained loans for investing in off-farm sources of income reported that their livelihoods were improved by having diverse income sources, better food security, and increased household assets. This suggests that access to credit enabled the adoption of more profitable income-generating activities for subsistence farmers, leading to a gradual improvement in livelihoods. Farmers in Prey Kdouch Village, Tramkak District, emphasised this point: "It is not always true that loans from *angkar* (MFIs) cause people in the community to become even poorer ... It depends on the way people use the loans. If they borrow money to buy a motorbike, or spend carelessly, they will certainly become poorer. Many people took a loan for pig raising, collecting rice to sell, or starting a small business, and they were successful. They have better lives now."

The impacts of credit on semi-commercialised farmers were uncertain. These farmers depended solely on WS and DS rice production for income. Without other livelihood activities, they were vulnerable to crop failure. The increasing number of MFIs operating in the province enabled farmers to access capital to invest in rice farming, transforming their production systems from subsistence to semi-commercial farming. Most farmers accessed loans from multiple MFIs. The impact of this use of credit could be positive or negative depending on farm profit, which was largely determined by weather conditions and the price of rice. Farmers felt they could be profitable if the price of rice was KHR 1200/kg or above. With this price, farmers could repay their loans and pay for the fertilisers and pesticides used during the season. In this scenario, income from rice farming contributed to increased household well-being. However, in 2012, the rice price was about KHR 700/kg, reducing the rice gross margin to KHR 1.3 million (USD 328) per ha. Many farmers fell deeper into debt because of the low price. The common coping strategies used were reducing household consumption of both food and non-food items and, in some cases, selling household assets to repay debts.

Credit for commercial rice farmers clearly contributed to improved household livelihoods. Farmers reported increasing rice yields from around 3 t/ha to almost 6 t/ha, largely due to access to capital, irrigation, mechanisation, and an export market. The presence of MFIs contributed to the acceleration of agricultural commercialisation in these areas. Farmers also reported increased household assets, including better housing and more machinery, better education for their children, and greater food security. However, the increasing cost of production and fluctuations in the price of rice may reduce farmers' incomes and capacity to service loans. In addition, there were reports of more frequent and extensive pest damage and increasing pollution due to the intensive use of pesticides. These trends could undermine the productivity of the farming systems and affect welfare directly through impacts on health.

CONCLUSION

The study found several challenges affecting viable credit use by rice farmers in Takeo Province. High and increasing farm input costs, the low quality of fertilisers, and the rising cost of mechanisation due to the price of fuel continued to hinder long-term agricultural growth, not just in rice production. While these trends implied greater capital needs, hence increased demand for credit, the squeeze on farm profits was reducing the viability of agricultural loans, especially for small farmers engaged primarily in rice production. The lack of a formal land title to use as collateral also remained a barrier to credit use for many farmers. The Land Management and Administration Project (LMAP) has accelerated the process of issuing rural land titles, but the needs of small farmers must be further prioritised.

The need to ensure high repayment rates by carefully assessing potential clients meant that MFIs effectively screened out poor farmers with little collateral and high vulnerability. Nevertheless, poor farmers often resorted to loans from formal and informal sources when subject to livelihood shocks, despite the high risk of falling even deeper into indebtedness. Some MFIs such as CREDIT and Vision Fund had developed a special loan package (subsidised by donors) for this category of farmers. Increased investment in the government's social protection programme would also help reduce the vulnerability of the poor and protect them from falling into burdensome debt.

A challenge facing all farmers was the high interest rates for formal credit. At the time of the study, MFIs were charging around 30% p.a. (2.5% p.m.) to cover operational costs and maintain financial sustainability. Many depended on outside sources of capital, while low domestic savings remained a barrier to lowering the interest rate. However, MFIs have reduced interest rates over time due to increasing competition and local saving. MFI representatives interviewed claimed to be aiming for an interest rate of 1% p.m. In 2017, the National Bank of Cambodia (NBC) introduced a ceiling of 18% p.a. for MFIs, or 1.5% p.m. Some farmers have recently reported paying interest at rates less than this, though not yet as low as 1% (Moeun Nhean 2017).

Demonstrating improved cultivation methods to farmers could help increase productivity and farm income and make farm investment more viable. With training in business skills, farmers would be better able to use MFI loans successfully, whether for rice farming or other livelihood activities. Rural development NGOs could better integrate these measures into their extension programmes. Long prioritised by the government to improve productivity and reduce production and market risks, further investment in infrastructure is also needed to help transform subsistence farming. Greater private sector involvement in the rice sector through contract farming may also be a means of facilitating and financing smallholder development (see Chap. 17).

Increasing regional demand for rice plus government policy to promote rice exports is driving growth of the commercial rice sector. Government efforts to improve agricultural infrastructure, such as irrigation facilities and the road network, and increased private sector investment in rice milling and storage will boost production, facilitate trade, stabilise market demand for rice, and help smooth rice price fluctuations. All these will increase the importance of small-scale agricultural credit. Provided the vulnerability of poor farmers is addressed in the ways suggested above, this expansion of credit use will have a largely positive impact on rural livelihoods.

NOTES

1. Moeun (2017) quotes a villager from Kandal Province who said: "Most of my villagers know that MFIs are part of the private sector, but we have called them Angkar since a very long time ago." Angkar, meaning "organisation", was the term used in the Khmer Rouge era to refer to "the government".

2. Urea cost about KHR 135,000 (USD 34) per sack if paid for in cash. If a farmer bought the urea on credit, the cost increased to KHR 165,000 (USD 41). However, the farmer could obtain as much urea as he needed and was only required to repay the merchant when the harvest was finished. Normally, the period for wet-season rice production was six months. This implied a monthly interest of about 3% if the urea was bought at the start of the season or 5% if it was bought four months before harvest.
3. This works out to be about USD 580–770 per ha, presumably referring to gross revenue (cf. Table 15.2).
4. Harvesting was also increasingly undertaken by Vietnamese traders using combine harvesters; hence, the harvested paddy was of high moisture content and was transported directly to rice mills in Vietnam with drying capacity.

REFERENCES

ACI, 2005. *Final Report for the Cambodia Agrarian Structure Study*. Prepared for the Ministry of Agriculture, Forestry and Fisheries, Royal Government of Cambodia, the World Bank, the Canadian International Development Agency (CIDA) and the Government of Germany/Gesellschaft für Technische Zusammenarbeit (GTZ). Bethesda, MA: Agrifood Consulting International.

ADB, 2001. *Financial Sector Blueprint for 2001–2010*. Phnom Penh: Asian Development Bank.

CMA, 2011. *Microfinance Information Exchange*. Cambodia Microfinance Association. Available at https://cma-network.org/en/microfinance-information-exchange.

CMA, 2015. *Annual Report 2014*. Phnom Penh: Cambodia Microfinance Association.

CMA, 2017. *Microfinance Information Exchange*. Cambodia Microfinance Association. Available at https://cma-network.org/en/microfinance-information-exchange.

Moeun Nhean, 2017. Misuse of MFI loans leaves locals in debt. *Phnom Penh Post*, 21 March 2017.

Ovesen, J., Trankell, I. B., Heng, K., and Sochoeun, C., 2012. *Rice Farming and Microcredit in Takeo*. Phnom Penh: Uppsala University and Intean Poalroath Rongroeurng Ltd.

Phlong, P., 2009. *Informal Credit Systems in Cambodia*. Master's Thesis, Northern Illinois University, DeKalb, Illinois.

World Bank, 2007. *Sharing Growth: Equity and Development in Cambodia*. Phnom Penh: World Bank for the Cambodia Development Cooperation Forum.

CHAPTER 16

Contract Farming of High-Quality Rice in Kampong Speu

Nou Keosothea and Heng Molyaneth

INTRODUCTION

Contract farming is seen as one of the policies to overcome current imped-
iments to commercialisation in the Cambodian rice sector. The Angkor
Kasekam Roongroeung Co. Ltd. (AKR) was the first agribusiness firm to
implement contract farming of rice, beginning in 1999 with about 100
farmers; it currently claims to have over 50,000 contracted farmers in four
provinces (AKM 2015). The approach was later adopted by other development organisations, such as the Cambodian Centre for Study and
Development in Agriculture (CEDAC). A study by Cai et al. (2008), for
the Asian Development Bank Institute (ADBI), on the impacts of AKR's
rice contract farming scheme on farmers' performance provides some useful insights. In general, however, little is known about rice contract farming

N. Keosothea (✉)
National Committee, Economic and Social Commission for Asia and the Pacific,
Ministry of Foreign Affairs and International Cooperation,
Phnom Penh, Cambodia

H. Molyaneth
Faculty of Development Studies, Royal University of Phnom Penh,
Phnom Penh, Cambodia

© The Author(s) 2020
R. Cramb (ed.), *White Gold: The Commercialisation of Rice
Farming in the Lower Mekong Basin*,
https://doi.org/10.1007/978-981-15-0998-8_16

327

in Cambodia in terms of its contractual arrangements, inclusiveness, benefits, and challenges.

To help fill the knowledge gap, this study aimed to examine three aspects of AKR's rice contract farming: (a) inclusion of smallholder farmers and contractual arrangements, (b) benefits of contract farming for farmers, and (c) challenges faced by farmers and agribusiness firms. AKR was selected as a case study because the company operates the largest scale of rice contract farming in Cambodia. Findings of this study will hopefully contribute to policymaking on how to make rice contract farming more developmental. In this chapter, we first review the existing literature on the effects of contract farming, then outline the methods used in the study. This is followed by a presentation of research findings on the inclusion of smallholders, the nature of the contractual arrangements, and the benefits and challenges of contract farming, and a discussion of policy options based on these findings. A summing-up concludes the chapter.

Understanding the Effects of Contract Farming

Although one of the purported benefits of contract farming is to help smallholder farmers integrate into global agri-food supply chains, smallholders have not always been included because working with them incurs high transaction costs and a high risk of producer defaults (Key and Runsten 1999). On the other hand, smallholders are preferred in some cases to minimise the negative effects of crop failure, enable a flexible production portfolio, enhance the quality of produce, and reduce the drop-out rate of members (Birthal et al. 2005). The experience of various contract farming schemes in Thailand suggests that, where production requires large amounts of capital, medium- and large-scale farmers are chosen; but when hard work and commitment are more important, small-scale farmers have a better chance to participate (Sriboonchitta and Wiboonpoongse 2008).

Findings on the effects of contract farming on farmers and agribusiness firms are mixed and inconclusive. Farmers might enjoy some benefits—increased profitability and income; better access to production inputs such as machinery, seeds, fertilisers, infrastructure, and credit; a guaranteed and stable price; a reliable and secure market; and improved technical farm management skills (Setboonsarng et al. 2005). At the same time, they can experience losses due to the failure of agribusiness firms to comply with agreed terms and conditions of the contract, whether intentionally or unintentionally. The benefits could also be negatively affected by the farmers' limited ability to apply required farming technology, resulting in failing to attain defined levels of productivity and quality. Moreover, farmers could be in debt because of their over-reliance on easily accessible credit provided by the contracting agribusiness firm.

Agribusiness firms can similarly have both positive and negative experiences. On the positive side, they could secure sources of supply with required quality and standards, reduce production and transaction costs, transfer production risks to farmers, and get more agricultural support from government, such as credit and subsidies. On the other hand, some firms experience losses because farmers break the contract by selling to third parties when the price increases. This practice of extra-contractual marketing is an often-reported problem facing agribusiness firms involved in contract farming schemes.

RESEARCH METHODS

The study was based on interviews with key informants in Kampong Speu Province over two periods: May 2012 and June 2013. According to Cai et al. (2008), more than 80% of contract farmers were in Kampong Speu Province. All but two interviewees were from Prey Khmeng Commune and Chom Sangker Commune in Phnom Srouch District. This district was an ideal site for the study due to AKR Company's long history there. A total of 20 key informants were interviewed—ten farmers, four village heads, one commune clerk, three staff of the Society for Community Development in Cambodia (SOFDEC), a local NGO, and two staff of the AKR Company (whose office was in Angsnoul District). Farmer interviewees were selected through snowball sampling with the support of SOFDEC staff. Since village heads were also farmers, a total of 14 farmers were interviewed, comprising 11 former contract farmers, 2 current contract farmers, and 1 non-contract farmer. All farmers interviewed cultivated a single crop of rainfed rice in the wet season.

All interviews were done in a semi-structured manner. There were four different interview guides—for farmers, representatives of the commune association, the staff of the local NGO, and representatives of the AKR Company. Some common questions were asked of former contract farmers, current contract farmers, and non-contract farmers, but there were also specific questions for each type of farmer. All interview guides covered reasons why farmers and agribusiness firms cooperated under a contract, and the terms and conditions, costs and benefits, and challenges of working under a contract.

Inclusion of Smallholders

AKR considered several factors in deciding where to start contract farming in the early stages of its operation. The foremost factor was the agronomic conditions. The company started with *Pkar Malis* (a type of aromatic rice), a variety that is selective in terms of agronomic conditions. AKR had examined agronomic conditions in several provinces and chose four to start its rice contract farming: Kampong Speu, Kandal, Takeo, and Kampot. Agronomic conditions remained a critical factor when deciding on the specific locations within the province. Not all communes, villages, and households had the conditions suitable for *Pkar Malis* rice, hence some were excluded from the contract farming scheme from the outset.

Another criterion for selecting villages was the degree of concentration of interested farmers. Due to a strong requirement for varietal purity, villages having interested farmers who were geographically dispersed were not eligible for the contract. AKR staff interviewees explained that when *Pkar Malis* rice is grown next to other varieties, pollen of other rice varieties can reduce the varietal purity of the *Pkar Malis*. To avoid this, AKR only selected villages where many farmers were interested in participating in contract farming and farmed close together.

The size of landholding was not a condition for selecting farmers when the company began its operation. Nevertheless, in 2000, the share of farmers contracted with less than a hectare of land was only about 5% of AKR's total number of contracted farmers. Three main reasons explain this low representation of very small holdings. First, very poor farmers could not spare their land for the production of commercial rice. Second, even if they wanted to join the scheme, if their agronomic conditions were not suitable, the company did not accept them. Third, in some cases, farmers had a large area of land but the land that could successfully grow *Pkar Malis* rice was less than a hectare, hence the company only accepted the suitable land.

A minimum of one hectare of suitable land was enforced after a few years of operation because the company found it difficult to work with farmers owning less than this. Often the urgent need for money had forced poorer farmers to sell their rice to informal traders at the farm gate. Sometimes farmers consumed all the grain produced. Even with these breaches of contract, the company could not take measures against the farmers because they were too poor to be held responsible for their actions. Therefore, despite a few exceptions, the company decided to exclude farmers with less than a hectare of land. Exceptional farmers were those who were committed to the company, hardworking, and strongly recommended by village heads.

A subsequent significant change in the buying policy of AKR made the area of land an irrelevant condition. Due to some challenges (see below), AKR changed the policy of buying rice from individual farmers to buying collectively from each village. As long as farmers had rice to sell to AKR, they could sell through the village, regardless of farm size. Data from interviews with all the four village heads confirmed this practice. Such collective purchase prevented the company from knowing the extent of participation of very small-scale farmers in their contract farming scheme.

CONTRACTUAL ARRANGEMENTS

In order to gain villagers' trust and as a more efficient way to manage contract farming, AKR established "commune associations". Each association comprised the head and deputy head of the commune and the village heads. The associations had various roles, beginning with helping AKR persuade and select the contract farmers. After one year of attempting to introduce the concept of contract farming directly to farmers, the company realised that it was difficult to gain farmers' trust in this way. This led them to seek the support of the local authority at commune and village levels in explaining the idea to farmers. Commune associations then assisted AKR in evaluating the suitability of farmers in terms of their agronomic conditions and commitment. The company delivered quality seeds and technical advice to contracted farmers through these associations. During the production stage, commune associations were obliged to monitor their contracted members and report to AKR on the production process, progress, and challenges. In exchange for the services of the commune associations, AKR provided incentives to the commune and village heads at the rate of KHR 30 and 40, respectively, for each kilogramme of rice sold by members of their association.

The "resource-providing" type of contract adopted by AKR seems to have worked well in the Cambodian context, where the market for farm inputs remains underdeveloped. Farmers in general often faced problems of limited access to necessary production inputs such as seeds, fertilisers, credit, and extension services. They also had difficulty selling their products in markets at reasonable prices. By signing a contract with AKR, farmers had access to quality seeds, extension services, a secure market, and competitive prices. The company advanced seed to farmers without interest.

However, the contract was not prepared in a participatory manner. The company drafted the contract and asked farmers to sign it. AKR and members of the commune association held a village meeting to explain the concept of contract farming and the terms and conditions of the contract to farmers. Village heads recorded the names of interested farmers and, together with AKR's technical team, examined their agronomic conditions. If the land was suitable, the company invited farmers to its office and explained the contract again to ensure farmers' proper understanding before having them sign the contract.

Although the contract specified a number of necessary clauses, it lacked several important aspects. It mentioned the amount of seed borrowed by farmers but did not indicate whether the company would provide seed to farmers every year. It described the obligations of AKR to provide contract farmers with fees for transporting paddy rice to the company and to pay members of the commune associations for their services. It also included conditions under which the company would buy paddy rice from farmers. The penalty clause specified the consequences for farmers who breached the contract but stated nothing about the consequences for the company if it was to breach the contract. The contract failed to mention the date on which farmers needed to return the seed, the duration of the contract, and how each party could end the contract.

Benefits of Contract Farming

Access to Market

Access to an export market with a competitive price was the first and most important reason why farmers were interested in joining the contract farming scheme and was the major expected benefit for their participation. The price provided by AKR was competitive in two ways. First, it was much higher than the prices for ordinary varieties grown by farmers before AKR came. In 1999, the market for *Pkar Malis* rice had not been developed in Phnom Srouch District. Farmers grew ordinary varieties for household consumption. When in urgent need of money, farmers sold their paddy rice for KHR 200–300 per kg. The AKR was the first to introduce *Pkar Malis* to farmers and the buying price was KHR 500–700 per kg. Second, the price was competitive when compared to the price offered by informal traders for the same type of rice. After the market of *Pkar Malis* rice was established, traders and CEDAC also bought this variety. However, AKR always bought rice from contract members at a higher price than other buyers.

With a well-established market for the *Pkar Malis* variety, contract farmers saw the importance of the price provided by AKR less in terms of its value and more in terms of insurance. AKR's higher price came with many production challenges (see below), which lessened its attraction for some farmers. Others, however, maintained their relationship with AKR or CEDAC, despite the production challenges, in order to keep reaping the benefit of the consistently high market price.

Access to Quality Seeds

The second most important benefit was access to quality seeds. Although CEDAC also bought *Pkar Malis* rice, it did not follow AKR's policy to advance quality seeds to farmers. Farmers increasingly appreciated this benefit. Information from interviews with former AKR contract farmers suggested that, in the early phase of contract farming when rice farming was only for the domestic market and household consumption, farmers cared less about the varietal purity of harvested paddy rice. Their main reason for participating in the contract farming scheme of either AKR or CEDAC was the access to markets with a competitive price provided by the two institutions. With the establishment of an export market, farmers were more concerned about the availability of quality seeds as a primary reason for contracting to supply AKR. These seeds produced high yields with excellent varietal purity, which was one of the conditions of the high-end export markets.

A current AKR contract farmer explained that, although she had already joined the contract farming scheme of CEDAC, she still contracted with AKR to receive new quality seeds because her old seeds were no longer pure after several years of farming. A village head related that, in 2006, only 30 out of 159 households in the village expressed their interest in contract farming with AKR because they had not realised the necessity of quality seed. By 2011, 98 households had registered with AKR to obtain new foundation seed, but the company did not advance seed to them.

Access to Technology

The third benefit of contract farming with AKR was access to quality extension services. In order for farmers to produce grain with the required standards in terms of varietal purity and yield, AKR delivered training and ongoing technical support to its contracted farmers. However, the company did not exclusively provide this benefit because CEDAC and SOFDEC also offered extension services to farmers.

AKR only delivered training on production techniques to its members during the first year of contract farming. The company trained members of commune associations who would further train their contract farmers. The content of the training covered the whole production process. The benefits of the training seem to have continued even after the termination of contract farming. For example, a former contract farmer of AKR appreciated the training since he could apply the production technology when he grew other rice varieties after quitting the AKR scheme.

Access to Credit and Other Benefits

Despite their irregular provision, other secondary benefits received from AKR included access to credit with a low interest rate, fees for the services provided by the commune associations, fees for transporting rice to the AKR office, and the use of trustworthy scales to weigh their crop.

The company originally provided loans to its members without interest, but the policy at the time of the study was to charge 1.4% per month. This compares with an average monthly interest rate charged by microfinance institutions in Cambodia of around 3% for loans in riels (see Chap. 15).[1] In 2011, the company gave loans to about 500 households.

In terms of the fee for the services of the commune association, AKR was not consistent in issuing this payment as stipulated in the contract. Only in the early stages of the operation did it pay the associations, although they were still working for the company by collecting paddy rice from farmers.

AKR also did not consistently pay farmers the transportation fee as stated in the contract, supposedly because of the varying volumes delivered. When farmers had low yields, they were not able to sell the required amount, resulting in the company not being able to fulfil its export orders. This loss was partly made good by withholding transport fees.

Finally, farmers pointed out that AKR used reliable and trustworthy scales when weighing their paddy rice. This was another benefit compared

Table 16.1 Estimated costs and returns for a one-hectare rice farm, by type of farmer

	Former contract farmer	Current contract farmer	Non-contract farmer
Yield (kg/ha)	2500	2000	2000
Price (KHR/kg)	1350	1450	1350
Gross revenue (KHR/ha)	3,375,000	2,900,000	2,700,000
Variable costs (KHR/ha)	1,230,000	986,000	1,000,000
Gross margin (KHR/ha)	2,145,000	1,914,000	1,700,000

Source: Interviews with key informants

with selling to local traders, who not only offered a lower price but, farmers claimed, always under-weighed their paddy using doctored scales.

Increased Profit

The above-mentioned benefits enabled contract farmers to increase their profit from rice farming. Based on information provided in the interviews, with a yield of 2 t/ha, contract farmers could generate a gross revenue of around KHR 2.9 million per ha, which was lower than the former contract farmers' KHR 3.4 million per ha but moderately higher than non-contract farmers' KHR 2.7 million per ha (Table 16.1). The same ranking was observed in gross margin per ha, with former contract farmers netting KHR 2.1 million, contract farmers KHR 1.9 million, and non-contract farmers KHR 1.7 million. This finding confirms the estimates given by Cai et al. (2008). The implication is that entering into contract farming increases the profitability of rice farming, but that farmers who "move on" from contract farming achieve even higher returns.

CHALLENGES OF CONTRACT FARMING

Contract farming can be regarded as successful when the agribusiness firm and the contracted farmers are both satisfied with the benefits they receive and thus maintain their business relationship. The lower revenue and gross margin of contract farmers compared to former contract farmers suggest that there were problems with the contract farming scheme in this case that made it less profitable, resulting in farmers withdrawing. The interviews provided insights into the challenges faced, how these were addressed, and the support still needed.

High Requirement for Varietal Purity

One of the great challenges was the requirement for high varietal purity of the paddy rice produced. Farmer interviewees expressed different attitudes towards the difficulties inherent in the purification process. Some former AKR contract farmers raised the purification problem as one of the main reasons they left the scheme, despite the high price. However, some current farmers did not see the requirement as too difficult to meet, just needing some extra effort on their part. The policy of AKR specified that paddy rice had imperfect varietal purity if there were three or more grains of the wrong variety in every 100 sample grains. Different levels of varietal purity were reflected in the different prices that farmers received. Thus, contract farmers ran the risk of receiving a lower price if they had not made enough effort in purification. Some contract farmers avoided the challenge by leaving the AKR contract farming scheme. Although informal traders offered a lower price, they attached no conditions to their purchase.

AKR started contract farming to fulfil export requirements in terms of quality and quantity. Varietal purity was one of the quality criteria, especially for the high-end market. The company did not consider the requirement too high for contract farmers. Instead, they attributed the inability to fulfil this condition to farmers' low commitment. Such attribution resulted in selective discontinuity in the business relationship between AKR and contract farmers. To maintain high varietal purity, AKR changed the improved foundation seed for their contract members every two to three years. The company based the decision to distribute new foundation seed on the farmer's past purity levels.

Strict Requirement of Moisture Level

AKR contract farmers faced a problem with drying their paddy. One of the contract conditions was that paddy rice had to have a moisture level less than 16%. The company trained its contracted farmers on how to measure the moisture level, but it was not easy for farmers to dry their paddy rice to the required level due to unfavourable weather conditions and their reliance on sun-drying. Former and current contract farmers explained that, to get down to 16% moisture, they needed to dry their paddy for about two consecutive days under the hot sun, but sun-drying was unreliable. If the dried grains were exposed to rain, they were likely to germinate, yellow, or rot.

AKR accepted paddy rice with a slightly higher moisture level than required but reduced its price accordingly. Farmers did their own mental calculation and reported that sometimes it was not profitable to sell to AKR. They felt that, no matter how hard they tried, the company could reduce the price due to imperfect varietal purity and/or excess moisture. Former contract farmers preferred selling their paddy rice to informal traders, who put no conditions on their purchase. Instead of purifying and drying paddy, former contract farmers chose to spend their time on other income-generating activities.

The strict requirement for moisture level posed a difficulty not only for the farmers but also for the company. The AKR staff observed that the company was successful in contract farming in terms of price but not in terms of flexibility when compared to informal traders. The company was able to pay a higher price, but contract farmers needed to produce very pure and dry paddy rice. Since the company operated on a very large scale, it was unable to buy wet paddy rice from farmers and sun-dry it in its facilities. Unlike the company, informal traders could buy wet paddy and dry it themselves in local drying yards. Given their credit constraints, farmers were inclined to sell to informal traders right after harvesting without any drying.

Farmers' inability to fulfil the moisture requirement had resulted in AKR not being able to satisfy export demand. To meet its export orders, the company had resorted to buying paddy from other sources. All sellers had to fulfil the requirements of purity and moisture level, though to different degrees and for different prices. Current contract farmers received the highest price because the quality of their paddy was also the highest (Table 16.1). At the time of the study, the company bought about 60% of its exported volume from traders because it could not get enough paddy from its farmers. Due to insufficient capital, AKR was not able to get to the root of the problem of unfilled export orders. The company realised that, if it could buy wet paddy from farmers and dry it, it would be able to collect larger quantities of paddy. However, acquiring high-capacity drying machines was beyond the company's financial capacity.

Limited Access to High-Quality Seed

Another challenge for contract farmers and a factor affecting low varietal purity was limited access to high-quality seed. New quality seed could produce higher and purer yields, with the capacity to retain seed for the next two to three crop seasons. The continued use of retained seed beyond this period would result in lower yields and mixed varieties. AKR was the first and only agribusiness firm to advance quality seed to their contract farmers without interest. Current contract farmers in one of the selected villages expressed concern about the purity of their retained seed stocks. AKR had provided quality seed of *Angkong Seouy* to them in 2010 and they had already used their retained seed stocks in 2011 and 2012. As the company had not provided new seed for them in 2013, they continued to use their retained seed for another year, despite running the risk of lower yield and producing mixed varieties.

Farmers adopted diverse strategies to deal with the shortage of high-quality seed. Some non-contract farmers turned to AKR for new seed, but they were disappointed because the company advanced seed neither to them nor to the current contract farmers in 2013. Some former AKR contract farmers and current and former contract farmers with CEDAC were willing to join SOFDEC to obtain new seed of the *Pkar Roumdoul* variety. However, the seed provision scheme of SOFDEC had only just started in the year of the study, and the variety provided was not one that was purchased by AKR or CEDAC. Moreover, the scheme was not large enough to cover all farmers, resulting in a considerable number missing out, and in any case, SOFDEC only advanced seed to farmers but did not contract to buy the harvested rice.

The staff of AKR raised farmers' low commitment to the company as a reason why the company could not continue to provide quality seed to all participants. The company advanced seed to all its contract farmers in 2000 and 2001, but it was no longer the practice. The staff observed that, due to both drought and lack of commitment, contract farmers could not produce high yields of sufficient purity, causing the company a great loss. Learning from this experience, the company advanced seed only to a few communes whose farmers were committed to the company.

Breaches of Contract

Ordinary contract farmers and contract farmers who were also members of commune associations suffered from AKR's irregular payment for transportation of paddy to the company and for the services of the associations. As mentioned above, the contract specified that AKR would pay these fees. In reality, this was not consistently implemented, resulting in participating farmers losing some of their expected revenue from selling rice to

AKR. Also, despite its promise to pay each commune and village head KHR 30 and KHR 40, respectively, for every kilogramme of paddy sold by their members, the company only occasionally adhered to this commitment. Even though the contract was legally binding, farmers had no ability to hold the company accountable.

A long-standing problem for AKR was the extra-contractual marketing undertaken by contracted farmers. AKR staff explained that the company knew if farmers broke their contracts in this way but was not able to take any measures against them in the way commercial banks or microfinance institutions could. Interviews with contract farmers revealed that none had been fined for extra-contractual marketing. The only solution the company saw was to explain to farmers the costs and benefits of selling paddy to the company and to traders.

Another example of contract breach by farmers was the misuse of their membership cards. There were reports that some farmers had rented their membership card to traders or other non-contract farmers, enabling them to sell paddy to the company at the highest price. As noted above, the company did buy from other sources but reserved the best price for holders of current cards. A former AKR contract farmer complained that he and other farmers still wanted to continue with contract farming since it improved his livelihood, but the company had already withdrawn from his village. He suspected this was due to some farmers in the village engaging in this practice of renting out their cards.

Credit Constraints

Credit constraints represented a serious challenge for both contract farmers and AKR. Farmers with credit constraints were under pressure to sell their paddy quickly to informal traders or were not willing to sell paddy to AKR on credit (i.e., with delayed payment). This contributed to farmers' extra-contractual marketing. Informal traders made selling to them convenient for farmers by not placing any conditions in terms of moisture content or varietal purity and by paying farmers immediately. On the other hand, AKR used to buy from their contract farmers on credit, resulting in a large number of farmers quitting the scheme. The company was able to improve its financial position in 2010 and paid cash on delivery to its current members, but it was unknown whether the company could sustain this practice.

In addition to its past inability to pay contract farmers immediately, credit constraints prevented AKR from investing in large-scale paddy driers. Acquiring high-capacity driers would have significantly improved the company's ability to purchase wet paddy rice from farmers, increasing their export volume and making life easier for their contract farmers.

Rainfall Variability

Variability in rainfall during the growing season had a direct negative impact on contract farmers and an indirect negative impact on AKR. When drought affected their crop, farmers could not produce a high yield, resulting in their inability to supply the amount of rice stipulated in their contract. For example, in one of the study villages, contract farmers were not able to sell any surplus rice to AKR in 2011 and 2012 due to drought.

The AKR staff reported frequent losses due to drought. In the early years of their operation, the company experienced dramatic losses since farmers could not return the advanced seed, which was very expensive. The company terminated contracts with several communes because of drought. Despite otherwise favourable agronomic conditions, the company still withdrew because the communes were drought-prone.

POLICY OPTIONS FOR CONTRACT FARMING

Raising Awareness

Raising farmers' awareness of the costs and benefits of contract farming could help increase their commitment to the company. As the study revealed, one of the conditions that AKR considered when terminating contracts with any village was the overall level of commitment of the farmers in that village. On the other hand, despite their limited landholdings, the poorest farmers could participate in the contract farming scheme as long as they were highly committed to the policies of the company. The Ministry of Agriculture, Forestry and Fisheries (MAFF) could provide education on weighing up the costs and benefits of participating through its extension service, or make use of existing commune associations created by the AKR to conduct the training. Such intervention would need to take the stance of an independent adviser, however, to avoid seeming to persuade or coerce farmers to enter into contracts reluctantly.

Rice-Drying Technology

Removing technical constraints for paddy drying would create more benefits for both parties. One possible measure is to improve farmers' knowledge of new drying technology. The MAFF could collaborate with the International Rice Research Institute (IRRI) which has been working in Cambodia on adaptive technology to deal with post-harvest losses, including paddy drying. Farmer representatives could also attend a training course on agricultural mechanisation at the Don Bosco Technical School, which has received technical assistance from IRRI. However, in general, small-scale village-based driers have not been successful.

Another possible measure is to encourage the private sector, for example rice millers, to invest in drying technology. According to the 2013 report of an ADB-IRRI training course, only a large-scale rice miller and a farmer cooperative had so far provided drying services to farmers in Cambodia.[2] This practice needs to be expanded to reach farmers across the country.

Yet another measure would be to provide agricultural credit to AKR directly to invest in high-capacity drying machines. AKR would be able to buy a larger volume of wet paddy rice from farmers if the company had such drying capacity. Hence, there may well be a business case for financing this investment.

High-Quality Seed

Improving farmers' access to high-quality seed could be achieved by accelerating the implementation of the current rice policy. High-quality seed determines the production volume as well as the quality, including the level of varietal purity. The government has already included this issue as a "quick-win" measure in its policy paper, *The Promotion of Paddy Production and Rice Export* (RGC 2010). The implementation of this seed policy was observed during fieldwork. However, the varieties distributed by the local authority, for example *Sen Pidor* and *Chulsar*, were for household consumption rather than for commercial purposes. As a policy measure, the government could coordinate with rice exporters on the varieties to be exported and distribute seed accordingly.

Access to Credit

Due to credit constraints, farmers could not store their paddy long enough to sell to AKR or CEDAC for high prices, or survive the subsequent waiting period until receiving payment. The urgent need for cash pushed farmers to engage in extra-contractual marketing, undermining the viability of the contract farming scheme. Improving farmers' access to credit should be able to reduce this extra-contractual marketing. On the other hand, increasing access to credit for agribusiness firms could help overcome their current capital constraint to paying farmers on time (for paddy delivered as well as service fees) and investing in drying equipment. The above-mentioned policy paper specifies measures to alleviate the credit constraints facing farmers and agribusiness firms, mainly through the expansion of microfinance institutions (see Chap. 15). However, there may be a need for an Agricultural Bank as in Thailand to increase the flow of credit for profitable investments for farmers and the agribusiness sector.

Contract Enforcement

The study found that contract farmers and AKR both experienced breaches of contract but were unable to take any legal measures. There was no institution to oversee compliance with the contract on the part of both parties. MAFF could consider implementing Article 7 of Chapter 2 of the Sub-Decree on Contract Farming regarding the establishment of a coordination committee. As stipulated in the Sub-Decree, the Coordination Committee for Agricultural Production Contracts (CCAPC) "shall intervene or reconcile arguments or conflicts that might occur from the implementation of the contract farming". While the Sub-Decree indicates that the CCAPC would function at the national level, the government should consider establishing provincial-level committees for easier access by farmers.

CONCLUSION

The study found that the rice contract farming scheme of Angkor Kasekam Roongroeung Co. Ltd. (AKR) was inclusive of poor farmers with small farms, even those with less than a hectare. With access to several important benefits of the scheme, contract farmers were able to increase their returns from rice farming. However, some flaws in the contractual arrangements and the requirement to deliver high-quality rice for the export market posed a number of challenges to both AKR and the participating farmers, some of which could be addressed through policy interventions. Overcoming these challenges will enhance the benefits of contract farming for both farmers and agribusiness firms and thus contribute to further commercialisation of the rice sector and rural poverty reduction.

NOTES

1. MFTRANSPARENCY Case Study on Lending Interest Rate in Cambodia. Available at http://www.mftransparency.org/wp-content/uploads/2012/05/MFT-BRF-302-EN-Outlawing-Flat-Interest-in-Cambodia-2011-10.pdf (accessed 5 July 2013).
2. Cambodia: Postharvest project assesses outcomes. Available at http://irri-news.blogspot.com/2013/06/cambodia-postharvest-project-assesses.html (accessed on 17 July 2013).

REFERENCES

AKM, 2015. Angkor Rice website. Available at http://angkorrice.com/farmer-member/ (viewed 23 June 2017).

Birthal, P. S., Joshi, P. K., and Gulati, A., 2005. *Vertical Coordination in High-Value Food Commodities: Implications for Smallholders.* MTID Discussion Paper No. 85. Washington DC: International Food Policy Research Institute (IFPRI).

Cai, J., Ung, L., Setboonsarng, S., and Leung, P., 2008. *Rice Contract Farming in Cambodia: Empowering Farmers to Move beyond the Contract toward Independence.* ADBI Discussion Paper 109. Tokyo: Asian Development Bank Institute (ADBI).

Key, N., and Runsten, D., 1999. Contract farming, smallholders, and rural development in Latin America: the organization of agroprocessing firms and the scale of outgrower production. *World Development* 27(2): 381–401.

Royal Government of Cambodia (RGC), 2010. *Policy Paper on the Promotion of Paddy Production and Rice Export.* Phnom Penh: Council of Ministers.

Setboonsarng, S., Leung, P., and Cai, J., 2005. *Contract Farming and Poverty Reduction: A Case of Organic Rice Contract Farming in Thailand.* Tokyo: Asian Development Bank Institute (ADBI).

Sriboonchitta, S., and Wiboonpoongse, A., 2008. *Overview of Contract Farming in Thailand: Lessons Learned.* ADB Institute Discussion Paper No. 112. Tokyo: Asian Development Bank Institute (ADBI).

PART V

The Overflowing Rice Bowl

Trends in Rice-Based Farming Systems in the Mekong Delta

Nguyen Van Kien, Nguyen Hoang Han, and Rob Cramb

INTRODUCTION

In this and the next three chapters, the focus shifts to the Mekong Delta in Vietnam, which accounts for 15% of the Lower Mekong Basin by area and 27% of the total paddy area but in 2015 produced 48% of the Basin's total rice output and about 60% of its rice exports (Fig. 17.1). Thus, it is indisputably the "overflowing rice basket" of the region. It is also overflowing in the sense that much of the Delta is naturally flooded in the wet season and has been subjected to major hydraulic works to permit inten-

N. Van Kien (✉)
An Giang University, Long Xuyen, Vietnam

Fenner School of Environment and Society, Australian National University, Canberra, ACT, Australia
e-mail: nvkien@agu.edu.vn

N. Hoang Han
Charles Sturt University, Wagga Wagga, NSW, Australia

R. Cramb
School of Agriculture and Food Sciences, University of Queensland, St Lucia, QLD, Australia
e-mail: r.cramb@uq.edu.au

© The Author(s) 2020
R. Cramb (ed.), *White Gold: The Commercialisation of Rice Farming in the Lower Mekong Basin*,
https://doi.org/10.1007/978-981-15-0998-8_17

347

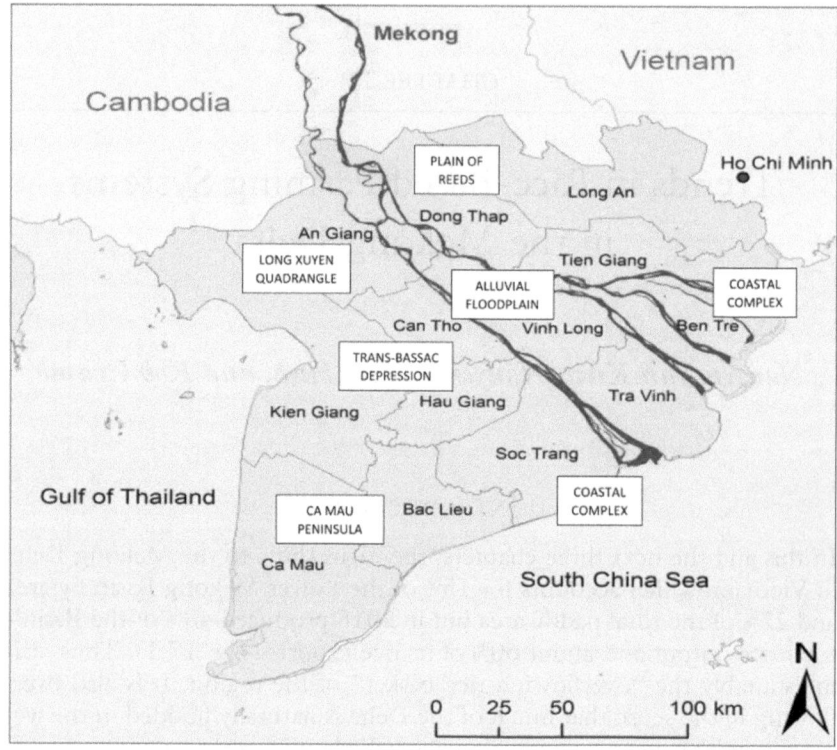

Fig. 17.1 Mekong Delta showing provinces and agro-ecological zones. (Source: Base map by CartoGIS Services, College of Asia and the Pacific, Australian National University)

sive rice farming throughout the year. It is also under threat of sea-level rise. This chapter reviews the trends in rice-based farming systems in the Delta as a whole while subsequent chapters report on field studies in An Giang and Hau Giang Provinces in the upper and middle Delta, respectively. These studies examined the domestic rice value chain from input suppliers to consumers (Chap. 18) and the cross-border trade from Cambodia (Chap. 19). Chapter 20 examines the more specialised cross-border trade in sticky rice between Savannakhet Province in Central Laos and Quang Tri Province in the North Central Coast region of Vietnam (thus complementing the analysis of rice marketing in Savannakhet in Chap. 9).

GEOGRAPHY OF THE MEKONG DELTA

As a geographical unit, the Mekong Delta comprises a triangle of almost 50,000 km² of mostly fertile alluvial and marine deposits extending from Phnom Penh in south-eastern Cambodia through southern Vietnam to the South China Sea and the Gulf of Thailand. About 39,000 km² or 78% of the total area lies within Vietnam. We use the term "Mekong Delta" in this chapter to refer to the Vietnam portion of the Delta. In this portion, canal construction, first for transport, then for irrigation and drainage, has been undertaken over the past two centuries, accelerating under French rule in 1910–1930 and again since the end of the Indochina War in 1975 (White 2002). The Delta now has over 10,000 km of canals and 20,000 km of dykes, profoundly altering the hydrology and agroecosystems of the region. About 90% of the cropland is now irrigated.

The climate of the Mekong Delta is similar to that of the Lower Mekong as a whole—a tropical monsoonal climate with distinct wet and dry seasons (Fig. 17.2). The wet season occurs from June to October, when monthly rainfall averages over 200 mm, and the dry season from December to April. Variation in temperature combined with the seasonality of rainfall gives rise to three rice-cropping seasons: (1) a cooler wet season from July to October (the main or "autumn crop"); (2) a cooler, late wet/early dry

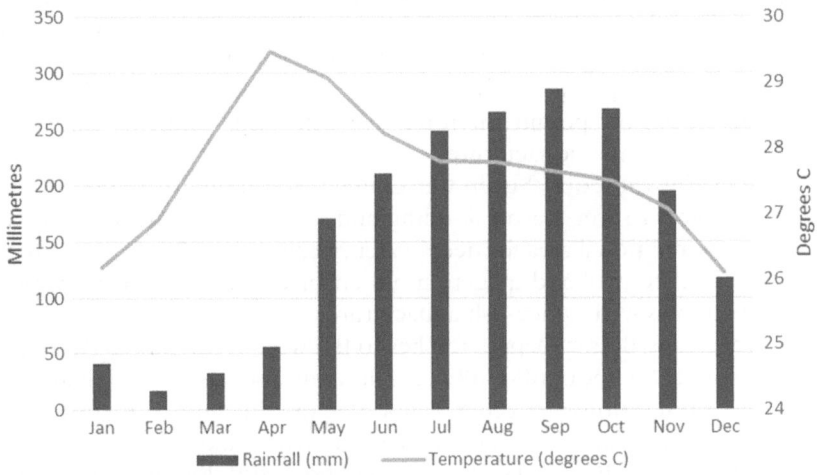

Fig. 17.2 Mean monthly rainfall and temperature at Can Tho. (Source: Climatic Research Unit, University of East Anglia)

season from November to February (the "winter crop"); and (3) a hot, early wet season from March to June (the "spring crop").

Wet-season flooding has been the dominant constraint in the upper Delta, with flood depths of more than four metres, while saline intrusion in the dry season is the major factor affecting land use in the lower Delta, limiting rice production to one crop per year (White 2002). Fertile alluvial soils make up about 30% of the Delta, mainly along the banks of the Mekong and Bassac Rivers, but acid sulphate soils occur in broad depressions over 40% of the area—about half the Delta is subject to saline intrusion. The Delta comprises six distinct agro-ecological zones with different potentials for rice-based farming systems (Nguyen, D. C. et al. 2007; Biggs 2015; Biggs et al. 2009; Fig. 17.1):

- The Alluvial Floodplain is the freshwater zone along the Mekong and Bassac (Hau Giang) Rivers, accounting for about 900,000 ha. The rivers and canals are tide-affected in the middle reaches of the floodplain, enabling farmers to irrigate and drain their land with the tides. This zone has fertile alluvial soils, and farmers practise intensive rice farming with two or three crops per year, while many have diversified into orchards, vegetables, and rice-fish aquaculture.
- Away from the main rivers, there are four large depressions. The Plain of Reeds in the north, accounting for about 500,000 ha, is the lowest part of the Delta at 0.5 metres below mean sea level. This zone floods in the wet season and features acid sulphate soils. Farmers traditionally planted deep-water rice in this zone but flood-control measures now permit intensive rice cultivation and rice combined with freshwater aquaculture.
- Similarly, the Long Xuyen Quadrangle, accounting for 400,000 ha, is subject to wet-season flooding and has acid sulphate soils. It was also a traditional area for deep-water rice, but since 2000, investment in flood control and irrigation has enabled more intensive rice production as well as rice-fish aquaculture.
- The Trans-Bassac Depression lies to the west of the Bassac River and accounts for about 600,000 ha. This zone also has acid sulphate soils but is not seriously affected by flooding or saline intrusion, providing good conditions for intensive rice production and other field crops.
- The Ca Mau Peninsula encompasses about 800,000 ha at the southernmost part of the Delta where the Delta is actively growing due to sediment deposition. This zone is subject to dry-season saltwater

intrusion, limiting rice production to a single wet-season crop. Large parts of this zone have been developed for shrimp farming.

- The Coastal Complex includes about 600,000 ha of coastal flats and sand ridges, much of which is subject to saline intrusion, though coastal dykes have altered the hydrology. Along with the Ca Mau Peninsula, this has been the major zone for the expansion of brackish-water shrimp farming.

About 17.5 million people live in the Delta, including Kinh (90%), Khmer (6%), Hoa (2%), and Cham (2%) ethnic groups, accounting for one-fifth of Vietnam's population. However, the population growth rate is only 0.3–0.5% due to out-migration (CGIAR 2016). Rice forms the basis of livelihoods for the millions of smallholders in the Delta, both as their staple food and as a major source of income. In 2016, Vietnam as a whole had 3.8 million ha of paddy land, producing over 40 million tons of unhusked rice, half of which came from the Delta. In the same year, Vietnam exported 4.5 million tons of milled rice worth USD 2 billion, 90% of which was produced in the Delta (Demont and Rutsaert 2017; Thang 2017). The planted area and yield of rice have increased over the past 20–30 years as irrigation and flood control have increased and as farmers have adopted high-yielding varieties (HYVs), increased fertiliser use, and small-scale mechanisation. In many parts of the Delta, farmers now cultivate three crops of rice per year.

However, the rice sector faces problems of low farm incomes and increased environmental hazards. Diversification of the farming system is now seen by both farmers and the government as a way to address these challenges. In 2000, the government issued the first of a series of policies to encourage farmers to diversify their production. Farmers responded by planting more non-rice crops on paddy land in the dry season, such as maize, vegetables, and watermelons, as well as combining rice with aquaculture. Fruit trees were also extensively planted on flood-protected upland areas (Nguyen, D. C. et al. 2007). In this chapter, we review the intensification and growth of rice production and assess the development of these more diversified rice-based farming systems.

PROFILE OF RURAL HOUSEHOLDS

A household survey conducted in the mid-2000s gives a snapshot of rural livelihoods in the different agro-ecological zones of the Delta (Nguyen, D. C. et al. 2007). Some key characteristics of the average household in

Table 17.1 Characteristics of households in the Mekong Delta by agro-ecological zone, 2005

Variable	Agro-ecological zone					
	Alluvial Floodplain	Plain of Reeds	Long Xuyen Q.	Trans-Bassac D.	Coastal Complex	Ca Mau Peninsula
Mean area owned (ha)	1.36	1.83	1.40	1.75	2.51	2.08
Mean no. in household	4.93	4.24	5.76	5.47	5.01	5.00
Mean no. aged 16–55	3.11	2.33	3.79	3.95	3.43	3.41
Mean no. in off-farm work	0.64	0.05	0.42	0.28	0.29	0.34
Mean no. in non-farm work	1.88	1.05	2.71	2.50	2.05	2.12
Edcn. of head (% scndry.)	34	37	21	58	43	64
Gender of head (%F)	16.2	2.4	5.3	0.0	6.3	4.9
% owning pump	74.7	76.2	78.9	50.0	91.4	27.4
% owning boat	38.0	76.2	78.9	50.0	91.4	27.4
% owning sprayer	73.6	57.2	68.4	63.3	2.9	0.0
% receiving loan	44.2	42.9	71.1	71.7	85.5	90.2

Source: Nguyen, D. C. et al. (2007)

each zone are presented in Table 17.1. The average farm size was uniformly small, but lowest in the more productive Alluvial Floodplain (1.4 ha) and largest in the Coastal Complex (2.5 ha). Household size averaged around five members throughout the Delta, with an average of two to four members of working age. The proportion of working-age members engaged in off-farm labouring was highest in the Alluvial Floodplain (20%) but mostly below 10% in other regions. However, the proportion engaged in non-farm work was high across the Delta, ranging from 45% in the Plain of Reeds to over 70% in the Long Xuyen Quadrangle. The education level of the household head was similar across zones, with one to two-thirds having completed lower secondary school. Almost all household heads were men, but in the Alluvial Floodplain 16% of households were headed by a woman. The ownership of key equipment varied across the zones depending on variations in livelihoods. Pumps were an essential item for most households (75–90%) except in the Ca Mau Peninsula where there was less reliance on irrigation. Similarly, ownership of a boat was essential in the flood zones of the deep depressions and the Coastal Complex. Sprayers were owned by 60–70% of households in the major rice-growing areas but were less common in the coastal shrimp zones. The use of credit varied from around 40% in the Alluvial Floodplain to 90% in the Ca Mau Peninsula.

The different patterns of household land use are indicated in Fig. 17.3. Paddy land dominated in the Alluvial Floodplain and surrounding depressions, averaging between 1.0 and 1.5 ha. Orchards of 0.2–0.3 ha were also a significant feature of land use in these zones. However, shrimp and fish farms were the dominant land use in the Coastal Complex and the Ca Mau Peninsula, averaging 2.0–2.4 ha.

The land-use patterns were partly reflected in the sources and levels of income (Table 17.2). Farmers in the Alluvial Floodplain and the adjacent depressions (Plain of Reeds and Long Xuyen Quadrangle) obtained 80–90% of their farm income from rice and field crops, earning USD 1300–1500 in 2005. In contrast, farmers in the Coastal Complex earned 85–90% of farm income from shrimp and fish, with coastal farmers averaging USD 1700 from this source. Non-farm income was more significant for households in the Alluvial Floodplain, indicating greater livelihood diversification in this more accessible and densely populated zone. Overall, these farmers and those in the Coastal Complex obtained the highest incomes per household (USD 2100–2300) and per capita (USD 420–470). Those in the Ca Mau Peninsula were the poorest, with about half the mean income of the more prosperous zones.

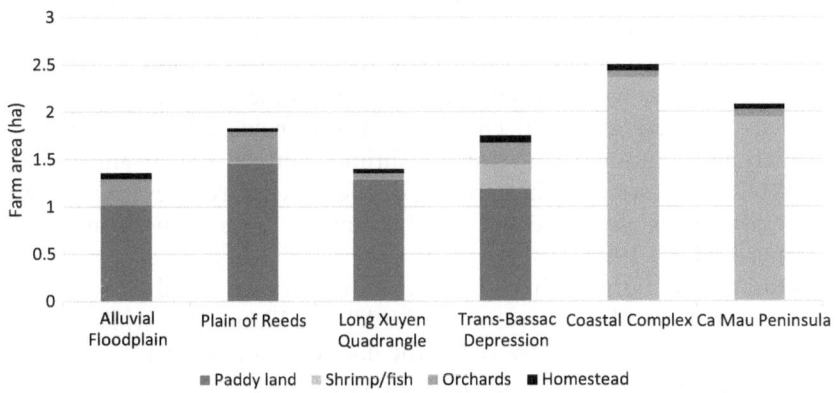

Fig. 17.3 Average farm area in Mekong Delta by land use and agro-ecological zone, 2005. (Source: Derived from data in Nguyen, D. C. et al. (2007))

Table 17.2   Mean household income by source and agro-ecological zone, 2005 (USD/year)

Source of income	Agro-ecological zone					
	Alluvial Floodplain	Plain of Reeds	Long Xuyen Q.	Trans-Bassac D.	Coastal Complex	Ca Mau Peninsula
Agriculture	1652	1592	1482	1052	1932	937
Rice-based farming	1321	1470	1318	731	0	0
Fish/shrimp farming	0	0	0	0	1598	673
Capture fisheries	22	16	44	90	69	163
Livestock	281	59	54	209	50	13
Off-farm work	28	47	67	22	216	89
Non-farm income	548	0	198	241	144	234
Remittances	96	0	56	258	48	0
All sources	2296	1592	1736	1551	2124	1171
Income per capita	466	375	301	283	424	234

Source: Nguyen, D. C. et al. (2007)

THE INTENSIFICATION OF RICE FARMING

Khmer farmers cultivated rice in the Delta for perhaps 2000 years (Chap. 1). From the eighteenth century, Vietnamese farmers occupied the Delta under the expanding Nguyen dynasty and began to extend the paddy area (Xuan and Matsui 1998; Le Coq et al. 2001). Traditionally, farmers settled along the river levees and along sand ridges in the coastal zone. The typical farm comprised a homestead (usually with livestock), ponds (used for aquaculture or as wild fish refuges and for domestic water use), dykes and gardens for trees and cash crops, and paddy fields for rice cultivation (sometimes combined with fish or shrimp culture) (Xuan and Matsui 1998). In some systems, farmers dug ditches, used as refuges for wild fish, and raised beds alongside for growing annual crops such as sugarcane (Nguyen, H. C. 1994).

During the eighteenth and nineteenth centuries, local rice varieties were cultivated in the wet season on the Alluvial Floodplain, while floating or deep-water rice was cultivated in the flooded zones such as the Plain of Reeds and the Long Xuyen Quadrangle (Le Coq et al. 2001; Biggs et al. 2009).[1] After harvesting floating rice in December, farmers planted field crops such as mung bean, sweet potato, and maize, harvested in February or March (Xuan and Matsui 1998; Nguyen, H. C. 1994). Traditional, photosensitive varieties were cultivated in most areas of the Delta (Nguyen, H. C. 1994; Xuan and Matsui 1998). Early maturing varieties were cultivated in the Coastal Complex to permit harvesting before November when saline water intruded into the paddy fields. Medium-maturing varieties were grown in the tide-affected Alluvial Floodplain where water levels were difficult to control. Late-maturing varieties were cultivated in low-lying areas at risk of flooding. Floating or deep-water rice was mostly cultivated in the depressed zones where flooding in the wet season was inevitable—the Plain of Reeds and the Long Xuyen Quadrangle (Nguyen, H. C., 1994; Nguyen and Howie 2018).

The development of the Delta for rice-based farming systems can be divided into three stages: (1) adapting to existing conditions, (2) semi-control, and (3) total control (Le Coq et al. 2001; Kakonen 2008; Biggs et al. 2009; Vormoor 2010). For the first 200 years of settlement by Kinh farmers, there was little or no infrastructure for rice farming, other than the canals that farmers progressively constructed. Farmers cultivated only one rice crop per year where conditions were suitable, whether traditional wet-season rice or floating rice. They also harvested wild fish in the paddy fields during the wet season and planted dry-season vegetable crops (Vo 1975; Nguyen, H. C. 1993).

In the second stage, from the mid-1970s, irrigation infrastructure was developed, permitting the intensification of rice production using modern varieties (Le Coq et al. 2001; Kakonen 2008; Biggs et al. 2009). A single crop of local rice continued to be cultivated under rainfed conditions on higher land, while single- or double-cropping of modern varieties was introduced across the Delta where low dykes and irrigation infrastructure had been constructed. Many farmers accessed irrigation water using locally developed, portable, axial-flow (or "shrimp-tail") pumps (Nguyen, V. K. et al. 2016; Biggs et al. 2009).

In the third stage, following a 1996 government decision, increased investment in raising dykes and extending internal irrigation canals enabled widespread triple-cropping of rice (Yasuyuki 2001; Biggs et al. 2009; Nguyen, V. K. et al. 2016).[2] For example, farmers in Cai Lay District of Tien Giang Province began triple-cropping in the late 1990s (Berg et al. 2017). An Giang and Dong Thap Provinces also started to raise dykes to enable triple-cropping, with the greatest progress after 2000 (Nguyen et al. 2016). The central government offered direct financial support to provinces, districts, and communes to build dykes to increase the extent of triple-cropping even in provinces subject to wet-season flooding. The area of irrigated land increased from 52% of the Delta in 1990 to 91% in 2002 (CGIAR 2016).[3]

With the expansion of irrigation and increased cropping intensity, the area of rice planted in the Delta increased from 3.2 million ha in 1995 to 4.3 million ha in 2016, an annual increase of 1.4% (Fig. 17.4). Most of this increase came from the main wet-season (autumn) crop, the share of which increased from 44% to 55%. Farmers tended to replace the dry-season winter crop with an HYV crop in spring. The winter crop fell to 8% of the total area while the spring crop increased from 32% to 36%, with the main increase occurring in the period 1995–2000. Farmers shifted from winter rice to other dry-season crops such as maize, sweet potatoes, cassava, and vegetables, or left the land idle in this season. In coastal areas, farmers shifted to brackish-water shrimp farming in the dry season.

The total production of paddy rice has doubled from 12.8 million tons in 1995 to 24.2 million tons in 2016, an annual increase of 3.2% (Fig. 17.5). The increase in production was about equally due to the increase in planted area and an increase in yields. The yield of the main autumn crop increased from 3.8 t/ha to 5.3 t/ha, that of the spring crop from 5.2 t/ha to close to 7.0 t/ha, and that of the winter crop from 2.9 t/ha to 4.3 t/ha (Fig. 17.5). The yield increase in all seasons was attributable to better water management, use of HYVs, and greater use of fertilisers.

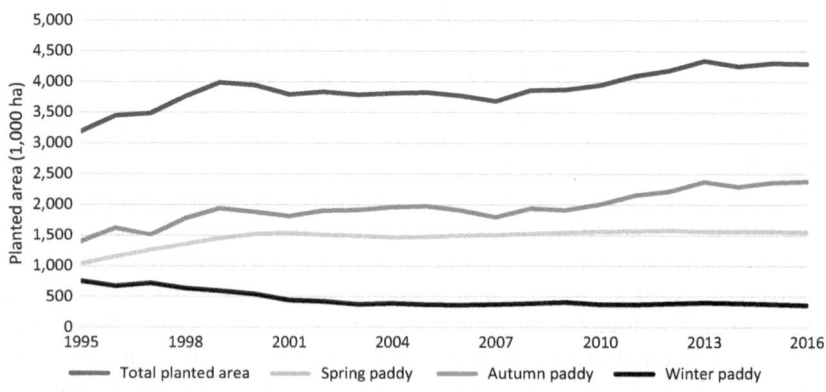

Fig. 17.4 Planted area of rice in the Mekong Delta by season, 1995–2016. (Source: General Statistics Office 2016)

Despite this growth in area, yields, and production in the Delta, the rice sector faces several challenges (Nguyen, D. C. 2011). The focus has been on producing high-yield but low-quality rice, especially for the export market, with consequent low farm-gate prices (Demont and Rutsaert 2017). The use of inputs has increased, resulting in increased yields, but the net returns to rice farmers remain low (Berg et al. 2017). The use of fertiliser increased from 40 kg per ha in 1976–1981 to 120 kg/ha in 1987–1988 (Xuan and Matsui 1998) and reached over 600 kg per ha in 2015 (Nguyen et al. 2018). The application of pesticides increased three to six times from 2000 to 2015. The cost of these inputs has also increased, adding to the cost-price squeeze farmers are facing.

Rice production in the Delta also faces a series of interconnected environmental problems. The intensification of rice cultivation in the last 15 years has led to an increase in soil and water pollution from the overuse of agricultural chemicals (Nguyen, D. C. et al. 2015; Berg et al. 2017), and reduction in wild fish supply (Nguyen et al. 2018). Moreover, the Mekong Delta is highly vulnerable to the effects of climate change, including sea-level rise, increased flooding, and saline intrusion, particularly in the Coastal Complex and the Ca Mau Peninsula (Dasgupta et al. 2007; Phạm

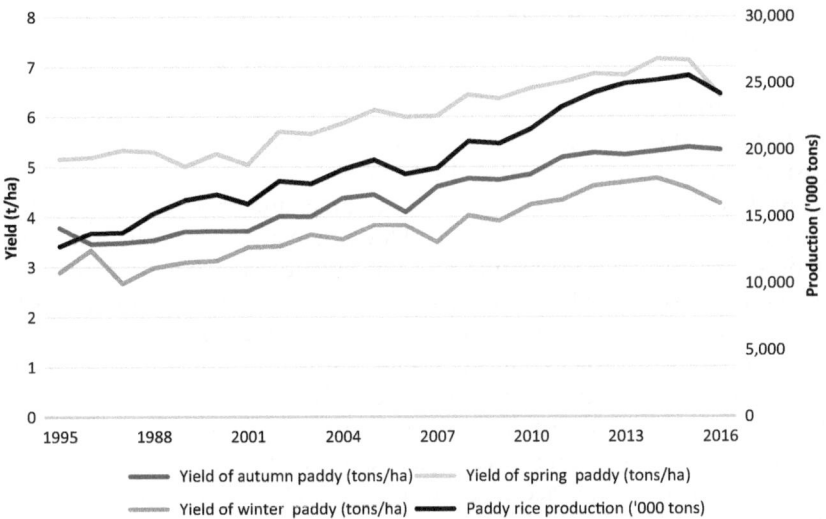

Fig. 17.5 Paddy yields by season and total paddy production in the Mekong Delta, 1995–2016. (Source: General Statistics Office 2016)

and Furukawa 2007; MONRE 2009). Recent evidence shows that saline intrusion is having adverse impacts on both rice and rice-shrimp farming systems (Mainuddin et al. 2011, 2013; Ling et al. 2015; Thuy and Anh 2015; USAID 2016; Leigh et al. 2017; Stewart-Koster et al. 2017).

DIVERSIFICATION OF RICE-BASED FARMING SYSTEMS

Farmers in the Delta have always combined other livelihood activities with rice production, giving rise to a range of rice-based farming systems in the different agro-ecological zones. With the trends in rice production described above and official encouragement to diversify production, a range of farming systems have been developed since the 1990s (Bosma et al. 2005; Tong 2017). The Delta now has seven dominant rice-based farming systems (Xuan and Matsui 1998):

- rice-rice-rice
- rice-rice
- rice-upland crops
- rice-livestock
- rice-wild fish
- rice-freshwater aquaculture
- rice-saline aquaculture

In addition, there has been a widespread conversion of paddy land to orchards in the Alluvial Floodplain. Trends in farming systems over the past 40 years are summarised in Table 17.3.

As described in the previous section, double- and triple-cropping of rice have expanded mostly in the Alluvial Floodplain and in the adjacent broad depressions. Systems alternating rice and upland crops have expanded on the natural levees and back swamps of the upper floodplain and on coastal sand ridges. Rice-livestock farming systems, in which the paddy fields are used for rice while ducks, chickens, pigs, cattle, or buffaloes are raised in the home yard, are found in most zones of the Delta (Xuan and Matsui 1998). The rice-wild fish farming systems are found in the Ca Mau Peninsula and in the Long Xuyen Quadrangle and Plain of Reeds. Farmers cultivate only one local rice crop from May to December and harvest wild fish from paddy fields and canals during the cropping season. The rice-freshwater aquaculture (fish and freshwater prawn) systems have mostly expanded in Tien Giang and Can Tho Provinces in the tide-affected

Table 17.3 Trends in farming systems in major landform units, 1976–2016

Landform unit	Landform sub-unit	Farming system	Trend
Upper floodplain	Natural levees	**House-garden + winter-spring upland crops**	Increased
	Back swamp	**Winter-spring rice + summer-autumn rice**	Increased
		Winter-spring rice + summer-autumn jute	Decreased
		Winter-spring upland crops + summer-autumn rice	Increased
	Open floodplain	Floating rice + upland crops	Decreased
		Winter-spring rice	Decreased
		Pineapple + cashew nut	Decreased
	Closed floodplain	**Summer-autumn rice + winter-spring rice**	Increased
		Floating rice + yam + local rice	Decreased
		Melaleuca + bees + fish	Decreased
Tide-affected floodplain	Natural levees and back swamp	**Home compounds and orchards**	Increased
		Local rice + upland crop	Decreased
		Summer-autumn rice + winter-spring rice	Decreased
		Summer-autumn rice + local rice	Decreased
		Winter-spring rice + spring-summer rice + summer-autumn rice	Increased
	Freshwater broad depression	Local rice	Decreased
		Sweet potato + autumn-winter rice (HYVs)	Decreased
		Summer-autumn rice + winter-spring rice	Increased
Broad depressions	Broad depression	Summer-autumn rice + local rice	Decreased
		Local rice	Decreased
		Autumn-winter rice (HYVs)	Increased
		Melaleuca trees + fish + bees	Decreased
Coastal complex	Coastal flat	Wet-season rice (HYVs)	Decreased
		Local rice mixed with HYVs	Decreased
		Autumn-winter rice (HYVs) + shrimp	Increased
		Summer-autumn rice after ridge flushing	Decreased
		Coconut garden + fish + shrimp	Decreased
		Upland crops + orchards on ridges	Decreased

Alluvial Floodplain (Berg et al. 2017; Håkan 2002; Xuan and Matsui 1998). From 2000, with greater flood control, the rice-freshwater shrimp system was extended into the deep flood areas of An Giang and Dong Thap Provinces (the Long Xuyen Quadrangle and the Plain of Reeds) (Nguyen, V. K. 2014). Rice-brackish water shrimp farming systems have evolved over 80 years and are found in the coastal provinces of the Delta (Xuan and Matsui 1998; USAID 2016). Farmers cultivate one local rice crop in the wet season, when rainfall and freshwater flows enable salinity to be flushed out of the paddy fields, and modify the fields to culture brackish-water shrimp in the dry season. The main rice-based systems are discussed in more detail below.

Rice and Upland Crops

As described above, farmers traditionally planted field crops in upland plots in or near the homestead. In recent years, farmers have begun to allocate paddy land to these crops in some seasons, thereby increasing and diversifying their incomes. This trend has recently received encouragement from the government, which was previously intent on retaining all paddy land in the Delta for rice production. The Ministry of Agriculture and Rural Development (MARD) announced a land-use plan for 2014–2020 that mandates more flexible use of paddy land (Table 17.4). The plan encourages provinces to shift from rice to maize, soybean, sesame, vegetables, flowers, animal feed, and aquaculture. In total, the plan envisages a reduction of 316,000 ha of rice (7% of the 2013 rice area), mainly in the dry season, and an equivalent increase in non-rice crops, over half of which is to be taken up by maize (83,000 ha) and vegetable and flower crops (87,000 ha). More importantly, the government issued a resolution in 2017 to develop strategies for sustainable and climate-resilient development in the Delta.[4] This resolution provided the basis for reducing the extent of triple-cropping of rice and further diversifying cropping systems.

Farmers have developed appropriate rice-based cropping systems depending on local hydrological, soil, and topographical conditions. Two common patterns that have emerged are (1) one crop of rice followed by one crop of maize or sweet potatoes or several crops of vegetables and (2) two crops of rice followed by one crop of maize or sweet potatoes or one crop of vegetables. From 1995 to 2016, the area of maize has increased from just over 20,000 ha to nearly 35,000 ha, and the area of sweet potatoes has doubled from about 10,000 ha to 20,000 ha (General Statistics

Table 17.4 MARD land-use plan for paddy land in Mekong Delta for 2014–2020 period

Crop	Area in 2013 (ha × 10³)	Planned change in land allocation, 2014–2020 (ha × 10³)				Area in 2020 (ha × 10³)
		Spring	Autumn	Winter	Total	
Rice	4338	(160)	(129)	(28)	(316)	4022
Maize	40	29	52	1	83	123
Soybean	2	17	4	0	21	23
Sesame	29	14	11	0	25	54
Vegetables and flowers	254	50	34	3	87	341
Animal feed	7	17	4	3	24	31
Rice and aquaculture	174	5	8	42	54	228
Other	53	13	9	0	22	75

Source: Approved land-use plan for changing cropping systems on rice land in the 2014–2020 period, MARD 31 July 2014

Office 2016). The area of all types of vegetables increased sharply from under 20,000 ha in 2000 to over 45,000 ha in 2011.

Rice and Livestock

Pigs are an integral part of farming systems in the Delta. Small-scale pig raising is very common—about 70% of smallholders own a pigpen, raising several pigs—while some operations raise several thousand head (Huynh et al. 2007). This activity creates employment for household members and provides a major source of income (Huynh et al. 2007). Farmers face market and disease risks, causing the number of pigs to fluctuate as prices vary and outbreaks of diseases such as foot-and-mouth disease or swine influenza occur. Nevertheless, total numbers in the Delta increased from 2.4 million in 1995 to 3.8 million in 2016.

Farmers traditionally raised chickens and ducks inside the homestead for both meat and eggs (Xuan and Matsui 1998). They were fed with rice, food waste, and local aquatic animals such as fish and snails. Each household raised small numbers of chickens and ducks for home consumption, or up to several hundred for sale. Some specialised, large-scale farmers raised up to several thousand head. However, the bird flu epidemic in the mid-2000s[5] had a negative impact, with many small-scale farmers ceasing to raise poultry. The number of poultry in the Delta dropped sharply from 51.5 million in 2003 to 31.4 million in 2005. Nevertheless, medium- and large-scale poultry farming has increased dramatically since 2005, with total numbers reaching 64.7 million in 2016 (General Statistics Office 2016).

Before the 1980s, cattle and buffaloes were used for draught power, both ploughing and transportation (Xuan and Matsui 1998). However, following the *doi moi* reforms, most farmers have replaced buffaloes with two-wheeled tractors imported from Japan and China. Consequently, the number of buffaloes in the Delta has declined dramatically, from 113,000 in 1995 to 40,000 in 2001, with slower decline thereafter to 31,000 in 2016 (General Statistics Office 2016). In contrast, the number of cattle increased sharply, from 150,000 in 1995 to 680,000 in 2006, and continues to hover around 700,000. The primary use of cattle is now for commercial beef production, with high demand in nearby Ho Chi Minh City. Rice-growing households fatten up to ten cattle in sheds in the house compound (Fig. 17.6). Farmers grow forage grasses or use rice straw as forage, and also buy imported soybean cake.

Fig. 17.6   Cattle being fattened for sale in a farmyard shed in My An Commune, Cho Moi District, An Giang Province. (Source: Nguyen Van Kien, September 2017)

Rice and Aquaculture

Four main rice-based farming systems in the Mekong Delta incorporate the capture and/or rearing of aquatic species (Xuan and Matsui 1998):

- rice-wild fish capture
- rice-aquaculture (freshwater fish, e.g., *Pangasius* spp.)
- rice-aquaculture (freshwater shrimp, e.g., *Macrobrachium rosenbergii*)
- rice-aquaculture (brackish-water shrimp, e.g., *Penaeus monodon*)

Wild fish capture has long been practised in the transplanted rice zones of the Delta, producing about 190,000 tons/year. Farmers dug ponds or ditches in the paddy fields to create refuges for fish during the rice-growing season. After the rice harvest, fish moved to the ponds where farmers could harvest them for home consumption. This system could yield 2–3 tons of rice/ha and 150–200 kg of fish/ha (Xuan and Matsui 1998). The average yearly consumption of fresh fish products was estimated at 21 kg per capita in 1995 (Rothuis 1998). Harvesting wild fish for sale provided additional income for most households. This system dominated before the rapid spread of HYV rice in the Delta (Xuan and Matsui 1998), significantly reducing wild fish capture. For example, the total fish catch in the upper delta decreased by one-third from 1995 to 2016.

While traditional wild fish capture has declined, fish aquaculture has markedly increased (Fig. 17.7). This involves rearing fish in pens or floating cages and, increasingly, in ponds along the main rivers and canals using pelleted feed. Local catfish (*Pangasius* spp.) are the main species reared. The catfish industry began in the late 1990s in An Giang and Dong Thap Provinces in the upper Delta, and within a decade involved 800,000 farmers managing 6000 ha of ponds to produce 1.5 million tons, much of it exported to the US and European Union (EU). In the last decade there have been trade disputes with the US and concerns over quality in the EU, causing fluctuations in demand. Nevertheless, the area of freshwater fish aquaculture has continued to increase. For example, in An Giang Province, the area has increased from 1465 ha in 1995 to 1690 ha in 2016 (Fig. 17.8).

Systems combining rice with freshwater shrimp are mainly found in low-lying areas of the Delta. In the 1990s, farmers in Phung Hiep District, Hau Giang Province, in the Trans-Bassac Depression, began double-cropping with short-medium duration HYVs integrated with giant fresh-

Fig. 17.7   Rice-fish-poultry system in My Phu Dong Commune, Thoai Son District, An Giang Province. (Source: Nguyen Van Kien, September 2017)

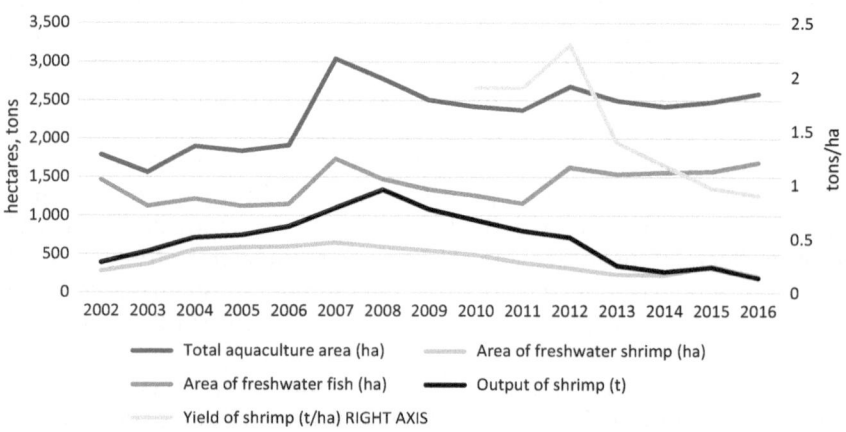

Fig. 17.8   Area, yield, and production of freshwater shrimp and fish in An Giang Province. (Source: Statistical Office of An Giang Province (2016))

water shrimp or fish such as snakehead and climbing perch (Xuan and Matsui 1998). The rice-shrimp farming system was introduced to An Giang Province in the Long Xuyen Quadrangle during the flood season of 2000. Tu Xang in Phu Thuan Commune of Thoai Son District cultured several hectares of shrimp and obtained a high economic return, thanks to good yields and a high price.[6] In 2002, farmers in Chau Phu District followed this practice and cultured shrimp over 282 ha in the flood season, rotated with HYV rice in the dry season (Fig. 17.9). Nguyen (2014)

Fig. 17.9 Freshwater shrimp farming in Vinh Thanh Trung Commune, Chau Phu District, An Giang Province. (Source: Nguyen Van Kien)

Table 17.5   Economic returns to rice-freshwater shrimp farming in An Giang, 2006

Parameter	Units	Rice	Shrimp	Total
Yield	Tons/ha	6.76		
Price on sale	VND/kg × 10^3	2.269		
Total benefit	VND/ha × 10^3	15,338	102,540	117,878
Total cost	VND/ha × 10^3	7437	57,790	65,117
Net benefit	VND/ha × 10^3	7901	44,750	52,761
Net benefit	USD/ha	339	1920	2263

Source: Nguyen, V. K. (2014)

Note: USD 1 = VND 23,000

found that the net return from one rice-shrimp cycle in 2006 was USD 2263, 85% of which was from the shrimp activity. This was much higher than the net return from double- and triple-cropping rice in An Giang (Table 17.5). Consequently, the provincial government formulated a policy to promote "flood-based livelihoods" through rice-shrimp systems (An Giang People's Committee 2006). Farmers in Dong Thap Province in the Plain of Reeds followed those in An Giang, beginning shrimp culture in the flood season of 2004. In 2006 there were 146 ha of ponds, producing 230 tons. This increased to 700 ha producing 1200 tons in 2010.

The area and output of freshwater shrimp increased markedly up to about 2008. However, production has declined significantly in the past decade due to chemical pollution from neighbouring HYV rice paddies, reduction in flood levels (the flood peak in 2015 was the lowest in 100 years), and unstable market prices. The area of shrimp ponds in An Giang peaked at 650 ha in 2007 and fell to 214 ha by 2016, while the average yield fell sharply from over 2 tons/ha in 2012 to be less than 1 ton/ha in 2016 (Fig. 17.8 above). With this decline in area and yield, total production in An Giang fell from 1334 tons in 2008 to just 194 tons in 2016. The same trends have occurred in both An Giang and Dong Thap Provinces.[7]

Rice combined with brackish-water shrimp was observed in five coastal provinces as early as the 1930s (Nguyen, H. C. 1994). In 1984, there was about 5000 ha of rice-shrimp farming (Xuan and Matsui 1998). At this time, the yield of shrimp averaged 640 kg/ha and the yield of rice ranged from 3.5 to 4.0 tons/ha (Xuan and Matsui 1998). In 2000, the total area of rice-shrimp was 71,000 ha, distributed across five coastal provinces: Ben Tre, Soc Trang, Bac Lieu, Ca Mau, and Kien Giang. This had increased to 153,000 ha in 2014 (USAID 2016). These semi-intensive rice-shrimp systems include one crop of wet-season rice and one or two crops of tiger shrimp or white-leg shrimp (USAID 2016). This system is very common in Soc Trang, Ca Mau, Kien Giang, Ben Tre, and Tra Vinh Provinces. Total brackish-water shrimp production was 65,000 tons with yields ranging from 300 to 500 kg/ha (USAID 2016). Preston and Clayton (2003) found that farmers' incomes had improved significantly from adopting this system.

The rice-shrimp systems have several technical problems that threaten their sustainability. Nutrient use in rice-shrimp systems is less efficient than in dedicated shrimp grow-out ponds, causing low shrimp survival rates and low production (Dien et al. 2018). Leigh et al. (2017) found water temperature and salinity were too high in the dry season and dissolved oxygen was too low, causing low survival rate and low shrimp production. The rice crop was also affected adversely by high salinity levels (Leigh et al. 2017). Although rice-shrimp systems are economically and environmentally viable, farmers have tended to switch to intensive shrimp production systems (Preston and Clayton 2003).

Climate change poses new risks to rice-shrimp farming systems. Early saline water intrusion in November negatively affects rice yields while contributing to an accumulation of soil salinity over time (Preston and Clayton 2003; ACIAR 2016). The impacts of climate change have increased in recent years. Saline intrusion due to drought events occurs more frequently. The historical drought event in 2015 caused severe damage to rice, vegetables, flowers, fruit trees, livestock, buffaloes, cattle, and small fishponds. The coastal provinces were most affected, with rice, fruit, and aquaculture taking the brunt of the impact.

CONCLUSION

Rice-based farming systems in the Mekong Delta have been transformed over the last three decades due to farmer initiatives and government policy. Facing an urgent need to boost rice production after 1975, the government increased investment in water control and irrigation and promoted the intensification of rice farming through green revolution technology, leading to widespread adoption of double- and triple-cropping systems. With the expansion of irrigation and increased cropping intensity, the area of rice in the Delta increased from around 2.0 million ha in the immediate post-war decade to 3.2 million ha in 1995 and 4.3 million ha in 2016. The yield of the main wet-season (autumn) crop increased from about 2 t/ha in 1975 to 3.8 t/ha in 1995 and 5.3 t/ha in 2016 due to better water management, use of HYVs, and greater use of fertilisers. Total production of paddy rice was only about 4 million tons in 1975, increasing threefold to 12.8 million tons in 1995 and doubling again to 24.2 million tons in 2016. The increase is attributable in equal measure to the increase in planted area and the increase in yields. From being a net importer of rice in the 1970s and 1980s, Vietnam exported 1.4 million tons in 1989 following the first phase of intensification and market reforms. In 2016, exports totalled 4.5 million tons worth USD 2 billion, 90% of which was produced in the Delta. By any standard, this has been an amazing economic transformation.

However, the focus on rice intensification has shifted since 2000 as the impacts on farmer livelihoods and the environment have become apparent. Locking farmers into producing low-quality rice for the export market has not provided them with adequate returns, especially as both domestic and global demand have shifted in favour of higher-quality rice and more diverse diets. Specialisation in continuous rice production has also restricted the dietary diversity of rural households. Intensive use of fertilisers and pesticides has led to soil and water pollution and reduction in wild food supply. Moreover, the "total management" of hydrology in the Delta has had major impacts on water flows, sedimentation processes, aquatic species, and land-use options. The Delta is also highly vulnerable to the effects of climate change, including sea-level rise, increased drought and flooding, and saline intrusion, the latter affecting up to half the surface area.

In response, the government has progressively relaxed its restrictions on the use of paddy lands—originally conceived to achieve food security and maintain export earnings. Hence, rice-based farming systems have become more diversified in the last two decades, with the increased use of paddy lands for non-rice field crops, orchards, and freshwater and brackish-water aquaculture. Irrigated dry-season horticultural crops and productive and profitable orchards now abound in the Alluvial Floodplain. While the traditional inland fish catch has declined, the production of freshwater fish in ponds, especially local catfish, has developed into a major export industry in the upper Delta. Freshwater shrimp, however, after a rapid increase since 1990, appears to be in decline. In the Coastal Complex and the Ca Mau Peninsula, brackish-water shrimp culture has had a longer and more successful history, though it is facing challenges due to disease outbreaks, market fluctuations, and climate change.

The most recent indication of the shift in policy was the November 2017 resolution in support of sustainable development strategies for the Delta. These strategies include (1) promotion of high-quality rice, (2) reduction in the area planted to rice, (3) further diversification of farming systems, and (4) promotion of agro-ecological and organic agriculture. Targets have been set to increase the quality rather than the volume of rice and to diversify rice-based farming systems to make the best use of each agro-ecological zone in the Delta. Even traditional floating rice is being encouraged in the remaining deep-flooding pockets. Reduction in the area planted with rice is intended to help counter the overuse of chemicals in the paddy field ecosystem while opening further opportunities for the diverse range of crop, livestock, and aquatic products that are increasingly in demand in Vietnam's cities. Promotion of integrated (or agro-ecological) rice-based farming systems is intended to provide the basis for more profitable and sustainable rural livelihoods in the Delta, with greater adaptability to changing markets and climate.

NOTES

1. Floating rice can elongate at rates of 20–25 cm/day as the floodwaters rise (Cummings 1978).
2. Decision No. 99/TTg, 9 February 1996.
3. By 2015, the area planted to floating rice had fallen to only 200 ha (Nguyen, V. K. and Pittock 2016). However, the floating rice system has been re-evaluated in recent years (Nguyen, V. K. and Huynh 2015), leading to a plan to expand to 500 ha in deep-water areas by 2030 (An Giang Department of Agriculture and Rural Development 2014).
4. Resolution No. 120/NQ-CP of 17 November 2017.
5. Highly Pathogenic H5N1 Avian Influenza Viruses. The Clade 1 viruses first detected in late 2003 continued to circulate until 2007 (Wan et al. 2008).
6. Personal communication with Tu Xang in December 2016.
7. Personal communication with leader of Tam Nong Department of Agriculture and Rural Development, Dong Thap Province.

REFERENCES

ACIAR, 2016. *Improving the Sustainability of Rice-Shrimp Farming Systems in the Mekong Delta, Vietnam.* Canberra: Australian Centre for International Agriculture Research.

An Giang Department of Agriculture and Rural Development, 2014. Land use planning for rice cultivation using high technology in An Giang Province until

2020 and vision for 2030. Long Xuyen: An Giang Department of Agriculture and Rural Development.

An Giang People's Committee, 2006. Chuong trinh khai thac loi the mua nuoc noi tinh An Giang giai doan 2002–2010 (Programs for exploitation of the benefits of the 'rising water season' of An Giang Province from 2002 to 2010). Long Xuyen: An Giang People's Committee.

Berg, H. K., Soderholm, A. E., Soderstrom, A.-S., and Tam, N. T., 2017. Recognizing wetland ecosystem services for sustainable rice farming in the Mekong Delta, Vietnam. *Sustainability Science* 12: 137–154.

Biggs, D., 2015. Promiscuous transmission and encapsulated knowledge: a material-semiotic approach to modern rice in the Mekong Delta. In F. Bray, P. A. Coclanis, E. L. Fields-Black, and D. Schäfer, eds. *Rice: Global Networks and New Histories*, chapter 5. Cambridge: Cambridge University Press.

Biggs, D., Miller, F., Chu, H. T., and Molle, F., 2009. The Delta machine: water management in the Vietnamese Mekong Delta in historical and contemporary perspectives. In F. Molle, T. Foran, and M. Karonen, eds., *Contested Waterscapes in the Mekong Region: Hydropower, Livelihoods and Governance*, pp. 203–225. London: Earthscan.

Bosma, R. H., Udo, H. M. J., Verreth, J. A. J., Visser, L. E., and Cao Quốc Nam, 2005. Agricultural diversification in the Mekong delta: farmers' motives and contributions to livelihoods. *Asian Journal of Agriculture and Development* 2: 49–66.

CGIAR, 2016. *Assessment Report: Drought and Salinity Intrusion in the Mekong River Delta of Vietnam.* CGIAR Research Centres in Southeast Asia, 25–28 April 2016.

Cummings, R., 1978. Agricultural change in Vietnam's floating rice region. *Human Organization* 37(3): 235–245.

Dasgupta, S., Benoit, L., Craig, M., David, W., and Yan, J., 2007. *The Impact of Sea-Level Rise on Developing Countries: A Comparative Analysis.* Washington, DC: World Bank.

Demont, M., and Rutsaert, P., 2017. Restructuring the Vietnamese rice sector: towards increasing sustainability. *Sustainability* 9: 325.

Dien, L. D., Hiep, L. H., Hao, N. V., Sammut, J., and Burford, M. A., 2018. Comparing nutrient budgets in integrated rice-shrimp ponds and shrimp grow-out ponds. *Aquaculture* 484: 250–258.

General Statistics Office, 2016. *Agriculture, Forestry and Fisheries.* Hanoi: General Statistics Office.

Håkan, B., 2002. Rice monoculture and integrated rice-fish farming in the Mekong Delta, Vietnam—economic and ecological considerations. *Ecological Economics* 41(1): 95–107.

Huynh, T. T. T., Aarnink, A. J. A., Drucker, A., and Verstegen, M. W. A., 2007. Pig production in Cambodia, Laos, Philippines, and Vietnam: a review. *Asian Journal of Agriculture and Development* 4(1): 69–90.

Kakonen, M., 2008. Mekong Delta at the crossroads: more control or adaptation. *Ambio* 37(3): 205–212.

Le Coq, J. F., Dufumier, M., and Trebuil, G., 2001. *History of Rice Production in the Mekong Delta.* Paper presented at Third EUROSEAS Conference, London, 6–7 September 2001.

Leigh, C., Hiep, L. H., Stewart-Koster, B., Vien, D. M., Condon, J., Sang, N. V., et al., 2017. Concurrent rice-shrimp-crab farming systems in the Mekong Delta: are conditions (sub) optimal for crop production and survival? *Aquaculture Research* 48: 5251–5262.

Ling, F. H., Tamura, M., Yasuhara, K., Ajima, K., and Trinh, C. V., 2015. Reducing flood risks in rural households: a survey of perception and adaptation in the Mekong delta. *Climate Change* 132: 209–222.

Mainuddin, M., Kirby, M., and Hoanh, C. T., 2011. Adaptation to climate change for food security in the lower Mekong Basin. *Food Security* 3(4): 433–450.

Mainuddin, M., Kirby, M., and Hoanh, C. T., 2013. Impact of climate change on rainfed rice and options for adaptation in the lower Mekong Basin. *Natural Hazards* 66: 905–938.

MONRE, 2009. *Climate Change and Sea-Level Rise Scenarios for Vietnam.* Hanoi: Ministry of Natural Resources and Environment.

Nguyen Huu Chiem, 1993. Geo-pedological study of the Mekong Delta. *Southeast Asian Studies* 31(2): 158–186.

Nguyen Huu Chiem, 1994. Former and present cropping patterns in the Mekong Delta. *Southeast Asian Studies* 31(4): 345–384.

Nguyen, D. C., 2011. *Transformation of Farming Systems in Coastal Mekong Delta: Seeking for Better Management and Sustainability.* Paper presented at 6th International Symposium on Structural Transformation of Vietnamese Agriculture and Rural Society, Kagoshima University, Japan, 14–16 March 2011.

Nguyen, D. C., Duong, L. T., Nguyen, V. S., and Miller, F., 2007. Livelihoods and resource use strategies of farmers in the Mekong Delta. In T. B. Tran, T. S. Bach, and F. Miller, eds. *Challenges to Sustainable Development in the Mekong Delta*, pp. 66–98. Bangkok: Sustainable Mekong Research Network.

Nguyen, D. C., Sebesvari, Z., Amelung, W., and Renaud, F. G., 2015. Pesticide pollution of multiple drinking water sources in the Mekong Delta, Vietnam: evidence from two provinces. *Environmental Science and Pollution Research.* https://doi.org/10.1007/s11356-014-4034-x.

Nguyen, V. K., 2014. *An Economic Evaluation of Flood Dike Construction in the Mekong Delta: Cost-Benefit Analysis of Flood Protection Dike Construction in An Giang Province.* Lambert Academic Publishing.

Nguyen, V. K., and Howie, C. (eds.), 2018. *Conservation and Development of the Floating Rice Based Agro-Ecological Farming Systems in the Mekong Delta.* Hanoi: Agricultural Publishing House.

Nguyen, V. K., and Huynh, D. N., 2015. Conserving the benefits of floating rice in Vietnam. Available at http://www.mekongcommons.org/conserving-the-benefits-of-floating-rice-in-viet-nam/ (accessed 28 March 2015).

Nguyen, V. K., and Pittock, J., 2016. Scoping floating rice-based agro-ecological farming systems for a healthy society and adaptation to climate change in the lower Mekong Region and Myanmar. Canberra: Australian National University.

Nguyen, V. K., Dumaresq, D., and Howe, C., 2016. Dike compartments: case studies in water governance, farming systems, and adaptation to water-regime changes in An Giang Province, Mekong Delta, Vietnam. In *Dynamics of Water Governance in the Mekong Region*. Mekong Program on Water, Environment and Resilience (M-POWER) Series, vol. 5. Kuala Lumpur: Strategic Information and Research Development Centre.

Nguyen, V. K., Dumaresq, D., and Pittock, J., 2018. Impacts of rice intensification on rural households in the Mekong Delta: emerging relationships between agricultural production, wild food supply and food consumption. *Food Security* 10: 1615–1629.

Phạm Thị Thúy Hạnh, and Furukawa, M., 2007. Impact of sea-level rise on coastal zone of Vietnam. *Bulletin of the Faculty of Science, University of the Ryukyus* 48: 45–59.

Preston, N., and Clayton, H., 2003. *Rice-Shrimp Farming in the Mekong Delta: Biophysical and Socioeconomic Issues*. Canberra: Australian Centre for International Agricultural Research (ACIAR).

Rothuis, A., 1998. *Rice-Fish Culture in the Mekong Delta, Vietnam: Constraint Analysis and Adaptive Research*. Leuven, Belgium: Katholieke Universiteit.

Stewart-Koster, B., Anh, N. D., Burford, A., Condon, J., Qui, N. V., Hiep, L. H., et al., 2017. Expert-based model building to quantify risk factors in a combined aquaculture-agriculture system. *Agricultural Systems* 157: 230–240.

Thang, T. C., 2017. *Current Status of Vietnam Rice Export Quality*. Taipei: Food and Fertilizer Technology Center for the Asian and Pacific Region.

Thuy, N. N., and Anh, H. H., 2015. Vulnerability of rice production in Mekong River Delta under impacts from floods, salinity and climate change. *International Journal on Advanced Sciences Engineering and Technology* 5(4): 272–279.

Tong, Y. D., 2017. Rice intensive cropping and balanced cropping in the Mekong Delta, Vietnam: economic and ecological considerations. *Ecological Economics* 132: 205–212.

USAID, 2016. *Development of Rice-Shrimp Farming in Mekong River Delta, Vietnam*. Washington, DC: United States Agency for International Development.

Vo, X. T., 1975. Rice cultivation in the Mekong Delta: present situation and potentials for increased production. *Southeast Asian Studies* 13(1): 88–111.

Vormoor, K., 2010. *Water Engineering, Agricultural Development and Socio-Economic Trends in the Mekong Delta, Vietnam.* ZEF Working Paper Series No. 57. Bonn: Center for Development Research (ZEF), University of Bonn.

Wan, X.-F., Nguyen, T., Davis, C. T., Smith, C. B., Zhao, Z.-M., Carrel, M., et al., 2008. Evolution of Highly Pathogenic H5N1 Avian Influenza Viruses in Vietnam between 2001 and 2007. *PLoS ONE* 3(10): 1–12.

White, Ian, 2002. *Water Management in the Mekong Delta: Changes, Conflicts and Opportunities.* Technical Documents in Hydrology No. 61. Paris: UNESCO.

Xuan, V. T., and Matsui, S. (eds.), 1998. *Development of Farming Systems in the Mekong Delta of Vietnam.* Ho Chi Minh City: Ho Chi Minh City Publishing House.

Yasuyuki, K., 2001. Canal development and intensification of rice cultivation in the Mekong Delta: a case study in Cantho Province, Vietnam. *Southeast Asian Studies* 39(1): 70–85.

CHAPTER 18

The Domestic Rice Value Chain in the Mekong Delta

Dao The Anh, Thai Van Tinh, and Nguyen Ngoc Vang

INTRODUCTION

Due to the almost continuous growth of yield and a much smaller growth in cultivated area, rice production in Vietnam has increased fourfold from 11.6 million t in 1980 to a peak of 45.1 million t in 2015, dropping back to 42.8 million t in 2017 (Table 18.1). About 18% of milled rice production is exported, valued at USD 2.7 billion in 2017 and USD 2.2 billion in 2018, making Vietnam the third largest rice exporter globally after India and Thailand. The Mekong Delta accounts for about 56% of the total of 7.7 million ha cultivated, produces about 50% of total rice output, and contributes over 90% of rice exports. The export orientation of the

D. The Anh (✉)
Vietnam Academy of Agricultural Sciences, Hanoi, Vietnam

T. Van Tinh
Centre for Agricultural Policy, Institute of Policy and Strategy for Agriculture and Rural Development, Ministry of Agriculture and Rural Development, Hanoi, Vietnam

N. Ngoc Vang
An Giang University, Long Xuyen, Vietnam

© The Author(s) 2020
R. Cramb (ed.), *White Gold: The Commercialisation of Rice Farming in the Lower Mekong Basin*,
https://doi.org/10.1007/978-981-15-0998-8_18

375

Table 18.1 Number and type of value chain actors interviewed

Actors	No. interviewed
Rice-growing households	300
Commune authorities where rice is grown	20
Millers in the production region	70
Traders in the production region	60
Wholesalers in urban areas	50
Traditional retailers	85
Modern retailers	45
Input and service providers (land preparation, seed, fertilizer, extension, post-harvest) and provincial policymakers	14

Delta is further highlighted by the fact that 70% of rice produced there is channeled through the export value chain.

There have been several studies of the rice export value chain (Tran 2010; Vo and Nguyen 2011), but the domestic rice value chain, representing 82% of production nationally and 30% of production in the Delta, has been understudied. This chapter reports a study of the domestic value chain in the Mekong Delta. The study aimed to (1) describe the rice value chain in the Delta, focusing on the domestic chain; (2) conduct economic analysis of the actors in the rice value chain; and (3) examine the impact of government policies on the rice value chain (Fig. 18.1).[1]

Methods

The approach used in the study followed Kaplinsky and Morris (2000). The production area studied comprised 20 communes in An Giang and Hau Giang Provinces, with an average of 7922 rice producers per commune.[2] The combined production capacity of these provinces in 2012 was 5.12 million t of paddy, accounting for 21.1% of the total output of the Delta region (GSOV 2013). The consumption area studied included Can Tho and Ho Chi Minh Cities. These were the two largest cities in the region, with an average demand of 1.18 million t of rice per year.

We collected information using structured questionnaires for all the actors involved in the value chain (Chen et al. 2013). Actors were randomly selected in the research areas to ensure representativeness. The main actors were classified according to the scale of their operations. Farmers were classified as small (<1 ha), medium (1–2 ha), and large

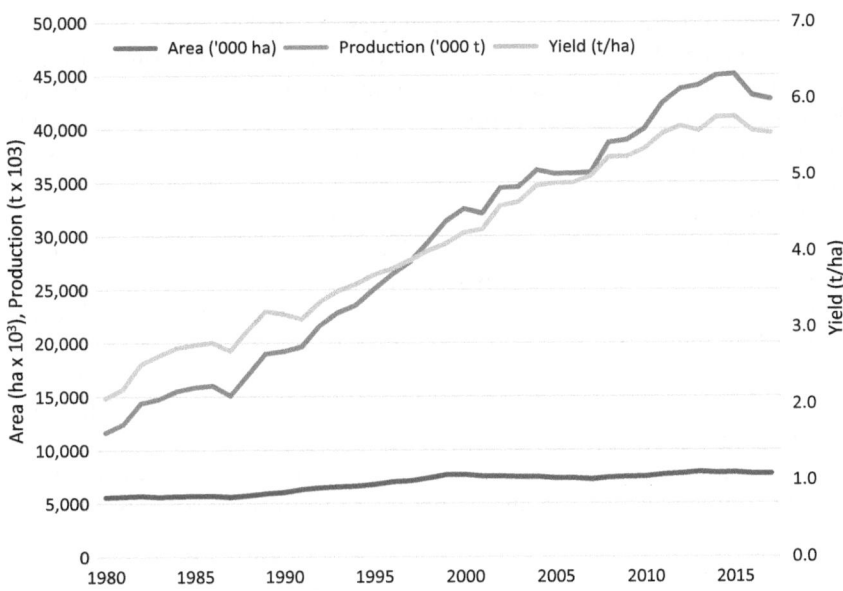

Fig. 18.1 Area, yield, and production of rice in Vietnam, 1980–2017. (Source: FAOSTAT)

(>2 ha). Rice mills were classified as small (<1 ton/hour); medium (1–5 t/hour); large (>5 t/hour); and milling/polishing plants. Descriptive statistics were used to compare means and proportions for each category of actors.

OVERVIEW OF THE RICE VALUE CHAIN IN THE MEKONG DELTA

In 2012, there were about 1.46 million rice farmers in the Mekong Delta cultivating about 4.1 million ha per year (given that rice is cropped 2–3 times per year in the Delta). Production of paddy was 24.6 million t, including short-term aromatic varieties (such as Jasmine 85, VD20, and ST5), short-term non-aromatic varieties (such as IR50404, VND95-20, and OM 576), and medium-term varieties (such as IR29723, IR42, and traditional local varieties). Farmers obtained production inputs such as fertilizers, pesticides, and farm equipment through a region-wide system of agricultural material stores and agencies, distinct from the traders and processors who handled the harvested product. There are now more than 100,000 millers and polishers operating in the region, of which up to 150 have been certified as rice exporters. Traders, wholesalers, and retailers operate through many different distribution channels in a widespread market (VFA 2012).

As illustrated in Fig. 18.2, the export value chain accounts for 70% of rice production in the region. This chain includes three channels: (1) a direct channel, in which paddy is sold by farmers to the exporting firms for milling and polishing, accounting for only 4% of exports; (2) a two-tier channel, in which paddy is sold by farmers to traders who bring it to the exporters for milling and polishing, accounting for 81% of exports; (3) a three-tier channel, in which traders sell paddy to the mills who sell rice to exporters as loose rice, accounting for 15% of exports.

The domestic rice value chain accounts for 30% of rice produced in the region. Wholesalers and retailers obtain rice from three sources: (1) from

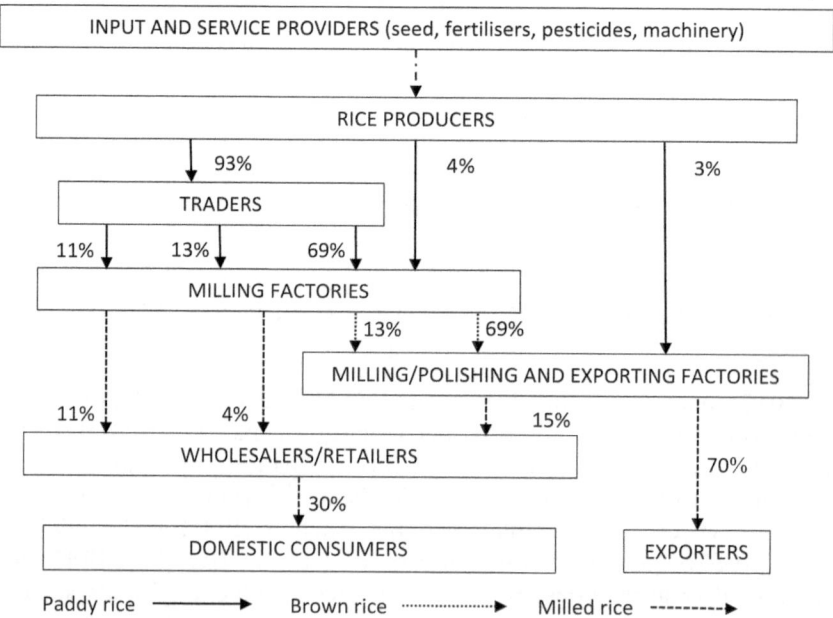

Fig. 18.2 Rice value chain in Mekong Delta

traders who buy paddy from farmers and have it milled for local consumption, mainly in the vicinity of the mill (37% of rice supplied to domestic consumers); (2) mills that supply rice for urban areas (13%); and (3) large milling and polishing firms that supply the cities (50%). In 2010, about 1.86 million t of paddy were imported from Cambodia into the Mekong Delta area (of which 90% was the high-quality Mien variety), mainly for the domestic market (Purcell 2010). This stream is not considered here.

ANALYZING THE ACTORS IN THE RICE VALUE CHAIN

Input and Service Providers

(1) *Production supplies.* In the Mekong Delta, as noted above, production inputs are provided through a system of agricultural stores and agencies. According to the 2012 survey, each commune had on average four input supply stores for farmers. This delivery system was highly organized, with large agencies distributing inputs to smaller shops which in turn distributed inputs throughout the communes and villages. The large agencies were also a conduit for technical advice to input suppliers and farmers.

The producer survey found that 100% of respondents purchased inputs from a store near their homestead (Table 18.2). The main reasons for their choice of supplier were the quality of the inputs and that they could defer payment, implying the provision of short-term store credit. This was

Table 18.2 Status of input use by farmers in study area (n = 300)

	Small farms	Medium farms	Large farms	All
% buying from store near house	100.0	100.0	100.0	100.0
Reason for selecting input supplier				
Regular customer	15.7	42.7	86.3	48.2
Short distance	31.0	31.3	28.8	30.3
Lower prices	31.5	15.7	11.3	19.5
Quality guaranteed	100.0	100.0	100.0	100.0
Payment can be delayed	100.0	100.0	98.5	99.5
Satisfaction with input supply				
High	89.4	86.5	87.2	87.7
Moderate	10.6	13.5	12.8	12.3
Low	0	0	0	0

Source: Producer survey, 2012

related to their reasons that the store was in close proximity and they were regular customers. Farmer satisfaction was high (about 88%) across all size classes. However, only 25–45% of the farmers interviewed used certified rice seeds: most used seeds retained from their previous crop.[3]

(2) *Machinery services.* All stages of rice production in the Mekong Delta from land preparation to post-harvest operations were mechanized to a degree. In particular, land preparation and harvesting were entirely mechanized. Commune-level data for the 20 communes in the survey showed that the two-wheel tractors used for land preparation (and other purposes) were available at an average density of three tractors per 100 ha (Table 18.3). Given a capacity of 1.5 ha/day, these tractors could complete land preparation for a region in 22 days on average. Transplanters were not used in the region, even though transplanting seedlings is a critical, labor-intensive activity; farmers preferred to save labor by broadcasting seeds rather than transplanting. Combine harvesters had spread throughout the Delta and were provided on a contract basis. The average density was two harvesters per 100 ha. With a capacity of 3 ha/day, this number of machines could complete the harvest in 17 days. As the window for harvesting is 7–10 days, harvesters had to be contracted from other provinces to augment the local supply of this service.

Considering the Mekong Delta as a whole, the Ministry of Agriculture and Rural Development (MARD) reported that there were 12,300 harvesters in 2011, including 8600 combine harvesters (obviating the need for threshers). Given a harvesting capacity of 3–5 ha/day, there were clearly too few harvesters to service the 1.5 million hectares of rice land—at 4 ha/day

Table 18.3 Availability of machinery services in study communes in 2011 (*n* = 20)

Indicator	Hau Giang	An Giang	Average
Mean no. of households per commune owning two-wheel tractors	387	456	421.5
No. of two-wheel tractors per 100 ha	2.3	3.7	3.0
Mean no. of transplanters per commune	–	–	–
No. of transplanters per 100 ha	–	–	–
Mean no. of harvesters per commune	271	345	308
No. of harvesters per 100 ha	1.5	2.5	2.0
Mean no. of dryers per commune	3.1	2.6	2.9
No. of dryers per 100 ha	0.3	0.2	0.2

Source: Survey of staff in 20 communes, 2012

the harvest would take at least 30 days, not counting movement between fields and breakdowns. The state has supporting policies for acquiring agricultural machinery, especially interest rate support. However, the technical and financial capacity of machinery manufacturers to expand in Vietnam is limited, hence so is the supply of more efficient and affordable harvesters.

The cost to farmers for harvesting services was very high and varied with the stage of the harvest and the type of harvester. The normal fee was VND 260,000–400,000 per 1000 m², but at the end of the harvest, when farmers were more desperate, the fee was VND 350,000–450,000 per 1000 m². The lack of harvesting services in high season in a given location meant that producers hired services from other localities, but they had to access these services through an intermediary who charged VND 15,000–20,000 per 1000 m².

The provision of drying services for the harvested paddy was the most limiting constraint in the production system in the study area, especially with the spread of combine harvesters. According to the survey of 20 communes, there were only 2.9 dryers per commune on average, giving a density of only 0.2 dryers per 100 hectares. At an average yield of 6 t/ha, this implied a total harvest of 3000 t/dryer. Yet most of these dryers were owned by the local rice mills with a very low capacity of 15 t/day, requiring 200 days for one crop. This had led to a situation in which, during peak season, many farmers had to sell "wet paddy" at a discount because they did not have access to a dryer or drying yard.

(3) *Agricultural extension and training.* Extension and training for rice farmers in the Delta region almost all take place through farmer groups or cooperatives. Farmers do not pay for the training because they are funded by the state and the private sector; in the survey, 21% of farmers were only trained by a private-sector actor, 7% were only trained by a state agency, and 72% were trained by both. The survey of villages and communes showed that 17.7% of communes had a cooperative and 14.4% had a cooperative or farmer group related to rice production (Table 18.4). The average number of extension officers in each commune was 0.6, meaning that many communes did not have regular access to this source of technical advice.

Producers

The survey sampled rice producers from three different size classes (Table 18.5). Interestingly, household size and the number of family

Table 18.4   Extension offices and cooperatives in communes

Content	Hau Giang	An Giang	Average
Average no. of extension offices per commune	0.5	0.7	0.6
% of communes with farmer cooperatives	15.3	20.2	17.7
% of communes with cooperatives dealing in rice	10.3	18.4	14.4

Source: Surveys, 2012

Table 18.5   Characteristics of rice producers in study area (n = 300)

Item	Farm size category			
	Small (<1 ha) (n = 87)	Med. (1–2 ha) (n = 124)	Large (>2 ha) (n = 89)	All (n = 300)
Household size (persons)	5.3	4.0	3.0	4.0
No. of workers/household	3.2	2.2	1.5	2.3
Paddy area (ha)	0.67	1.42	5.20	2.45
Area rented (ha)	0.50	1.32	2.53	1.45
Paddy yield (t/ha)	6.0	6.2	6.2	6.1
Farm-gate price (VND × 10^3/kg)	5211	5226	5258	5232
Gross revenue (VND × 10^3/ha)	31,110	32,140	32,600	31,915
Production cost (VND × 10^3/ha)	22,543	21,057	20,750	21,450
Net income (VND × 10^3/ha)	8567	11,083	11,850	10,465

Source: Producer survey, 2012
Note: USD 1 = VND 22,727 (11 August 2017)

workers decreased with increasing farm size, perhaps reflecting outmigration from the larger, more prosperous farm-households which also were more mechanized and employed hired labor. Obviously, the paddy area increased with farm size, with small and medium farmers renting in most of their paddy land (75% and 93%, respectively) while large farmers rented in under half their paddy land on average. Despite these differences, the productivity of the three groups did not differ greatly. The yield of the medium and large farmers was only slightly higher than that of the small farmers. This, combined with a slight upward trend in farm-gate price with farm size, perhaps reflecting the production of higher-value varieties on larger farms, meant that gross revenue also increased slightly with farm size. With a slight decreasing trend in production cost per ha with farm size, the net income per ha showed a more significant increase across the size classes, with medium farms earning 30% more than small farms, and large farms earning 38% more.

Farmers sold their paddy in different forms according to the time of sale. At harvest time, 75% of farmers sold wet paddy, that is, not dried to the required moisture content, and 25% sold dried paddy. Although the government encourages farmers to sell dried paddy to increase their profits, the cost of investing in drying equipment is large. Most producers had to accept the loss of profit from selling wet paddy, incurring a price difference of VND 900–1000/kg. In the months between harvests, all paddy sold was dried, this paddy coming from households with higher storage capacity.

Traders acquired 93% of the farmers' paddy (Fig. 18.2). However, the relationship between traders and farmers was quite loose. Trading through paddy brokers, who acted as local collectors, accounted for 55% of purchases. Only 4% of the paddy produced was purchased directly by millers, who operated in the same locality as the farmers from whom they bought. The polishing/exporting firms purchased directly only 3% of paddy produced. In acquiring paddy from farmers, 85% of buyers paid a deposit at the rate of 20–25% of the total value of paddy acquired, 10% made a "definitive purchase" (i.e., paid in full at the time of acquisition), and 5% paid only after the paddy was delivered to the mill. Thus, the majority of paddy sold was subject to flexible arrangements between farmers and traders.

Traders

Traders were a key link in the value chain as 93% of paddy produced was sold to these actors (Fig. 18.2). The traders surveyed were mostly small, multi-enterprise businesses without warehouses or shops (Table 18.6). They transported paddy by boat, with an average capacity of 26 t (Fig. 18.3). On average, each trader purchased 113 t of paddy per month from farmers. Only 8% of traders interviewed represented a particular rice mill; the majority of the traders operated independently and were not bound to particular farmers or buyers. There was no overarching organization of traders and no state policy that directly impacted on them.

About 13% of paddy bought by traders was sold to rice mills in the region (Fig. 18.2). Another 11% was put through the mills for milling and polishing and then sold as finished rice to wholesalers and retailers.

Table 18.6   Characteristics of rice traders in study area (*n* = 60)

Characteristic	Value
Average number of employees	2.2
Number of years in operation (years)	9.7
Traders who began as farmers (%)	20
Traders linked to a single rice mill (%)	8.3
Traders with storehouse, shops (%)	1.7
Traders owning boats/ships (%)	100
Average number of boats	1
Average tonnage of boats	25.9
Average value of boat (VND × 10³)	173,000
Average paddy volume purchased (t/month)	113

Source: Trader survey, 2012
Note: USD 1 = VND 22,727 (11 August 2017)

Fig. 18.3   Trader transporting paddy in Can Tho Province. (Photo: Dao The Anh)

However, most paddy (69%) was put through first-stage rice mills and sold in bulk to large export firms for polishing, bagging, and shipment overseas.

In performing these transactions, 71% of traders sold through "rice intermediaries" who linked them to the factories. The appearance of such intermediaries between farmers, traders, and millers helped the marketing system to operate, but it created a long chain, adding to costs and reducing the direct links between actors. About 40% of traders received an advance deposit from the factories in order to buy paddy and 60% only received payment after the paddy was delivered and so had to provide their own working capital.

Processors

Processing factories could be involved in any combination of de-husking, removing the rice bran, and polishing. However, as shown in Table 18.7, the processes were distributed quite differently among the four types of processor in the Mekong Delta. Small mills mainly produced white rice for local consumption on a daily basis, whereas medium and large mills were mainly engaged in the de-husking process, supplying brown rice to the large polishing factories, though 25% of the large mills performed all the processes through to polishing (Fig. 18.4). The large export firms mainly acquired de-husked or de-branned rice from the mills for polishing (92.5%).

The owners of the processing firms did not differ greatly in age or education, though the miller-polishers had more years of education on average (Table 18.8). The larger businesses had more experience in the industry (averaging 11–15 years) than the small millers (6 years). All of

Table 18.7 Types of rice processor in study area (*n* = 70)

Process	Product	Small mills (n = 10)	Med. mills (n = 15)	Large mills (n = 4)	Miller-polisher (n = 41)
		% of processors in each category			
De-husking	Brown rice	0	72.7	75.0	0
Polishing only	Polished rice	12.5	0	0	92.5
De-husking, de-branning	White rice	87.5	27.3	0	0
De-husking, de-branning, polishing	Polished rice	0	0	25.0	7.5
Total		100	100	100	100

Source: Processor survey, 2012
Note: USD 1 = VND 22,727 (11 August 2017)

Fig. 18.4 Large rice mill in Can Tho Province. (Photo: Dao The Anh)

Table 18.8 Characteristics of rice processors in study area ($n = 70$)

Characteristic	Small mills (n = 10)	Med. mills (n = 15)	Large mills (n = 4)	Miller-polisher (n = 41)
Age of owner (years)	42	49	45	46
Education of owner (years)	8.4	9.9	9	12.7
Years of business	6	11	15	12
Private firm (%)	100	100	100	75.6
State joint-stock firm (%)	0	0	0	12.2
Private joint-stock firm (%)	0	0	0	12.2
Area of factory (m^2)	87	1550	2500	4825
Capacity of factory (t/hr)	0.2	3.8	9.6	20.8
Value of factory (VND × 10^3)	111,000	2,420,000	4,000,000	2,975,610
Operating capital (VND × 10^3)	4612	1,353,346	1,325,000	46,471,073
Rice throughput (t/month)	48	2268	6768	13,791

Source: Processor survey, 2012
Note: USD 1 = VND 22,727 (11 August 2017)

the rice millers were private firms, whereas a quarter of the miller-polisher businesses were private or state joint-stock companies. The different functions of the four types of processor corresponded to different scales and operating capacities, as seen in the area and capacity of the factory and the fixed and operating capital tied up in the business (Table 18.8). Hence, the throughput of rice varied from 50 t/month for the small rice mills to 14,000 t/month for the large milling, polishing, and exporting firms.

The rice mills mainly focused on providing a de-husking service for the rice traders, accounting for 80% of their output (Fig. 18.2). Purchasing paddy for milling accounted for 17% of millers' output, including 13% sold in bulk to exporters and only 4% sold to wholesalers and retailers. The polishing/exporting businesses were the main actors supplying rice to the domestic wholesale and retail market (15% of total rice output from the region, or half the domestic supply) and all the export market (70% of total rice output) through export contracts. However, not all polishing factories had the right to export under the government's Decree 109 (only 150 enterprises in Vietnam have an export license for rice). Such factories either sold rice to firms that were able to export or bought rice from these enterprises to enable them to export.

WHOLESALERS AND RETAILERS

Focusing on the domestic market, the main actors supplying rice to consumers were the wholesalers, traditional retailers, and modern retailers.

The wholesalers were on average medium-sized enterprises with about 60 m² of storage space and a throughput of 68 t/month, requiring working capital of around USD 25,000 (Table 18.9). They bought half their supplies from the polishing factories, 37% from traders, and 13% from rice millers. Most of their sales (85%) were to traditional retailers.

Table 18.9   Characteristics of rice wholesalers and retailers in study area ($n = 180$)

Characteristic	Wholesalers (n = 50)	Traditional retailers (n = 85)	Modern retailers (n = 45)
Age of owner (years)	42	32	–
Number of employees	2.1	1.6	0.7
Length of operation (years)	8.7	8.9	6.3
Floor space for rice stocks (m²)	60.7	17.9	10.6
Operating capital (VND × 10³)	532,600	39,133	–
Volume of rice sold (t/month)	68	1.5	23.5

Source: Wholesaler/retailer survey, 2012
Note: USD 1 = VND 22,727 (11 August 2017)

Sales to other wholesalers accounted for 5% and direct sales to consumers and small shops for 8%. Very little was sold to modern retailers. About 92% of the wholesale product provided to retailers was sold in plastic packages of 50–55 kg. Of the packaged rice, 60% did not have clear information on the packaging, 26% had factory information and a brand mark, and 14% had only factory information.

Each traditional marketplace in the Delta had on average 6.7 traditional retailers operating small stores with about 18 m² of storage space for rice (Table 18.9). Thus, retailers were spread widely across the region, each selling only 1.5 t/month on average, requiring working capital of under USD 2000. Almost all traditional retailers surveyed (99%) sold rice in loose form, providing customers with plastic bags at the point of sale. Although the rice was packaged and labeled in 50–55 kg packs when bought from the wholesaler, selling loose rice was a feature of traditional retailing, enabling consumers to better assess the product in the market.

Modern retailers (supermarkets, food stores) were central actors in the government's rice price stabilization policy. However, rice was not the main item for these retailers, so they did not exploit its full potential. On average, only 10 m² were allocated for rice stocks, and there were no employees dedicated to the rice product (Table 18.9). The modern retailers sold almost all their rice (92%) in plastic packages, including 2 kg, 5 kg, and 10 kg sizes. They also paid special attention to the brand mark and product information—97% of bags had this labeling. This meant that the price of the same type of rice from a modern retailer was much higher than from a traditional retailer, hence the number of consumers buying rice from modern retailers was low, most consumers still relying on traditional retailers. Nevertheless, modern retailers averaged sales of 24 t/month.

Marketing Margins in the Rice Value Chain

The costs of producing, processing, and delivering rice to domestic consumers in the Mekong Delta were assessed. The average cost of rice production was obtained from the farmer survey (Table 18.10). It can be seen that there was a little variation in the per-hectare cost of production between farm size classes. Small farmers incurred somewhat more expenditure for fertilizers and pesticides than large farmers, and large farmers paid more interest on working capital, but the distribution of cost items and the total costs per ha (both paid-out or cash costs and imputed costs) were not significantly different. On average, total production costs were about VND 21.5/million/ha (USD 950/ha) and VND 3.5 million/t of paddy (USD 155/t). At a milling conversion rate of 75%, the production cost per kg of milled, polished rice was VND 4688 (USD 0.20).

To analyze the costs and margins along the domestic and export value chains, selling and purchase prices were converted into the equivalent weight of rice and the value of one actor's output was taken as the input cost of the next actor in the chain (Table 18.11). For farmers, input costs were taken to be the costs of seed, fertilizers, and pesticides, while other paid-out and imputed costs were classified as "incremental costs". The total value added in the domestic value chain was VND 3303/kg (USD 0.15/kg), with nearly 70% of this total coming from the post-milling actors (polishers, wholesalers, and retailers). In the export value chain, the total value added was VND 2131/kg as the chain was not followed through to the foreign buyers.

In both chains, the margins obtained by each actor represented a return over operating costs of 5–10%, except for the millers, who achieved returns

Table 18.10 Cost of paddy rice production (VND × 10³ per ha)

Item	Small farm		Medium farm		Large farm		All	
	Mean	%	Mean	%	Mean	%	Mean	%
Seed	1619	7	1684	8	1489	7	1554	7
Fertilizers	8400	37	7710	37	6280	30	7460	35
Pesticides	1023	5	1051	5	889	4	987	5
Irrigation	185	1	195	1	180	1	187	1
Wages	1587	7	1877	9	1296	6	1589	7
Machine hire	505	2	505	2	454	2	488	2
Land rental	710	3	680	3	600	3	667	3
Marketing	1594	7	1494	7	1393	7	1494	7
Interest	1343	6	492	2	3511	17	1824	9
Paid-out costs	16,966	75	15,688	74	16,092	77	16,250	76
Family labor	2743	12	2865	14	2918	14	2841	13
Depreciation	2835	13	2505	12	1740	8	2361	11
Total costs	22,543	100	21,057	100	20,750	100	21,450	100
Yield (t/ha)	6.0		6.2		6.2		6.1	
Cost of paddy (VND × 10³/t)	3757		3396		3347		3516	
Cost of rice (VND/kg)	5009		4528		4463		4688	

Source: Producer survey, 2012
Note: USD 1 = VND 22,727 (11 August 2017)

Table 18.11 Costs and margins in domestic and export rice value chains (VND per kg rice)

	Farmers	Traders	Millers	Polishers	Wholesalers	Retailers	Total
Domestic market							
Selling price	5232	5925	6780	7994	10,118	12,700	
Input cost	2213	5232	5925	6780	7994	10,118	
Incremental cost	2534	246	754	533	1423	1694	
Total variable cost	4747	5478	6679	7313	9417	11,812	
Value added	485	447	101	681	701	888	3303
Value added/cost (%)	10	8	2	9	7	8	
% of total value added	15	14	3	21	21	27	100
Export market							
Selling price	5232	7019	6780	7994	9555		
Input cost	2213	5232	5925	6780	7994		
Incremental cost	2534	1345	754	533	1139		
Total variable cost	4747	6577	6679	7313	9133		
Value added	485	442	101	681	422		2131
Value added/cost (%)	10	7	2	9	5		
% of total value added	23	21	5	32	20		100

Source: Surveys, 2012
Note: USD 1 = VND 22,727 (11 August 2017)

of only 2% (Table 18.11). Farmers obtained the highest return of 10% and contributed 15% of value added in the domestic chain and 23% in the (truncated) export chain. Thus, there was no indication that any actor in the chain was realizing excessive margins, reflecting a large number of actors at each stage and a competitive market overall.

IMPACT OF STATE POLICIES ON RICE VALUE CHAIN

Since 1975, rice policy in Vietnam has mainly focused on increasing productivity through the use of short-term, high-yielding varieties and increased fertilizer use. As a consequence, rice yields in Vietnam as a whole increased from 2.5 t/ha in 1975 to 5.8 t/ha in 2015. Moreover, the cropping intensity in favorable regions such as the Mekong Delta has increased, such that about 27% of the total rice area is cultivated three times a year. However, increased productivity has resulted in the predominance of low-quality rice in the export market. Moreover, the incidence of poverty among small rice farmers remains high, because the value of rice production per unit area is very low and has not improved for some time (Jaffee et al. 2012; Thu 2013).

In 2010, the government issued Resolution No. 63/NQ-CP on ensuring food security, stipulating that farmers must be assured a 30% profit. This policy was intended to discourage diversification out of rice production by boosting farmer returns. However, few rice processing enterprises buy paddy directly from farmers. Traders dominate this stage (93% of paddy purchases, Fig. 18.2) and transmit prices from the mills, earning a return of 7–8% compared with the farmers' 10% (Table 18.8). Thus, it is infeasible to enforce this policy.

With regard to rice exports, Decree No. 109 of 2010 introduced regulations about the enterprises allowed to export rice, namely, those having a factory with a capacity of over 10 t/hour, storage capacity of over 5000 t, and reserves in circulation of over 10% of the volume of rice that they exported in the previous six months. This led to the formation of informal networks among firms as the exporters that did not meet the conditions had to buy additional rice from other enterprises. However, the decree did not provide any benefit to farmers. Export prices are not listed; hence an increase in prices mainly benefits the exporters as farmers do not have the information or means to increase their margins. Currently, the government is preparing to issue an alternative policy whereby storage regulations will be replaced by product quality regulations. This will remove the limit on the number of businesses allowed to export, encouraging small businesses to export high-quality rice. Exporters will also be encouraged to develop contract farming areas, promoting higher rice quality.

The price stabilization policy has not been clear or consistent. The state does not have the capital to purchase and store rice, so reserves are required to be held by exporters with the support of state-subsidized loans to ensure rice prices for farmers. However, this mechanism is not suitable for the exporters because they are forced to use their own capital for the purchase of stocks and temporary storage, increasing their costs and reducing their competitiveness in the export market. The price stabilization mechanism does not distinguish between the objectives of food security, price stability, and the profitability of the exporters, yet an effective policy requires the clear separation of objectives to ensure benefits to all parties.

Provincial policies in the Mekong Delta have mainly focused on advising farmers to cultivate three crops of rice a year and concentrate on varieties to improve rice quality (though the quality of rice in the third season is mostly low). This policy has run into problems because the focus is only on rice producers. When farmers produce better-quality varieties, the market (traders and processors) still demands the low-quality varieties that form the bulk of the export trade, resulting in a situation in which farmers cannot sell their rice. In addition, high-quality rice varieties tend to have lower yields but the price premium is only VND 200/kg over normal rice, reducing farmers' profit. A good example of this contradiction is in An Giang Province, where the government discourages the planting of IR50404. However, this variety has high yield, is easy to grow, has fewer diseases, and is in high demand in the export market, so it is planted on up to 107,000 ha, accounting for 17% of the total cultivated area in An Giang.

Improving rice quality is one of the central strategies of the government. Accordingly, in 2013, the Ministry of Agriculture and Rural Development (MARD) approved the National Rice Product Development Project with two key tasks—to improve the competitiveness of rice products and enhance the return to actors in the value chain. In addition, in 2016, MARD issued a plan to restructure the rice sector to 2020—"Improving the efficiency of rice production and trade in Vietnam" (Decision No. 1898/QD-BNN-TT). This project is intended to (1) improve rice quality; (2) upgrade organization, policies, and institutions to improve value chain operations; (3) improve harvesting and processing technology; (4) promote sustainable market development; (5) facilitate environmental protection and adaptation to climate change; and (6) ensure food and nutrition security.

CONCLUSION

The rice value chain in the Mekong Delta is a large and complex system, successfully linking about 1.5 million small-scale rice farmers cultivating over 4 million ha per year to large numbers of traders, processors, wholesalers, retailers, and exporters. About 30% of production enters the domestic market and 70% is exported, accounting for over 90% of national exports.

There are many intermediaries in the domestic rice value chain in the Mekong Delta. Input suppliers are widely dispersed in the region, providing seeds, fertilizers, and other inputs competitively to small, medium, and large farmers. Agricultural extension and training are provided by both public and private sectors through farmer groups. The region has seen rapid mechanization, with the spread of two-wheeled tractors and combine harvesters, the latter mainly provided through contract services, including from outside the region. Despite the high fees (up to USD 180/ha), the labor-saving benefit of mechanical harvesting has ensured almost universal adoption. However, there is limited availability of driers at the farm or commune level, meaning that most farmers sell wet paddy at a discount, which then has to be dried at the mills.

Almost all the harvested rice crop is sold to local traders at the farm gate. These are small, independent operators who transport paddy by boat to the rice mills. However, they are mostly linked to processors through intermediaries who frequently advance funds to the traders to buy paddy. Most paddy goes through small mills that produce white rice, some of which are sold directly to domestic wholesalers but most of which go to exporters for polishing and bagging.

Wholesalers are medium-sized enterprises, buying from polishing factories, traders (after they have arranged milling), and directly from millers. Most of their sales are to traditional retailers who are spread widely across the region, operating small stores. The rice is sold loose and packaged after purchase. Modern retailers sell pre-packaged and labeled rice at higher prices than traditional retailers, and their share of the domestic market is low.

None of the actors in the domestic value chain appears to gain an excessive margin, with returns on working capital mostly in the range 7–10%, though small-scale millers average a lower return. It is unlikely that market efficiency could be improved through any structural intervention, given the high degree of competition at each stage. Rather, better forms of credit to enable producers (perhaps as farmer groups), service providers, and processors to invest in improved technology may do more to improve the efficiency of the value chain. Government policies need to consider the whole chain rather than focusing on one class of actors, for example, by encouraging farmers to cultivate high-quality varieties that are not in demand.

In order to increase the value and competitiveness of the rice value chain in the Mekong Delta, the government should implement a policy to promote the quality of rice through contract farming between cooperatives and private enterprises based on quality standards. The export policy of Decree 109 based on the capacity of the mill was not successful because of a lack of focus on quality. A revised policy should open the export market to private enterprises that obtain export contracts based on quality.

NOTES

1. This chapter is based on research supported under the Asian Development Bank TA-7648 Regional—Research and Development Technical Assistance (R-RDTA). A fuller version of the survey results and analysis has been reported in *Rice Value Chain Study in the Mekong River Delta, Viet Nam* by Dao The Anh, Thomas Reardon, Kevin Chen, Thai Van Tinh, Vu Nguyen, Nguyen Ngoc Vang, Nguyen Van Thang, and Le Nguyen Doan Khoi and subsequently incorporated in *Rice Value Chains in China, India, Lao PDR, and Viet Nam: 2012 Survey Results, Interpretations, and Implications for Policy and Investment*, a report submitted by the International Food Policy Research Institute to the Asian Development Bank, 15 September 2013.
2. Department of Agriculture and Rural Development, An Giang Province, various years. *Report on Agricultural Activities in An Giang Province*; Department of Agriculture and Rural Development, Hau Giang Province, various years. *Report on Agricultural Activities in Hau Giang Province*.
3. This may mean that farmers purchased new seeds every few years and then retained seeds from several crops before replenishing their stock.

REFERENCES

Chen, K., Reardon, T., Dao The Anh, Wong, L., Huang, Z., Das Gupsta, S., and Wang, J., 2013. *Rice Value Chain in China, India, Lao PDR and Vietnam: 2012 Survey Results, Interpretations, and Policies Implications for Investment.* Final Report submitted by IFPRI for ADB (TA-7648 REG Project, Component 2). Available at http://www.ifpri.org/publication/rice-value-chains-china-india-lao-pdr-and-viet-nam-2012-survey-results-interpretations.

General Statistics Office of Vietnam (GSOV), 2013. *Statistical Yearbook of Vietnam, 2013.* Hanoi: Statistical Publishing House.

Jaffee, S., Nguyen Van Sanh, Dao The Anh, and Nguyen Do Anh Tuan, 2012. *Vietnam Rice, Farmers and Rural Development: From Successful Growth to Sustainable Prosperity.* Hanoi: World Bank.

Kaplinsky, R., and Morris, M., 2000. *A Handbook for Value Chain Research.* Ottawa: IDRC.

Purcell, T., 2010. *Rice Production in Cambodia—Trends in Production and Productivity and Opportunities for Improvement.* Phnom Penh: Agricultural Development International.

Thu Huong, 2013. Vietnam Rice Development Policy Shortcomings. *Vietnam Economic Times*, 8 April 2013.

Tran Tien Khai, 2010. Export policies in Vietnam and problems that need to be adjusted. In *Proceedings of the Mekong River Socio-Economic Sciences and Regional Development Conference*, Can Tho.

VFA, 2012. *Annual Activity Report of 2012.* Hanoi: Vietnam Food Association.

Vo Thi Thanh Loc, and Nguyen Phu Son, 2011. Rice value chain analysis in the Mekong Delta. *Cantho University Scientific Magazine*, pp. 96–108.

CHAPTER 19

The Cross-Border Trade in Rice from Cambodia to Vietnam

Dao The Anh and Thai Van Tinh

INTRODUCTION

There is a paradox in the rice value chain in the Mekong Delta. Vietnam has a total paddy area of about 3.9 million ha and exported 4.8 million t of rice earning USD 2.2 billion in 2016, making it one of the top three rice exporters. Over half of rice production and over 90% of exports come from the provinces in the Mekong Delta. Yet in recent years, large quantities of paddy have been imported from Cambodia across its south-eastern border. Vietnam Food Association (VFA) estimated that, in 2008–2011, an annual average of 0.8–1.0 million t of paddy were imported from Cambodia through the border provinces (VFA 2011). Purcell (2010) reported an even higher figure of about 1.86 million t in 2010, of which 90% was for the domestic market. At the Tinh Bien border gate, about 1000 t of Cambodian paddy are imported into Vietnam each day during

D. The Anh (✉)
Vietnam Academy of Agricultural Sciences, Hanoi, Vietnam

T. Van Tinh
Centre for Agricultural Policy, Institute of Policy and Strategy for Agriculture and Rural Development, Ministry of Agriculture and Rural Development, Hanoi, Vietnam

© The Author(s) 2020
R. Cramb (ed.), *White Gold: The Commercialisation of Rice Farming in the Lower Mekong Basin*,
https://doi.org/10.1007/978-981-15-0998-8_19

397

the peak season. This paddy is milled in Vietnam due to the lack of milling capacity in Cambodia. The cross-border trade in Cambodian paddy had had an impact on rice production and rice exports in the Mekong Delta.

An Giang Province has the largest rice area and output in the Mekong Delta, with a total cultivated area of 605,720 ha and production in 2011 of 3.86 million t (see Fig. 17.1 in Chap. 17). However, An Giang is also considered an ideal market for Cambodian rice, particularly from the adjacent Takeo Province (Chap. 12). This study aimed to describe the rice value chain from Takeo to An Giang, analyse the roles of the different actors in this chain, examine the impacts on rice production, processing, and consumption in An Giang, and explore ways of managing the chain for mutual benefit. The study used the value chain theory of Kaplinsky and Morris (2000) and the methodology of GTZ (2007). Fieldwork was conducted in 2012 and focused on the rice value chain through the Tinh Bien and Khanh Binh border gates in An Giang Province. A total of 238 actors were interviewed, including 3 producers, 60 traders, 70 processors, 85 retailers, and 20 consumers.

BACKGROUND

Vietnam and Cambodia share a border of 1137 km, stretching through ten provinces in Vietnam and nine in Cambodia. An Giang Province in Vietnam shares a border of 96 km with the Cambodian provinces of Kandal and Takeo. An Giang has two international border gates with Cambodia: Tinh Bien-Phnom Den and Vinh Xuong-Kas Omsano. The terrain in the border region is relatively flat with many channels, creating favourable conditions for cross-border trade.

Rice production in Cambodia has experienced significant growth in recent years (Chap. 10). Rice is produced mainly in the wet season, which accounts for about 75% of total annual rice production. Most of the varieties grown in Cambodia are long-term varieties with good quality and are preferred by consumers. With total rice production reaching more than 9 million t in 2014, and with a population of over 15 million people, Cambodia had a surplus of up to 4 million t of paddy for export. The Food and Agriculture Organization (FAO) reported exports of 3.61 million t of paddy in 2013. While Cambodia has preferential access to many markets (e.g., Cambodia has unlimited duty-free access to the European Union market under the Everything But Arms arrangement for least-developed countries), its capacity to produce rice exports has been limited by constraints

Table 19.1   Area, production, and yield of paddy in An Giang Province in 2011

Crop season	Area (ha)	Yield (t/ha)	Production (t)
Winter-Spring	235,482	7.51	1,768,783
Summer-Autumn	232,987	5.58	1,301,992
Autumn-Winter	131,835	5.81	766,341
Summer	5398	4.32	23,334
Total	605,702	6.37	3,860,000

Source: DARD (An Giang Province) (2011)

in storage, processing, transportation, port facilities, and marketing (IFC 2015), hence the attraction of cross-border trade in paddy to Vietnam.

An Giang Province straddles the upper part of the Mekong Delta, with a total area of 353,676 ha and an agricultural land area of 246,821 ha (Table 19.1). Rice cultivation accounts for 82% of the agricultural land, and the total cultivated area in 2011 was 605,702 ha, implying a cropping intensity of almost 3.0. Over 80% of the cultivated area was planted with certified varieties, mainly IR and OM varieties.[1] In particular, the IR50404 variety was planted on 107,000 ha, or 17% of the cultivated area. The average yield in 2011 was 6.4 t/ha and production reached 3.86 million t. Thus, An Giang is the largest rice-producing province, not only in the Delta but also in Vietnam as a whole.

Vietnam and Cambodia have signed a number of important agreements to provide a legal basis for the development of cross-border trade. The Government of Vietnam issued Decision No. 254/2006/QD-TTg on 7 November 2006 on the management of cross-border trade, establishing preferential tax policies for imports from Cambodia to Vietnam, with rice exempt from any tax. The cross-border trade between An Giang and Takeo-Kandal increased by an average of 19% per year in the period 2006–2010. In 2010, the total value of goods moving across this border was estimated to be over USD 1053 million. This figure had increased by 51% over the previous year and accounted for over 50% of the total value of trade between Vietnam and Cambodia.

With regard to paddy, though it is exempt from tax, import quotas are imposed by Vietnam on Cambodian paddy and rice, including general-purpose, aromatic, and sticky rice (MIT 2008). The combined quota was 250,000 t of milled rice-equivalent in 2010 and 2011, increasing to 300,000 t in 2012 and 2013. Despite this policy, it is estimated that between 0.8 and 1.0 million t of paddy (well over the quota) are sold

annually via the border gates in Dong Thap, Long An, and An Giang
Provinces (VFA 2011). This phenomenon is partly because some
Vietnamese farmers cultivate rice as share-croppers in Cambodia and bring
their paddy to Vietnam to sell. Thus, the amount of paddy imported from
Cambodia to Vietnam informally is large, consistent with the broader situ-
ation of unregulated cross-border trade between the two countries.

Overview of the Rice Trade Between Takeo and An Giang

The broad structure of the cross-border value chain is shown in Fig. 19.1.
There are three main channels by which paddy produced in Cambodia is
imported into An Giang through the border gates: (1) paddy produced in
Cambodia by Vietnamese farmers is taken across the border to sell to
Vietnamese traders in An Giang (5%); (2) paddy produced in Cambodia is

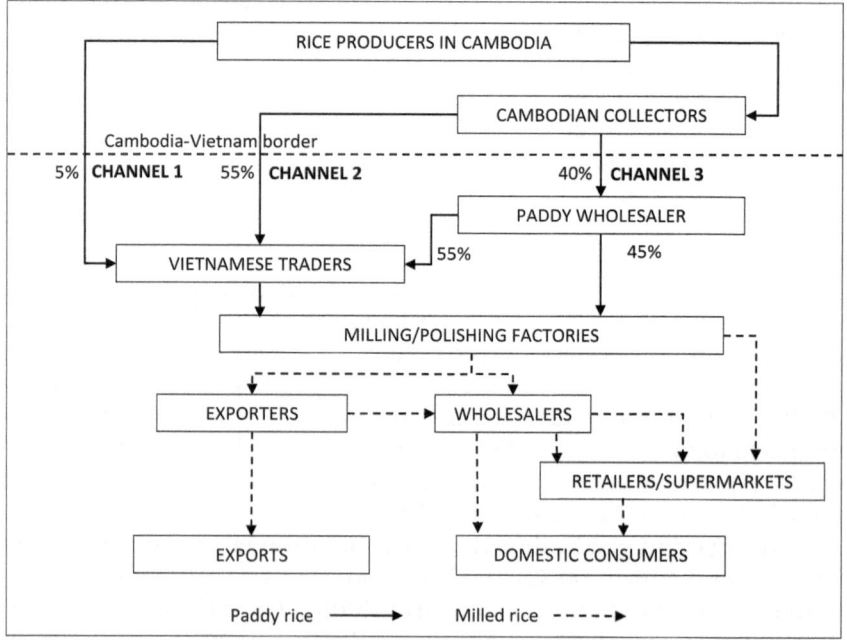

Fig. 19.1   Cross-border rice value chain between Cambodia and Vietnam

collected by Cambodian traders and sold to traders in An Giang Province at the border (55%); (3) Cambodian traders collect paddy and take it across the border to sell to paddy wholesalers with large granaries in An Giang (40%).

Two main types of paddy are imported from Cambodia to Vietnam: (1) Soc paddy, including Khaodak, Khaodakmali, and Jasmine, which accounts for 80–90%—domestic consumers in Vietnam prefer Soc paddy because it is of high quality and Cambodian farmers spray few chemicals[2]; (2) Than Nong paddy, or the low-quality IR50404 variety, which accounts for 10–20%. Once milled, this IR50404 rice is re-exported as low-quality Vietnamese rice.

There are five main reasons for the substantial flow of paddy from Cambodia into the Mekong Delta. First, Vietnamese farmers in the Delta mostly cultivate high-yielding, low-quality varieties such as IR50404, IR3217, OM1490, and OM1723 to sell to exporters. However, when production of this type of rice exceeds export demand, unsold rice is difficult to sell domestically. The domestic market, especially in the large cities, prefers high-quality aromatic rice. This is the main reason why paddy from Cambodia is imported to Vietnam through the cross-border trade.

Second, as noted above, farmers in the south-west region of Vietnam have rented land in Cambodia (in Takeo, Kandal, Prey Veng, and Svay Rieng Provinces) for rice cultivation. The cultivated area in Cambodia has expanded, and productivity has increased due to the application of intensive farming techniques from Vietnam. The cross-border trade is an essential outlet for rice cultivated by Vietnamese farmers in Cambodia. It is noteworthy that the rice produced by these farmers is not classified as either "Than Nong" or "Soc". Nevertheless, it consists of both IR50404 rice and aromatic rice varieties originating in Thailand.

Third, the difference in harvesting time between the two countries creates a demand for Cambodian rice later in the season. Vietnamese farmers use short-term varieties, so harvesting usually occurs 1–2 months earlier than in Cambodia. Sourcing paddy from Cambodia enables the rice mills in Vietnam to keep operating efficiently.

Fourth, seed of the standard IR and OM varieties has been brought from Vietnam to be planted in Cambodia. Although the quality of the paddy grown in Cambodia is lower (smaller grains, more cracked grains, and more chalkiness) due to poorer cultivation techniques, the price is also much cheaper than in Vietnam, hence Vietnamese traders and factories can benefit from dealing with this low-quality crop as well as the Soc paddy.

Finally, the differences in post-harvest technology between Cambodia and Vietnam create a demand for cross-border trade to process Cambodian paddy in Vietnam. According to unofficial statistics, Vietnam has more than 100,000 rice-processing factories and 150 enterprises which are allowed to export rice. Cambodia has around 200 rice processing factories, of which only 40 are capable of processing rice to the right standard for export to international markets (Purcell 2010). In addition, Cambodian rice mills lack storage capacity, hence paddy is exported to Thailand and Vietnam. Though Cambodian processing capacity is increasing rapidly, it still represents the major constraint on exports of processed rice from Cambodia (IFC 2015).

ANALYSING THE VALUE CHAIN ACTORS

Producers

The producers are mostly Cambodian farmers in Takeo, which is geographically part of the Mekong Delta (Chap. 1). These farmers traditionally cultivate the Soc paddy varieties under rainfed conditions. Yields are low, farm sizes are small, and farmers cultivate to meet family requirements, with little technical support. However, in 2002, the An Giang provincial government signed an agreement with the adjacent provinces of Takeo and Kandal to provide support in agricultural techniques. The An Giang Plant Protection Company was assigned to implement this programme by the An Giang People's Committee. Since then, the programme has organized many training events for provincial and district staff and Cambodian farmers to give them access to more intensive rice-farming techniques.

The Vietnamese producers in Takeo lease farming land from Cambodian villagers. They were previously traders who operated in the border areas and found untilled land which they requested to rent. Leases are annual or longer, up to 3–4 years. The farmers carry out land improvements and practise more capital-intensive cultivation. The result was a large increase in the supply of paddy to An Giang and the south-west border region.

Farmers use different means to transport farm inputs and outputs and consumer goods. Most (91%) use boats, while the remaining 9% use motorcycles and trucks. Most of the paddy is transported to Vietnam through informal trade channels using boats.

Table 19.2 Paddy prices in the border region between An Giang and Takeo

Actor	Type of paddy	Paddy price (VND/kg)			Paddy price in domestic market
		Channel 1	Channel 2	Channel 3	
Paddy producers	Soc	–	–	5200	–
	Than Nong	–	–	4700	5232
Traders/	Soc	–	5250	5280	–
collectors	Than Nong	5000	4740	4800	5330

Source: CASRAD survey, March 2012

Table 19.2 shows the price of paddy traded across the Takeo-An Giang border via the three channels described above and numbered in Fig. 19.1, compared with the price of paddy produced and traded within An Giang. For the Than Nong varieties of paddy produced on a large scale in An Giang, the price of imported paddy from Cambodia was much cheaper (e.g., VND 4700/kg if sold through Channel 3 or VND 5000/kg if sold through Channel 1, compared with VND 5232/kg in the domestic market). Even the preferred Soc varieties were cheaper than the domestically produced Than Nong varieties. Hence, traders and milling factories could use this cross-border supply to improve their profitability.

Traders

Traders are the most important link in the cross-border value chain. Most traders used to be rice farmers but have additional experience in small business. In the 2012 survey of 60 traders, 80% were men, their average age was 40, they averaged 9 years of education, and they had 10 years' experience in trading (Table 19.3). Their experience meant they knew how to evaluate product quality, how to deal with farmers in different locations and with different products, which factories were working efficiently, and what was happening moment-by-moment in the market. Most (82%) operated throughout the year but the cross-border trade was mainly in November–December, the main harvest period in Cambodia. They purchased paddy along both banks of the canals that cross the border and at gathering points along the border. In the flood season, traders took their boats directly to farmers' fields in Cambodia to purchase paddy. A small number (8%) only engaged in trading seasonally.

Table 19.3   Characteristics of Vietnamese traders engaged in cross-border paddy trade

Characteristic	Value (n = 60)
Mean age (years)	40.8
Gender (% male)	80
Education (years)	8.7
Working capital (VND × 10³)	149,370
Working capital owned by trader (%)	72
Mean length of experience in trading (years)	9.7
Proportion of traders operating throughout year (%)	82
Proportion of traders with boats (%)	100
Proportion of traders with large boats (> 33 t) (%)	15
Mean price of large boats (VND × 10³)	234,444
Proportion of traders with medium boats (13–33 t) (%)	73
Mean price of medium boats (VND × 10³)	193,289
Proportion of traders with small boats (<13 t) (%)	20
Mean price of small boats (VND × 10³)	85,833

Source: CASRAD survey, 2012

River transport was essential for traders, hence 88% of those surveyed owned boats and 12% rented boats (Table 19.3). However, the capacity of these boats varied with the capital resources and activities of the trader. Only 15% of traders had large boats (> 33 t), most (73%) had medium-sized boats (13–33 t), and 20% had small boats (< 13 t); note that some had more than one boat. The traders operating inside Cambodia had the larger boats and covered longer distances from farms in Takeo to mills in An Giang.

Traders bought paddy directly from farmers (5%), from Cambodian collectors (55%), and from large granaries along the border. From these purchase points, they transported the paddy for processing domestically at Vietnamese mills and polishing factories.

Paddy Wholesalers

As the informal cross-border trade in paddy has grown, particularly since 2009, the need for storage services at the border has increased. This has encouraged investment in storage granaries along the border. Once paddy is transported across the border, much of it is bulked in 6–7 paddy wholesale market places at the Tinh Bien and Khanh Binh border gates. A similar number of wholesale market places have collection stations along the

Mekong and its channels. The wholesalers have invested in large ware-houses and yards to facilitate the purchase and sale of paddy from Cambodia.

About 55% of the paddy acquired by these wholesalers is sold directly to traders and 45% is sold to milling and polishing factories through their agents. On average, over 1000 t of paddy is provided to traders each day in the peak season, who then transport this volume for processing before selling to domestic rice wholesalers. The method of transaction between Vietnamese traders and border collectors includes pre-ordering (44%) and spot transactions (56%).

Milling/Polishing Companies

An Giang is the largest rice producer in the Delta, so the processing sector is concentrated in this province. In 2012, there were 404 milling factories in the province with a capacity of 6.3 million t of rice per year. In addition, there were 236 polishing factories with a capacity of 2.4 million t of rice per year and storage for between 100 and 5000 t each. The output of milled rice from the province was 1.87 million t in 2008. The province had about 16 companies with the ability to export rice directly to other countries. One feature of the milling and polishing factories is that they only operate at full capacity during the harvest period. At other times, they lack the raw materials to maintain efficient throughput. The additional off-peak paddy supply from Cambodia enables the processing sector to optimize its operation when the domestic supply is less.

The milling operations varied greatly in scale and sophistication (Table 19.4). The owners/managers were mostly men in their 40s or 50s with 8–9 years of education and considerable experience, especially in the larger plants. The large rice milling and polishing factories had modern equipment and large warehouses, so the area of these factories was an order of magnitude larger than the small mills (2500–5000 m² compared with less than 100 m² for the small mills). The investment in equipment was also very different due to the differences in capacity and functionality, with the modern milling/polishing plants averaging USD 355,000 and requiring working capital of over USD 2 million. The mills surveyed were all privately owned and largely self-financed. However, a quarter of the milling/polishing plants producing rice for export were state-owned and had borrowed on average half the initial capital investment.

It can be seen that the investment in equipment and working capital to produce rice to export standard requires considerable experience, techni-

Table 19.4   Characteristics of milling/polishing factories

Characteristic	Size of mill			Milling-polishing plant	All
	Small	Medium	Large		
Mean age of owner/manager (yrs)	42	49	45	46	46
Gender (% male)	70	73	100	81	79
Education (yrs)	8.4	9.9	9.0	12.7	11.3
Business experience (yrs)	6	11	15	12	11
Private ownership (%)	100	100	100	76	95
Total operation area (m²)	87	1550	2500	4825	3344
Mean capacity (t/hr)	0.2	3.8	9.6	20.8	13.4
Loan capital at start-up (%)	12	0	8	51	32
% with complete white rice mill	20	0	0	24	17
Value of white rice mill (USD)	7180	0	0	418,062	359,365
% owning dryer	10	53	0	56	45
% using rubber roll huller	40	53	75	49	50
% using stone disc huller	30	40	0	46	40
% owning rice polishing machine	0	0	25	100	31
% owning discoloration machines	0	0	0	78	46
Mean value of mill (USD)	5313	115,845	191,479	355,000	244,453
Mean operating capital (USD)	221	64,784	119,674	2,224,561	602,310
% of operating capital borrowed	0	28.6	39.6	16.8	17.0
Grain processed (t/month)	48	2268	6768	13,791	7866

Source: CASRAD survey, 2012

cal knowledge, and capital. Such investment is occurring in Cambodia but is still quite limited. Hence, the processors in Vietnam are performing the milling and polishing for the hundreds of thousands of t of paddy imported from Cambodia every year.

Rice Wholesalers and Retailers

In Vietnam, rice wholesalers operate mainly in urban areas and less often in the production regions because the distance between the processors and the main consumer markets is typically very short. In fact, in production regions, much of the purchasing and selling of rice occurs directly between millers and retailers, without going through wholesalers. The characteristics of wholesalers surveyed in Ho Chi Minh (HCM) City are presented in Table 19.5. The wholesalers were mostly experienced and reasonably well educated. Most (86%) owned their own warehouse, averaging 63 m². They were almost all self-funded, with an average working capital of USD 25,500; hence it was difficult to gain entry to the wholesale market. In HCM City, 44% of wholesalers stocked rice originating from abroad, including Soc rice types from Cambodia. This is a significant number in a large rice-producing country such as Vietnam, reflecting the consumer demand for Cambodian rice.

The retailers in the study were younger than the previous actors (averaging 32 years), and there were as many women as men performing this role (Table 19.6). Despite being younger, the retailers interviewed were quite experienced, averaging almost 10 years in selling rice. About 40% had a market stall and 60% had a shop. The scale of operation for traditional retailers was generally small, with an average daily turnover of only 250 kg of all types of rice (modern retailers such as supermarkets do not sell Cambodian rice). However, they were widely distributed across the study area, with each traditional market having on average nine rice retail-

Table 19.5   Characteristics of urban rice wholesalers

Characteristic	Value (n = 50)
Mean age (years)	42.4
Gender (% male)	80
Education (years)	10.6
Mean working capital (USD)	25,500
% of self-funded working capital	97
Turnover period for rice stocks (days)	9.7
Contribution of rice to total sales (%)	100
% owning warehouse	86
% renting warehouse	33
Mean area of warehouse (m²)	63
% using trucks to deliver to retailers	8
% using motorbikes to deliver to retailers	100
Average number of employees (persons)	2.1

Source: CASRAD survey, 2012

Table 19.6 Characteristics of rice retailers in urban areas

Characteristics	HCM City (n = 48)	Can Tho City (n = 37)	All (n = 85)
Mean age (years)	32.5	31.0	31.8
Gender (% male)	49	52	51
Business experience in wet market (years)	7	11	9
Business experience outside wet market (years)	8	8	8
% with stall in wet market	38	49	42
% with shop in street	62	51	58
No. of years selling rice	7.4	10.0	8.9
Turnover of rice (kg/day)	115	188	154
Quantity of each purchase (kg)	678	1386	1001
Working capital (USD)			1873

Source: CASRAD survey, 2012

Table 19.7 Mode of selling rice by traditional retailers

Mode	%
Rice sold in bulk with plastic bag provided	99.2
Rice sold in sacks or plastic bags	0.8
Packed without information	99.2
Packed with information about factories and brand	0.8
Packed with only information about factories	0.0

Source: CASRAD survey, 2012

ers and many more selling rice in retail shops. The large number of sellers reduced the volume of sales in each business.

Almost all the traditional retailers sold rice in bulk and provided unlabelled plastic bags for customers to package the rice they bought (Table 19.7). While the variety, quality, and price of the rice could be assessed by the customers, there was no formal information provided about the origin, brand, and processing factory, hence no formal certification of the rice from Cambodia. Nevertheless, while domestic rice prices were in the range VND 11,000–13,000/kg, the price of Cambodian rice was VND 5500–13,000/kg higher (Table 19.8). About half (51%) of the consumers interviewed felt that Cambodian rice were of better quality than the domestically produced rice, and they preferred Soc products originating from Cambodia.

Table 19.8   Retail prices of Cambodian specialty rice in Vietnamese market

Type	Retail price (VND/kg)
Soc (traditional Cambodian varieties)	17,000–19,000
Khaodak	16,500–18,500
Jasmine	17,000–19,500
Phuong Hoang	23,000–26,000

Source: Survey 2012

IMPACT OF CROSS-BORDER TRADE ON THE RICE MARKET IN VIETNAM

Vietnam is one of the world's largest rice exporters. Increasing imports from Cambodia add to the supply of rice for the domestic market and hence increase the export capacity (as well as directly adding to rice exports). The import of Cambodian rice also helps to meet the demand for specialty rice in Vietnam as consumer preference for higher-quality products increases. There are many different kinds of rice in the market in Vietnam and rice imported from Cambodia has contributed to the diversification of products available in the domestic market.

However, there have also been some negative impacts. The increase in cross-border trade has pushed down the domestic price, affecting the income of rice farmers in the border areas. In fact, many collectors in the border region only deal in imported paddy from Cambodia because this gives a higher profit. Hence, some Vietnamese farmers cannot sell their paddy at the domestic price due to the pressure of competition from Cambodia. In Can Tho City and Ho Chi Minh City, 50% of rice retail stores sell varieties of rice originating from Cambodia. Though the price is higher than the price of domestically produced rice, the consumption of rice originating from Cambodia is very large and it is a competitive product with other specialty rice in the country. Vietnamese farmers also produce rice of high quality but the activities of marketing, product branding, and providing product information to domestic consumers are weak. This aspect of the value chain could be strengthened to improve the local product's position in the domestic market.

As noted above, large volumes of Cambodian paddy are imported into Vietnam via informal trade networks. Given that the price is lower than the domestic price, variations in the cross-border flow have led to price fluctuations in the Vietnamese market. Also, it has become difficult for the industry to identify and manage the supply of rice for the domestic market and for export. Coordination is needed between Vietnam and Cambodia to better manage paddy imports into Vietnam, limiting the large volume of paddy entering through informal trade (MIT 2008; Vo 2010).

CONCLUSION

The cross-border trade in paddy from Cambodia to Vietnam, especially between Takeo and An Giang Provinces, has increased substantially in volume, reaching an estimated 1000 t per day in the peak season of November–December. There are five main reasons for the growth of this trade: (1) the quality of the wet-season crop produced in Takeo meets domestic demand in Vietnam for this type of rice; (2) some An Giang farmer-traders rent land for cultivation in Cambodia and transport paddy into Vietnam for the domestic market; (3) the harvest in Takeo is later than in An Giang, helping maintain the throughput of the rice mills in An Giang; (4) Cambodian paddy is cheaper than Vietnamese paddy of the same type; (5) Cambodia does not have sufficient capacity or technology for processing and storage, leading to a surplus of paddy that is exported to Vietnam.

There are three channels in the cross-border rice value chain, in which traders and paddy wholesalers provide the critical link: (1) from producers directly to Vietnamese traders (5%); (2) from producers to Cambodian traders, who sell to Vietnamese traders (55%); and (3) from producers to Cambodian traders who sell to Vietnamese paddy wholesalers, operating large granaries along the border (40%). These granaries in turn sell to traders in Vietnam (55%) and directly to processors (45%). Processed rice is distributed to wholesalers in the cities or directly to numerous traditional retailers who operate market stalls or shops. While the Cambodian rice is identifiable to consumers, the mode of selling means there is no formal labelling or certification. Nevertheless, consumers pay a substantial premium for Cambodian rice—50–100% over ordinary Vietnamese rice.

The large volume of paddy imported from Cambodia increases the supply of specialty rice for the domestic market in Vietnam. Demand for high-quality rice is increasing in Vietnam as urban incomes increase. While domestic production cannot meet this demand, paddy imports from Cambodia are filling an important niche. They also free up more domestic rice for the export market. However, these cheaper paddy imports may also have a negative impact on domestic production and incomes, especially in the border areas. It may be necessary to develop joint policies to manage better the cross-border trade. This could perhaps entail establishing joint-venture companies to purchase paddy from Cambodian farmers and facilitate processing in-country to export standards; ensuring that Vietnamese farmers in border areas are not adversely affected by the cheaper imports; and improving the commercial value of specialty varieties in Vietnam to meet domestic demand and increase farmers' incomes.

NOTES

1. IR varieties have been bred by the International Rice Research Institute (IRRI) and OM varieties by the Cuu Long Delta Rice Research Institute at O Mon, Can Tho.
2. Cambodian Soc rice is a general name for any variety of traditional Cambodian wet-season (summer) rice, cultivated in an extensive manner.

REFERENCES

DARD (An Giang Province), 2011. *Report on Agricultural Activities in An Giang Province*. An Giang: Department of Agriculture and Rural Development.

GTZ, 2007. *ValueLinks Manual: The Methodology of Value Chain Promotion*. Eschborn: GTZ.

IFC 2015. *Cambodia Rice: Export Potential and Strategies*. Cambodia Agribusiness Series No. 4. Phnom Penh: International Finance Corporation.

Kaplinsky, R., and Morris, M., 2000. *A Handbook for Value Chain Research*. Ottawa: IDRC.

MIT, 2008. *A Report on the Vietnam Border Trade with Cambodia, 2001–6*. Hanoi: Ministry of Industry and Trade.

Purcell, T., 2010. *Rice Production in Cambodia: Trends in Production and Productivity and Opportunities for Improvement*. Phnom Penh: Agricultural Development International.

VFA, 2011. *Annual Activity Report of 2011*. Ho Chi Minh City: Vietnam Food Association.

Vo Hung Dung, 2010. Improving the Value Chain of Rice Exports of Vietnam (Report). Can Tho City: Vietnam Chamber of Commerce and Industry.

CHAPTER 20

Cross-Border Trade in Sticky Rice from Central Laos to North Central Vietnam

Dao The Anh and Pham Cong Nghiep

INTRODUCTION

Rice is the staple food in Vietnam, accounting for 78% of energy intake. In addition to the standard eating varieties consumed every day, sticky (glutinous) rice is also incorporated in many favourite dishes and is often consumed on special occasions. Sticky rice is produced in the North Central Region of Vietnam, but there is inadequate supply to meet domestic demand. Hence, every year the provinces in this region import sticky rice and paddy from Laos, where sticky rice is traditionally the staple, to add to local supply. The rice and paddy are imported mainly through Lao Bao Border Gate between Quang Tri Province and Savannakhet Province in Laos (Fig. 20.1). There has been little research on this cross-border value chain and its impact on rice production in Vietnam. The aim of this study was to describe and analyse the value chain and the commercial potential of this niche market in Vietnam.

D. The Anh (✉)
Vietnam Academy of Agricultural Sciences, Hanoi, Vietnam

P. Cong Nghiep
Centre for Agrarian Systems Research and Development, Vietnam Academy of Agricultural Sciences, Hanoi, Vietnam

© The Author(s) 2020   413
R. Cramb (ed.), *White Gold: The Commercialisation of Rice Farming in the Lower Mekong Basin*,
https://doi.org/10.1007/978-981-15-0998-8_20

Fig. 20.1 The Lao Bao International Border Gate between Quang Tri and Savannakhet. (Source: Bùi Thụy Đào Nguyên, https://commons.wikimedia.org/w/index.php?curid=17356460)

The methods included collecting the relevant secondary data, interviewing value-chain actors in Quang Tri Province, and analysing the cross-border trade via the Lao Bao Border Gate. Interviews were conducted with 60 actors, including 15 farmers, 7 collectors, 4 wholesalers, 7 retailers, 1 enterprise, 5 processors, 20 consumers, and 1 manager.

The Study Area

Quang Tri Province is located between 16° 18' and 17° 10' N and between 106° 32' and 107° 34' E. Quang Tri is bounded by Quang Binh Province to the north, Thua Thien-Hue Province to the south, the South China Sea to the east, and Savannakhet and Salavan Provinces in Lao PDR to the west. Quang Tri has a geographical advantage in that it straddles the East-West Economic Corridor from Myanmar, through Thailand and Laos, to the central Vietnamese ports of Cua Viet, Chan May, and Da Nang (ADB 2010). This corridor passes through the Northeast Region of Thailand (see Chaps. 2, 3 and 4) and Savannakhet Province in Laos (see Chaps. 5, 6, 7, 8, 9 and 10).

The terrain in Quang Tri descends from the Truong Son (Annamite) Range in the west to the narrow coastal plain in the east. Four types of terrain are recognised: (1) high mountainous terrain in the west, ranging from 250 to 2000 masl and with slopes of 20–30°, at high risk of erosion when cultivated and therefore suitable for forest, perennial crops, and live-stock; (2) low mountainous and hilly terrain, ranging from 50 to 250 masl, occasionally reaching 500 masl, and suitable for growing perennial crops like rubber, pepper, and fruit trees; (3) deltaic terrain, reaching no more than 25–30 masl, relatively flat, naturally fertile, and the key areas of food production, especially rice; and (4) coastal terrain, mostly flat, sandy areas where settlements are concentrated but not suited to cultivation due to the risk of flooding in some parts and drought in others.

Quang Tri has a tropical monsoonal climate with high annual rainfall, benefiting agricultural crops and forestry. However, the climate is considered rather harsh as it is influenced by strong, hot, dry southwest winds for about 45 days per year from March to September, often causing drought and having a major impact on agricultural production. From October to February, the region is influenced by the northeast monsoon, with heavy rains that can cause flooding. In addition, Quang Tri is influenced by tropical storms, especially from September to November, with strong winds and heavy rain that creates flash flooding and seriously affects agricultural production.

The average annual temperature is 24–25 °C in the deltas and 22–23 °C in the hinterland at elevations over 500 m. The cold season is from December to February, with the coldest month averaging 22 °C in the deltas and less than 20 °C over 500 m. The hot season lasts from May to August, when the average temperature is 28 °C, but in June and July, temperatures can reach above 40 °C.

The average annual rainfall is 2200–2500 mm, with 154–190 rainy days. Over 70% of the rainfall falls in September–November. The dry season lasts from December to August, with the driest month in July. This is the time of the southwest winds and the highest temperatures. In the wet season, intense rainfall often causes flooding; in the dry summer season, the lower rainfall often causes water shortages and drought. Humidity averages 83–88%. However, in April it averages only 22%, while in the wet season it averages 88–90%.

In general, the natural conditions of Quang Tri are not very favourable for rice production. The land that is suitable for sticky rice production is not extensive and the climate is unfavourable.

In 2010, the population of Quang Tri Province was 601,672, of whom 72% were living in rural areas. The population was growing at 1.1%. The average population density was 126 persons per km², much higher along the coast and much lower in the mountainous interior. About 92% of the population belongs to the majority Kinh ethnic group. Minority groups of the Katuic branch of the Mon-Khmer family (such as the Van Kieu or Bru and the Pako) generally occupy the mountainous zone, extending across the border into Laos.

Production and Consumption of Sticky Rice in Quang Tri

Production of sticky rice in Quang Tri is mainly for provincial consumption. There are two crops per year. The winter-spring crop is planted from the end of October and harvested in April. The summer-autumn crop is planted from the end of April and harvested in late September. According to the data obtained in 2011, the area planted to sticky rice (2213 ha) accounted for 5% of the total rice area. The yield averaged 4.4 tons/ha and total annual production was 9404 tons of paddy, equivalent to 6394 tons of milled rice.

The estimated disposal of this production is shown in Table 20.1. About 8% of total paddy produced is lost during harvest, amounting to

Table 20.1   Estimated production and use of sticky rice in Quang Tri Province in 2011

Item	Paddy (tons)	Rice (tons)[a]
Total output	9404	6394
Harvesting loss (7.6%)	715	486
Used for seed (49 kg/ha × 67% × 2213 ha)	73	49
Loss in trading process (7.1% of 5908 tons traded)	612	416
Consumption (12 kg of rice/person/year × 601,672)	10,706	7280
Other consumption (52 kg/household/year × 136,743)	10,517	7152
Net surplus/deficit	(14,481)	(8989)

Source: Estimated from survey data, 2011
[a]1 kg paddy = 0.68 kg rice

715 tons of paddy. Our survey showed that two-thirds of the cultivated area is planted the following season with saved seed and one-third is planted with purchased seed. Thus, about 73 tons of paddy are needed for seed. This leaves about 8616 tons of paddy or 5908 tons of rice that can be traded. However, losses also occur in the post-harvest stage, including milling, transportation, wholesaling, and retailing. Based on the estimates of the actors involved in these processes, the rate of loss is about 7% of the total rice in market circulation, that is, about 416 tons of rice. This leaves a total of 5492 tons of sticky rice available for consumption.

On average, regular consumption of sticky rice in Quang Tri averages 12 kg/year. With a population of 601,672, the province therefore needs 7280 tons of sticky rice or 10,706 of paddy. Households also consume sticky rice as rice wine, rice cake, and other products. The survey indicated that household consumption of sticky rice for these other purposes averaged about 52 kg/year. With 136,743 households in the province, an additional 7152 tons of rice was consumed in this way, equivalent to 10,517 tons of paddy (Table 20.1). Thus, the total demand for sticky rice in the province is about 14,481 tons.

The implied deficit amounts to 14,481 tons less 5492 tons, equal to about 8989 tons of rice or 13,219 tons of paddy. This shortfall is made up of imports from Laos via the Lao Bao Border Gate.

The Cross-Border Value Chain for Sticky Rice

Sticky rice and paddy are imported from Laos via the Lao Bao Border Gate, following both a formal path and an informal path (Fig. 20.2). The formal path, accounting for 95% of the trade by volume, involves food companies in Laos selling rice to import companies in Vietnam who in turn sell to wholesalers and retailers or to food processors who require sticky rice as an ingredient. The informal path, accounting for only 5%, involves traders in Laos selling paddy to millers in Vietnam or rice to wholesalers and retailers. Subsequently, the rice passes through wholesalers, food processors, and retailers in the proportions shown in Fig. 20.2.

The formal import of sticky rice is undertaken by accredited food importing companies. According to Decree No. 12/2006/ND-CP, enterprises involved in rice imports need to have an import permit from the Ministry of Industry and Trade (MIT). Rice should be checked for food safety and quality standards by a specialised state office before customs clearance and must have a certificate of product origin issued by the

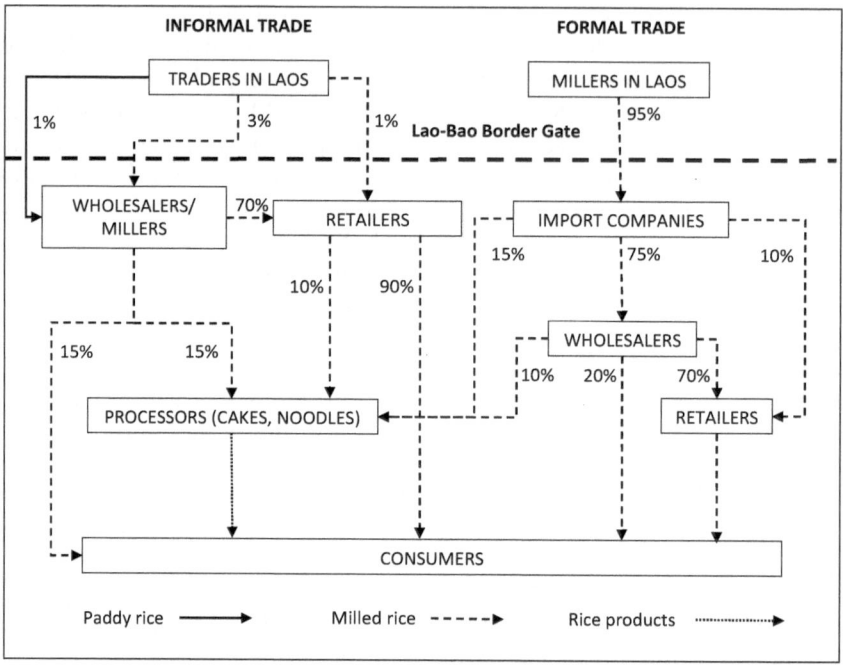

Fig. 20.2 Value chain for sticky rice and paddy imported from Laos through Lao Bao Border Gate

competent office in Laos. Rice imported from Laos to Vietnam within the formal quota incurs zero duty.

In 2010, within the formal trade quota, Vietnam imported 24,308 tons of Lao sticky rice and 3027 tons of Thai sticky rice through Lao Bao Border Gate. The quantity from Laos was more than double the previous year's total. Thai sticky rice imported through Laos also enjoys zero duty. However, because Thai sticky rice is of better quality and the price is higher than that of local rice in the North Central Region, the volume imported from Thailand has declined; the 2010 figure was less than half that of the year before.[1]

The informal trade in sticky rice is conducted by private traders on both sides of the border. In this case, there is no declaration, no quarantine, no payment of duty, and no formal contract, making it appealing to private actors. According to Vietnamese law, the value of purchases should not exceed VND 2 million per person per day, equivalent to 100–150 kg of sticky rice. However, Vietnamese traders often use their connections or ask other people to bring the rice across to increase the amount traded. Hence, one Vietnamese trader imports on average about 300 kg of sticky rice from Laos per day.

Lao traders also sell sticky rice and paddy at the border to wholesalers and retailers in Vietnam, each averaging about 1 ton/day. There are four of these traders who work about 320 days in the year, so the total amount of sticky rice imported from Laos through this channel can be estimated at about 1280 tons per year.

Thus, the total annual cross-border trade in sticky rice in 2010 was estimated to be 28,615 tons—including 24,308 tons of Lao rice and 3027 tons of Thai rice through the formal channel and 1280 tons of Lao rice through the informal channel. This compares with an estimated deficit of 8989 tons for Quang Tri Province, implying that almost 20,000 tons of the sticky rice imports were consumed in other provinces.

POLICIES AFFECTING THE PRODUCTION AND CONSUMPTION OF STICKY RICE IN QUANG TRI

The policies outlined in Chaps. 16 and 17 for rice production and exports in Vietnam as a whole have also indirectly supported sticky rice production and trade in Quang Tri Province. In 2009, focusing on national food security, the government set a target of keeping 3.8 million ha of paddy land to ensure producing 41–43 million tons of paddy each year to meet domestic demand and export 4 million tons. This policy has helped protect paddy land from conversion to other land use purposes.

The 2009 policy also aimed to ensure a minimum return to rice producers of 30% over the cost of production. Provincial People's Committees are to determine the local cost of rice production as a basis for determining the purchase price. Farmers can also access low-interest, long-term loans sufficient to purchase machinery, equipment, and materials for agricultural production. The domestic market for rice has also been freed up. Liberalisation of the rice market has reduced food shortages in rice-deficit areas. This policy has encouraged commercial actors to develop the rice supply chain, so consumers can easily access the products.

Before 2000, public investment in agriculture was mainly for irrigation systems, which accounted for 75% of total investment in agriculture during 1991–1995. After 2000, the government focused on investment in science and technology to produce new varieties and improve the quality of seeds.[2] In addition, the government has focused on human resource development and improving farmers' knowledge.

With regard to cross-border trade, the Ministries of Industry and Trade in Laos and Vietnam have signed bilateral trade agreements that regulate the trade in rice between the two countries. Commodities subject to regulated trade are exempt from duties. The quota for rice imported from Laos into Vietnam has continuously increased, from 40,000 tons in 2009 to 70,000 tons in 2012. Hence, there is increasing openness to cross-border trade in sticky rice.

However, according to traders and managers of food companies involved in importing sticky rice from Laos, the potential for this trade is limited. The volume of sticky rice imports is not likely to increase because Vietnamese sticky rice is milled to better quality and is cheaper. In the near future, sticky rice production in Quang Tri Province is expected to trend upwards because of a larger cultivated area and increased productivity through the wider use of varieties such as IRI352—a glutinous variety released by the International Rice Research Institute (IRRI) in 1990 and occupying 0.9 million ha in central and northern Vietnam in 2009 (Brennan and Malabayabas 2011). Hence, domestic production is likely to replace imports of Lao sticky rice. These informants also remarked that the trend in the sticky rice trade is towards more informal trade because of the complex procedures involved in complying with the formal path.

CONCLUSION

Sticky rice is widely consumed in Vietnam alongside conventional eating varieties, as well as in various processed foods. Though sticky rice is an established product in Quang Tri Province, the total output is currently not enough to supply the domestic demand within the province, let alone demand from towns and villages in other provinces. Given that, Quang Tri has access through the Lao Bao Border Gate to one of the largest rice-producing provinces in Laos, where most production is sticky rice, and through the East-West Corridor to Northeast Thailand, which also produces large quantities of sticky rice, it has become a conduit for cross-border trade in this commodity. Nearly 29,000 tons of sticky rice was imported through this gate in 2010, 90% of it produced in Laos and 10% in Thailand.

Almost all this volume (95%) came in through formal trade channels—labelled in Laos with a certificate of product origin, sold to registered enterprises in Vietnam with import permits, checked for food safety and quality standards, and subject to an annual quota. No duty was incurred on these quota imports. The remaining 5% entered in small lots via informal channels, avoiding all these requirements. It is thought by industry insiders that the proportion entering through the informal pathway might increase due to the lower costs.

Vietnam has many policies affecting the production and consumption of rice, which also have an impact on sticky rice production and distribution. These policies have three main purposes: ensuring supply to the domestic market, improving the incomes of rice farmers, and increasing foreign exchange earnings through exports. Hence, the policies involve retaining land in rice production, investing in improved varieties, training farmers, providing low-interest loans, attempting to set domestic prices to give rice farmers an adequate rate of return, and deregulating the domestic rice market to permit the emergence of more efficient supply chains. In this context, domestic production of sticky rice is likely to increase. Hence, the potential for expanding the cross-border trade from Laos is thought to be quite limited.

NOTES

1. Data from the Provincial Agriculture and Forestry Office, Savannakhet.
2. The government invests about 2% of gross domestic product in science and technology, a third of which is for agriculture.

REFERENCES

ADB, 2010. *Strategy and Action Plan for the Greater Mekong Subregion East-West Economic Corridor.* Mandaluyong City, Philippines: Asian Development Bank.

Brennan, J. P., and Malabayabas, A., 2011. *International Rice Research Institute's Contribution to Rice Varietal Yield Improvement in South-East Asia.* ACIAR Impact Assessment Series No. 74. Canberra: Australian Centre for International Agricultural Research.

PART VI

Conclusion

CHAPTER 21

Issues of Rice Policy in the Lower Mekong Basin

Rob Cramb

The commercialisation of rice farming in the Lower Mekong Basin has been at the centre of that region's remarkable journey out of poverty and food insecurity since the 1970s. Given that rice has long been the staple food in the region and that most of the population were rice growers, a development strategy that centred on opening up rice farming to productivity-enhancing investments had the double effect of increasing the incomes of large numbers of poor rural households while generating a marketable surplus to supply the rapidly growing urban population (and rice-deficit rural areas). The growth in export demand from elsewhere in Asia further added to the incomes of rice farmers in the more productive parts of the Basin (Northeast Thailand and the Delta), while initially providing a source of tax revenues and foreign exchange to support industrialisation strategies. Thus, a development pathway emerged that was driven by political necessity to be broadly based and inclusive, regardless of the nature of the political regime (Timmer 2008).

R. Cramb (✉)
School of Agriculture and Food Sciences, University of Queensland,
St Lucia, QLD, Australia
e-mail: r.cramb@uq.edu.au

© The Author(s) 2020   425
R. Cramb (ed.), *White Gold: The Commercialisation of Rice
Farming in the Lower Mekong Basin*,
https://doi.org/10.1007/978-981-15-0998-8_21

However, the very success of this pathway and the economic growth it has helped generate have created new policy issues and dilemmas, requiring adjustments in the long-term emphasis on the intensification of smallholder rice production. Most obviously, the rapid growth in production has contributed to a secular decline in market prices, particularly for bulk grades of rice, while the costs of fertilisers, fuel, and labour have been increasing, creating a classic cost-price squeeze on farmers' incomes. While a lower price of rice has been a major factor in reducing both urban and rural poverty (Warr 2015), it has led to pressure from surplus-producing farmers for price support, putting them at odds with downstream actors in the value chain and creating dilemmas for economic planners and policy makers. The volatility of export prices in what has always been a thin market but one which is increasingly interlinked with other global food and energy markets has added to the difficulty of managing domestic supplies to maintain low and stable prices for consumers.

These policy dilemmas are linked with an array of other issues, including the persistence of smallholdings in the face of a perceived need for larger production units to achieve new technical and marketing efficiencies; the growing preference of many rice farmers for more flexible and diversified farming systems, counter to long-standing rice intensification policies; and the role of state-owned and private enterprises in the processing and exporting sectors. These issues have arisen in the broader context of a changing physical environment—resulting from both hydropower development along the river and global climate change—that is creating particular challenges for rice production in the Basin (Friend et al. 2019). All of these developments raise questions about future investment in the research system that gave rise to the productive technologies underpinning the transformation of rice farming in the Basin.

This chapter briefly explores some of these policy issues and dilemmas using the categories in Chap. 1, Fig. 1.9—policies influencing *access* to resources (specifically, to land, water, and technology), the management of farm *activities* (whether specialised in production of high-quality rice or diversified into production of non-rice crops), and the *appropriation* of value (as determined by interventions in the marketing and pricing of paddy and rice).

ACCESS

The Scale Issue in Rice Farming

As outlined in Chap. 1, the commercialisation of rice farming in the Lower Mekong Basin has occurred in the context of (and has contributed to) economy-wide structural transformation that, notwithstanding sustained agricultural growth, has seen agriculture's share of national income and employment decline, following the general pattern of agrarian transition with economic development (Cramb and Newby 2015; Mellor 2017). Relatedly, there has been a dramatic fall in fertility and a slowing of population growth throughout the Basin. While agricultural employment continues to grow in most parts of the Basin, the rate of growth has significantly slowed; in Thailand, the absolute size of the agricultural workforce has been in decline since around 1990, including in the Northeast (Grandstaff et al. 2008). The slower growth in the total labour force, the movement of labour out of agriculture, and the ageing of the farm workforce have increased the demand for mechanisation in rice farming, as seen in all parts of the Basin, including Laos and Cambodia. This began everywhere with small, farmer-owned machines—two-wheeled tractors and portable pumps—but in Northeast Thailand, it has already evolved into medium-scale mechanisation with four-wheeled tractors and combine harvesters operated largely by contractors. Combine harvesters are also now widespread in the Delta and becoming more common in Cambodia and Laos.

The exodus of labour and the availability of mechanical technology might be expected to lead to an increase in average farm size, as commercial farmers with access to capital buy up the holdings of those who leave farming, following the historical pattern of the developed countries (Paarlberg and Paarlberg 2000; Rigg et al. 2016). Yet, as Rigg et al. (2016) have highlighted, smallholder farming has persisted in the Lower Mekong countries and elsewhere in Southeast Asia, especially in wet rice farming. In the Vietnamese Delta farm size averages 1.2 ha, but 38% of holdings are under 0.5 ha and there is a significant cohort of landless households. Even in Thailand, where rapid industrialisation began in the 1980s and the agrarian transition is most advanced, average farm size continued to decline into the 2000s, though it remained stable at 3.2 ha between the 2003 and 2013 agricultural censuses.

Rigg et al. (2016) point out some cogent reasons for this apparent lack of a trend to farm amalgamation, notably the "sheer competitiveness of small family farms, especially in rice-based systems" (p. 124); the micro-scale of mechanisation, well adapted to rice smallholdings, together with the emergence of contractors who can make the services of larger machines widely available; and the nature of rural livelihoods, such that many or most rural households have some members working in the urban sector while other, usually older members are able to maintain small rice farms that on their own would be "sub-livelihood." Cramb and Newby (2015) also emphasise the vital role of the traditional wet-season rice crop in underpinning the subsistence of even highly commercialised and diversi-fied farm households in Laos and Cambodia. Additional reasons men-tioned by Rigg et al. (2016) for holding on to small farms are the long-term social and economic value placed on land as an asset and the security it is perceived to provide in the face of the precariousness of much non-farm employment.

Rigg et al. (2016) also argue that the political economy of rice farming is such that government policy has in recent decades swung around from taxing to protecting rice smallholders, enabling them to persist longer than would be the case if technical and economic forces prevailed (see Chap. 2). The ill-fated rice-pledging scheme in Thailand is cited as evi-dence. However, this form of subsidy was clearly unsustainable and well in excess of the level of protection afforded by most governments in the Basin (Tobias et al. 2012). In any case, input and output subsidies are notoriously regressive, being appropriated mostly by larger farmers, increasing the incentive for them to expand their holdings (Poapongsakorn 2010).

Regardless of these arguments, it seems that governments are in fact increasingly concerned about the persistence of small rice farms and are looking for ways to encourage larger and more efficient operational units, though without upsetting the prevailing ownership structure. The motiva-tion is to increase both the quantity and quality of output through greater technical control over farming operations, thereby also enhancing farmers' persistently low incomes. This is being sought through local coordination to amalgamate adjacent paddies into larger fields more suited to mechani-cal operations, including realignment and land levelling.

In Northeast Thailand and Laos, such amalgamation of paddy fields is partly driven by contractors providing the services of four-wheeled trac-tors and combine harvesters who are seeking greater field efficiencies. Farmers can also benefit through land levelling that makes for more effec-tive irrigation and stabilisation of yields. A long-term study of a typical rice-growing village along the Chi River in Khon Kaen Province in Northeast Thailand found that while the total paddy area had hardly changed between 1981 and 2005 and average farm size had declined to 1.4 ha, the number of separate plots had been reduced by a factor of three (from 8401 to 2885) and the average size of plot had accordingly increased three times (from 663 to 1983 m^2) (Watanabe 2017). This dramatic consolidation of plots had proceeded with no government support. The beginnings of such a trend can be seen in some intensive rice-growing areas in Laos, with encouragement from local officials.

Notwithstanding these spontaneous developments, the military government in Thailand from 2014 to 2019 made "increasing efforts to tackle inefficiencies within small-plot farming" (FAO 2018: 6). The Agricultural Development Plan (2017–2021) includes a "Large Fields Scheme" or "area-based extension approach," intended to reduce production costs and increase farmers' returns. The approach involves consolidating many small paddy fields into a large farm while the ownership remains unchanged, setting up a farmers' organisation of the participating landholders, and appointing a farm manager for the enlarged farm. The Plan argues that this will facilitate site-specific government extension and support, economies of scale in production, and increased bargaining power for farmers (Pongsrihadulchai 2018).

In the longer term in Thailand, it is likely that, even in the absence of government intervention, the economic pressures for ageing smaller farmers to sell to larger entrepreneurial farmers will eventually prevail and the proportion of paddy land held in large holdings will increase over the coming decades, causing not only the size of paddies but the average farm size to gradually rise. In this context, Timmer (2015: 99) quotes Nipon Poapangsakorn of the Thailand Development Research Institute, who writes: "I think it is likely that the paddy farm size will easily reach 200–500 ha in the next ten years because the technology is there for a farm entrepreneur to manage such farm size."

In the Vietnamese Delta, with over a third of households having less than 0.5 ha of rice land, the scope for spontaneous land consolidation is much more limited. Rather, since 2010, the government has been encouraging a "Small Field, Large Farm" model to be coordinated by large agribusinesses such as the An Giang Plant Protection Company (now the Loc Troi Group) in a form of contract farming, encompassing as much as 10% of the paddy area in some provinces (Nguyen and Dao 2018).[1] In addition, a law issued in 2013 allows the accumulation of annual cropland (including paddy land) in the Delta to a maximum of 30 ha, specifically to encourage a more efficient and profitable scale of farming (Nguyen et al. 2017). Given the ongoing net outmigration from the Delta, farm size may soon begin to rise, even in this most densely populated part of the Basin.

As seen in Chap. 12, the land situation in southern Cambodia is similar to that in the Delta, with small farms and few plots per farm, these plots a result of land allocation in the 1980s to give households a share of different land types. Hence, spontaneous plot amalgamation is likely to be difficult. However, a similar contract-farming approach to that promoted in Vietnam has been tried here, as reported in Chap. 16. There are also some recent examples in Laos involving contracts between farmer cooperatives and rice milling companies.[2] Experience so far with this model is mixed as it requires an accumulation of trust between the parties and adequate capital on the part of the contractor, but there is likely to be continued policy emphasis in all jurisdictions on finding ways to manage small, independent landholdings to obtain the perceived benefits of larger operations.

River Management, Irrigation Schemes, and the Small-Pump Revolution

The Lower Mekong Basin is a vast catchment that is not only a productive rice bowl, producing a quarter of the world's rice exports, but also the world's largest inland fishery and a major source of hydroelectricity, powering the region's industrialisation. The trade-offs between these three functions of the river system are, however, becoming increasingly apparent, accentuated by climate variability and change, raising new issues for rice policy. Cosslett and Cosslett (2018: 111) identify three environmental factors having a major long-term impact on rice production in the Basin: "one, climate change and the outlook for global warming and associated sea level increases; two, El Nino and La Nina events that have been shown to be causative factors in some of the severe flood/drought weather cycles; and three, the construction and operation of China's mainstream dams on the Lancang River that have changed water flow regimes in downstream countries." Of these three factors, they consider that "the construction of dams on the Mekong River appears to pose the most imminent threat to both the near-term and the long-term sustainability of rice production in the Lower Mekong Basin" (Cosslett and Cosslett 2018: 111).

Of course, numerous hydropower dam projects have also been constructed along the tributaries of the Lower Mekong Basin itself, with at least 100 projects forecast for Laos by 2020 and new developments under-way in Northeast Cambodia (IFC 2017). Moreover, eleven large dams are planned for the main river in the Lower Basin, nine of them in Laos (Blake and Barney 2018); the controversial Xayaburi Dam on the main river within Laos is due to be operational in 2019. However, the total storage volume in all dams within Laos is estimated to be less than one-tenth the storage capacity of the Lancang cascade in China.[3]

The cumulative impact of all these dams on rice farming includes reduced delivery of upstream sediments and altered flow regimes (MRC 2017; IFC 2017; Hecht et al. 2019).[4] While the reduction in sediments is detrimental to the annual renewal of the Mekong floodplain and is likely contributing to the net erosion of the Delta since 2005, the general impact on flows has been mixed, with reduced flows (and flooding) in the wet season and increased flows in the dry season. The reduced wet-season flows may limit irrigation in that season (especially in drought years such as 2019), while the increased dry-season flows, though limiting the extent of traditional river-bank gardens, may increase the potential for pump irri-gation from both streams and groundwater (Hecht et al. 2019). The reduced extent of flooding is of course beneficial for wet-season rice in flood-prone areas of the Khorat Plateau, but reduces the potential for flood-recession rice in the Tonle Sap Basin and the Delta.[5] The moderat-ing effects of run-of-the-river dams can be contrasted with the effect of diversion schemes such as the Nam Theun 2 Dam in Laos, which can substantially reduce flows in one catchment while increasing the incidence of flooding in another (Blake and Barney 2018; Hecht et al. 2019).

Large-scale irrigation works, enabling farmers to supplement rainfall in the wet season and extend rice cultivation into the dry season, have also been implemented in the Lower Mekong Basin since Angkorean times, usually with more positive impacts on rice cultivation. Major public invest-ments in irrigation schemes occurred up to the 1990s, especially in Northeast Thailand and Laos (Hoanh et al. 2009). These were seen as an essential complement to the high-yielding seed-fertiliser technologies, which had been the experience in the earlier phase of the Green Revolution in Thailand's central plain, as well as in Indonesia and the Philippines. Yet, as Hoanh et al. (2009: 149) observe, "despite the great achievement in rice production in the LMB countries, there is a general consensus that irrigation systems have not lived up to expectations because of low perfor-mance in terms of control, water productivity, yields and quality of service delivery to farmers." Blake and Barney (2018) make similar observations specifically for Laos, particularly for schemes that are merely adjuncts to hydropower projects. Noble and Hoanh (n.d.: 7) conclude that "the provision of irrigation infrastructure has not proved to be a panacea." Notwithstanding this consensus, the further "greening of Isan" is still apparently a policy aspiration in Thailand, particularly under the military government (Molle et al. 2009; Blake 2019), and increased investment in irrigation is also espoused in Laos and Cambodia, though in practice the focus there is now more on rehabilitating, maintaining, and achieving fuller utilisation of existing irrigation works for dry-season rice production.[6]

Hence rainfed, wet-season rice farming remains the dominant production system in the Basin in terms of area, except in the Delta. Here there has been a much longer-term project of hydraulic engineering to achieve water control through dikes, canals, and irrigation works. This has proved critical to the expansion of double and triple cropping of rice in the central corridor of the Delta, utilising both tidal and pump irrigation systems (the latter accounting for 26% of the irrigated area). The rapid adoption of small, portable pumps has given farmers much greater flexibility in accessing and utilising this irrigation infrastructure, enabling a boost in dry-season cropping and increasing crop diversification, as described in Chap. 17. Yet as Biggs et al. (2009: 203) observe, "this *made* landscape, defined by ongoing canal-building enterprises and other works associated with a rapidly urbanizing human landscape, remains at constant risk of being *unmade* by the destructive and sediment-spreading natural effects of seasonal floods, erosion from daily tidal fluxes, storms and also the man-made effects from poorly placed dikes and other works." They add that "the nature of waterscape transformations is such that the state eventually has to cope with the maintenance of this hydro-agricultural 'machine'…," something they note that French engineers in the early twentieth century worried would become "a work without end" (Biggs et al. 2009: 216). They conclude that "…the financial implications of the need to maintain and protect the 'delta machine' are awesome…" (Biggs et al. 2009: 222). This "work without end" is being exacerbated by the growing threat of salinity and sea-level rise. In 2016, climate change combined with an El Nino drought resulted in the worst recorded salinity intrusion in the Delta, extending over 80 km inland and destroying at least 160,000 ha of crops. Chapman and Van (2018) report accelerated outmigration in those provinces worst affected by climate change and salinity.

While work continues on existing irrigation infrastructure throughout the Basin, in many ways a more significant trend has been the emergence of self-managed, on-farm irrigation through the construction of small ponds and shallow tube wells accessing groundwater, in both cases using the ubiquitous small-scale pumps. As noted in Chap. 2, digging of farm ponds adjacent to paddy fields began to take off in Northeast Thailand in the 1990s. This was a private initiative, subsequently supported through a revolving loan fund from the mid-2000s (Rambo 2017). The ponds took some land out of production but helped stabilise rice yields through supplementary irrigation in the wet season and permitted small-scale utilisation of paddy fields for non-rice crops (vegetables and field crops) in the dry season, as well as supporting the rearing of fish and livestock (Promkhambut and Rambo 2017). Farm ponds are also beginning to be recognised as a resource in the rainfed lowlands of Cambodia and Laos (Vote et al. 2019).

In addition, there has been a rapid increase in the private installation of tube wells in paddy fields, drawing on shallow alluvial aquifers in the Mekong lowlands (Johnston et al. 2013). For example, in Prey Veng Province in the south of Cambodia, the number of tube wells used for irrigation increased from 1600 in 1996 to 25,000 by 2005. The case studies in Chap. 13 show how on-farm irrigation enabled small-scale farmers to augment wet-season rice with an early wet-season rice crop and up to two short-term dry-season cash crops such as radish. While groundwater is available in most of the Tonle Sap Basin, "its sustainability as a resource is unclear" and "promoting extensive groundwater use before the resource is better defined is not recommended" (Johnston et al. 2013: 5–6). Nevertheless, the experience elsewhere in Asia suggests that farmers will continue to exploit this resource, even more so as rural electrification brings down the cost of pumping (Molle et al. 2003). Moreover, the independent control this gives farmers over water management will further underscore the trend to farm diversification discussed below, given the higher returns to irrigating non-rice crops. Nevertheless, research in Laos indicates that, even where sourcing groundwater for cropping is demonstrated to be technically feasible, dry-season production may be limited by the increasing cost of labour and energy (Clément et al. 2018).

Access to Biological and Mechanical Technology

Public investments in research and rural infrastructure (roads, canals, irrigation) were critical in giving smallholder rice farmers access to the improved varieties and complementary inputs, especially water and fertilisers, which have underpinned the growth in yields, output, and marketable surplus. The Green Revolution in the Lower Mekong came later and was more adaptive than the earlier input-intensive, high-yield package, based initially on IR8, which was implemented elsewhere in Asia. Long-term programmes of research and development were embedded in national institutions, enabling the selection, breeding, and dissemination of varieties that were suited to the preferences and circumstances of local farmers (Chap. 6). This can be seen in the impact of improved varieties of glutinous rice in Northeast Thailand and Laos, and of moderate-yielding but high-value local selections in Northeast Thailand and Cambodia. While IR8 and its high-input, high-yielding successors formed the basis of commercial growth in the Delta, here too policy is now favouring a shift to high-quality, high-value rice produced with less-intensive use of inputs.

Continuing collaborative research across the region has greater scope than ever to develop varieties adapted to local constraints of soil, pests, flood, drought, and salinity, partly a function of the altered hydrology of the river basin and of global climate change. Marker-assisted selection technology has incorporated pest and disease resistance in existing high-quality varieties such as KDML105, helping to stabilise yields.[7] Incorporating genes for drought and submergence tolerance will assist varietal development in Laos. For these farmers, the provision of more resilient wet-season varieties will continue to be a priority, underwriting diversification into dryland crops such as cassava, sugarcane, and rubber. In the Delta, breeders are using marker-assisted selection and crosses with wild rice to develop varieties more tolerant of salinity, submergence, acidity, drought, and heat (Bui and Nguyen 2017). These improved varieties will provide increased resilience not only for specialised, commercial rice producers but also for those farmers in the marginal rainfed lowlands, for whom the single wet-season crop is still an essential component of their more diversified livelihoods. Policies to develop improved seed systems will be important to enable farmers to capitalise on the continuing gains made by crop breeders (Chap. 8).

The small-scale mechanisation (especially pumps, tractors, and combine harvesters) that has evolved in response to the increasing scarcity and cost of farm labour has not required policy intervention, other than providing a favourable environment for commercial innovation and distribution. This development of mechanical technology is complementing the gains made through biological technology by enabling rural households to remain in rice production while diverting labour to more remunerative non-rice and non-farm pursuits, thus improving rural livelihoods. As discussed above, these technologies are creating economies of scale in rice production, favouring larger fields and, in the long term, larger farms.

Activities

Diversification of Rice-Based Farming Systems

The structural transformation of the Mekong economies discussed above is also having an impact on the domestic demand for rice and hence the profitability of conventional rice production. As urban and rural incomes rise, the importance of rice in the diet declines and the demand for fruit, vegetables, and meat increases (Timmer 2015). As noted in Chap. 1, Table 1.2, rice consumption per capita has declined markedly in Thailand and is following a similar trend in Vietnam (Nguyen 2013). More telling is that the share of rice in the per-capita consumption of calories in Vietnam declined from 61% in 1993 to 13% in 2014 for the urban population, and from 76% to 27% for the rural population (Nguyen et al. 2017: 20). The nominal farm-gate price of paddy has stagnated during the 2010s in all Lower Mekong countries (Fig. 21.1), while the cost of inputs has continued to increase.

These trends in the domestic economy, combined with the long-term decline in the world rice price, have translated into the decreasing

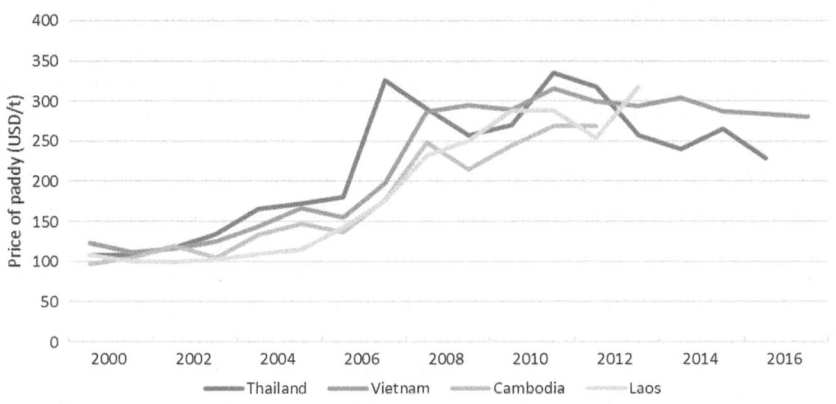

Fig. 21.1 Nominal farm-gate price of paddy in Lower Mekong Basin countries, 2000–2017 (USD/t). (Source: FAOSTAT)

profitability of specialised rice production and the increasing relative profitability of non-rice crops, hence a strong preference of rice farmers to diversify into these crops. Structural shifts in consumer demand in neighbouring countries are also increasing the relative profitability of non-rice crops such as maize and bananas produced within the Basin for cross-border trade (mainly to China, Thailand, and Vietnam). The dilemma for governments, given their long-standing food security goals, is that they want to maintain the area of paddy land and encourage its maximal utilisation for rice production. In particular, their investments in water control have been predicated on the intensification of rice production and provide little flexibility for diversification. Yet coercing farmers to grow the now less-profitable rice is counter to the espoused goals of rural development. These tensions are being worked out in different ways in the four Lower Mekong countries.

In Northeast Thailand, there has been a confusing succession of policies, at first seemingly designed to further intensify rice production (notably through price support), then designed to wean farmers away from rice production, including discouraging dry-season rice in irrigated areas. As mentioned in Chap. 2, in 2016, the military government initiated a scheme that paid farmers subsidies to stop planting rice in areas deemed to be unsuitable and to develop integrated farming systems instead, with on-farm irrigation, fish, and livestock. Official maps appear to designate some of the most profitable areas for jasmine rice production in the southern part of the Khorat Plateau as unsuitable for rice growing (Sunsuk 2016).[8] Yet markets continue to pay an ever-higher premium for this highly valued variety (Chap. 2, Fig. 2.4), offsetting the effect of lower yields, as shown in Chap. 3. Moreover, it would seem that farmers in the Northeast have already allocated land less suitable for rice to field crops such as cassava, sugarcane, and rubber.

In Laos, the current policy is to focus support on more productive areas, especially where irrigation schemes have been established along the main Mekong corridor, to ensure that these schemes are working effectively and that farmers have access to high-quality inputs (seed and fertilisers) and modern rice mills, capable of producing export-quality rice. The 2016 National Rice Policy proposes to prohibit the conversion of paddy land within these focal areas to other uses. At the other extreme, in the sloping uplands, government policy has long favoured the reallocation of land to non-rice crops such as maize, bananas, and rubber, both to eliminate shifting cultivation and increase farm incomes. However, the widespread leasing of land to Chinese investors for banana plantations in Northern Laos has extended to lowland paddy fields, prompting provincial authorities to ban this change in land use, though it is unclear whether this is to protect paddy land as such (including from pesticide pollution) or to assert official control over the banana boom.

In Cambodia, there are no land-use controls as such, with the availability of irrigation in the dry season being the key determinant of rice intensification. As outlined in Chap. 13, farmers in flood-recession areas are finding that dry-season cultivation of high-yielding varieties for export to Vietnam is a profitable option, whereas farmers in wet-season rice areas are more inclined to use the limited on-farm irrigation from ponds and tube wells to support more diversified dry-season cropping. Most farmers, of course, have neither of these options and rely more on non-farm sources of income to augment the rainfed wet-season rice crop.

In the Delta, ongoing investments in water control have enabled double and triple cropping of rice in formerly flood-prone areas, giving farmers in these districts little option but to specialise in rice (though fish farming within the canals and flooded paddies allows a degree of diversification, as described in Chap. 17). These rice-specialist districts have somewhat lower incomes and higher incidence of child malnutrition and poverty (Nguyen 2013). Cazzuffi et al. (2018), using panel data from 2008 to 2016 for the whole of Vietnam, found that households selling a higher proportion of the rice they produced did not have significantly higher income or food consumption than those selling less rice, though they accumulated more assets. At the same time, households earning a higher proportion of total income from rice had lower income and no significant difference in asset accumulation. These results suggest that, while increased selling of rice has improved the welfare of rice-producing households over time, it is diversification into non-rice and non-farm activities that is providing higher household incomes. Thus, even in the rice bowl of the central Delta, there has been a trend to rotating rice with non-rice crops (maize, soybean, sesame, vegetables, and flowers) and the establishment of permanent orchards of fruit trees on the natural levees and back swamps of the alluvial floodplain. The policy response so far has been to marginally reduce the paddy area designated exclusively for rice production—from 4.3 million ha in 2013 to 4.0 million ha in 2020 (Chap. 17). It is likely there will be increasing pressure to further relax land-use controls in the coming decade.

Even with these trends towards diversification away from rice, it will be important for ongoing public investment in research to maintain productivity and sustainability of especially the wet-season rice crop, which continues to provide the platform for diversified livelihoods throughout much of the Basin (Cramb and Newby 2015).

Specialisation in High-Value Rice Production

In those areas best suited for intensive rice cultivation, whether due to natural conditions or established infrastructure, the alternative to increasing farm incomes through diversification is to pursue specialisation in higher-yielding, higher-value, lower-cost rice production. In this regard, Northeast Thailand has led the way. Historically, the Green Revolution in this region differed in important respects from the high-input, high-yield model seen elsewhere in Asia in the 1960s and 1970s. As described in Chap. 2, plant breeding delivered a well-adapted glutinous variety (RD6) to secure subsistence supplies and a high-value fragrant variety (KDML105) for sale to domestic and export markets. Within the Northeast, many farmers in the southern provinces specialised in jasmine rice for sale, thereby creating a regional demand for a marketable surplus of glutinous rice produced mainly in the northern provinces. Hence, two groups of farmers benefited from specialisation in high-value, though not necessarily high-yield production.

Whereas Thai jasmine rice attracts a significant premium in export markets, there is growing domestic and export demand from environmental- and health-conscious urban consumers for organic rice, as well as rice with special characteristics such as nutritious, coloured, and local or native rice (Pongsrihadulchai 2018). As noted above, the Agricultural Development Plan (2017–2021) differentiates between "suitable" and "unsuitable" paddy land, with the former to be the focus of efforts to increase the value and reduce the costs of rice production, particularly through conversion to organic rice,[9] while in the latter areas farmers are offered incentives to change from rice to other activities (Pongsrihadulchai 2018). However, the official target for 60% of production to be organic by 2027 seems ambitious, driven as much by ideology as market analysis. The case studies in Ubon Ratchathani Province reported in Chap. 3 showed that the organic farming village obtained a somewhat higher average return to land but a much lower return to labour than the conventional farming village— in an increasingly labour-scarce economy. According to the SCB (Siam

Commercial Bank) Economic Intelligence Centre (2017), "organic farming remains a tiny niche industry in Thailand… just 0.3% of the country's agricultural land is certified as organic… fewer than 0.2% of Thailand's farmers practise organic agriculture… [but] domestic consumption of organic foods is likely to grow." In 2017 the military government introduced a new scheme to promote organic rice. Farmers receive THB 9000 per rai (USD 1700 per ha) over three years if they sign up to the scheme, almost equal to the average cost of production as reported in Chap. 3, hence it is unlikely to have wide coverage.

Cambodia's push into export markets since 2010 has centred on the promotion of its own high-quality rice as "white gold." As described in Chap. 19, over a million tonnes of paddy are exported annually from Cambodia to Vietnam, 80% of which is *soc* paddy, comprising traditional Cambodian wet-season varieties that are preferred by domestic consumers in Vietnam because of their high quality and the minimal use of agrochemicals by Cambodian farmers, and only 20% of which is *than nong* paddy, the low-quality IR rice that is milled and re-exported as standard Vietnamese rice. The export-promotion campaign has sought to increase the production and local milling of Cambodian fragrant rice (*phka malis*) and dry-season fragrant varieties for export not just across the border to Vietnam and Thailand but to Europe and increasingly to other destinations (IFC 2015). As noted in Chap. 11, the more profitable Cambodian aromatic varieties, mainly grown in the western provinces, now account for 10% of the annual cultivated area and 30% of total production (World Bank 2015) and by 2018 milled rice exports, though below the government's target, had risen to well over 600,000 t from only 40,000 t in 2010. To capitalise on the export potential of high-quality Cambodian rice, some miller-exporters are contracting with groups of farmers to provide pure seed, appropriate inputs, and credit in an effort to secure an adequate volume to fill export contracts, as described in Chap. 16. Though there were many issues with this project, such as credit constraints and side-selling, farmers who had participated achieved on average 12–25% higher gross margins. It is noteworthy that support for specialised, high-quality production in Cambodia is mainly coming from the agribusiness sector rather than from government agencies as in Northeast Thailand.

Laos is also looking to increase returns to rice farmers by promoting high-value production systems, mainly for export to China. There is a small demand for organic glutinous and jasmine rice and, as in Cambodia, miller/exporter companies have initiated contract-farming schemes with farmer cooperatives to provide seed and organic fertiliser on credit and purchase the paddy at an agreed price, up to 50% above that for standard glutinous rice destined for the domestic market. However, it has been difficult to meet the required export standards and thus to fill agreed quotas. In any case, the overall impact on farm incomes is not likely to be widespread.

In the Delta, despite specialist rice farmers achieving the highest yields and cropping intensities in the Basin, profitability remains low. Nguyen (2013) found that rice farmers in An Giang Province in 2009–2010 averaged a profit over three seasons of VND 3.8 million (USD 1,012) per household which, assuming the average household size of 4.4 persons, translates to VND 316,250 per person per month—below the poverty line of VND 400,000. Hence, to escape poverty, most households derived a considerable portion of income from non-farm sources. Based on a 2012 decree, the government offers direct financial support of VND 500,000 per ha per year for farmers on specialised paddy land and VND 100,000 per ha per year for farmers on other paddy land (not including uplands), but this has been difficult to implement and monitor and in any case has only a small impact on household income (Nguyen 2013). Other support includes a reduced land tax, subsidised credit, subsidised seed, and exemption from irrigation fees. More strategically, policy makers see the need to shift from the low-quality, low-priced rice that has formed the basis of competitive rice exports in previous decades in order to increase the returns to specialist rice producers.

As well as shifting to high-value varieties, there is a growing emphasis on increasing the sustainability of production systems in the Delta, both to reduce costs and to improve quality-based competitiveness. The Ministry of Agriculture and Rural Development's "One Must Do, Five Reductions" (1M5R) campaign promotes increased use of quality seed together with reduced seed rate, fertiliser use, water use, insecticide use, and post-harvest losses. Agribusiness groups such as Loc Troi are aiming to establish a certified sustainable value chain, using Vietnam's Good Agricultural Practice (VietGAP) standard. However, according to Demont and Rutsaert (2017), there is limited demand for VietGAP-certified product, limiting the price incentive for farmers to comply. Moreover, the fragmented nature of the supply chain, with numerous small collectors as the first point of contact with farmers, makes monitoring of sustainable practices problematic.

Appropriation

While rice policy in the Lower Mekong Basin has embraced a raft of interventions affecting farmers' access to resources and their production activities, it is price policy that has been seen as critical to the political legitimacy and survival of Mekong governments, both in the socialist one-party states of Vietnam and Laos and the quasi-democratic states of Cambodia and Thailand. Price policy determines who appropriates the "white gold" derived from the Basin's increasingly productive paddy fields. All regimes have manoeuvred to juggle the interests of net producers of rice and net consumers, with downstream actors in the value chain (traders, millers, wholesalers, and exporters) and various government departments (agriculture, finance, commerce) also caught up in what is an essentially political process. According to Timmer (2013), there is a historically deep-seated political imperative for each state to maintain food price stability within its borders.

The perennial dilemma has been to keep farm-gate prices high enough so that commercial rice farming remains profitable and rural protest is contained, while keeping the price to consumers low enough to avoid economic hardship and urban unrest. An array of tools has been employed, including mandating a minimum farm-gate price, intervening to purchase, store, and sell paddy and rice at administered prices, retail price controls, and subsidised provision of rice to vulnerable groups, and controlling exports through licensing, floor prices, quotas, and bans. The use of these tools has varied between jurisdictions and over time, with the long-term pattern conforming to Anderson and Hayami's (2019) finding for East Asia that, with economic growth and structural transformation, developing countries switch from taxing to protecting agriculture, particularly rice production, responding to the declining importance of rice in gross domestic product (GDP), employment, and consumer expenditure and the growing political power of the farm lobby, which can be mollified at lower budgetary cost. The conundrum is that there is "an urban bias in poor countries when farmers are a majority of the population, and a rural bias when urban consumers are a majority of the population" (Timmer 2013: 85).

Rice Price Policy in Thailand

Ricks (2018) has analysed the evolution of rice price policy in Thailand from a political economy perspective. From the 1950s to the 1970s, the authoritarian nature of Thailand's government meant that rural interests did not influence rice price policies, which were aimed at benefiting urban consumers and exporters while generating government revenue and foreign exchange. The tax burden on rice farmers was equivalent to a quarter of the total value of rice production in 1960. However, from 1979 to 2000, the voting system required individual politicians to develop local political networks, and rice millers and traders were in the best position to assist. Moreover, elected governments now had to placate increasingly vocal farmers. This led to measures to reduce the overall tax on rice farming and to support farmers and millers through early forms of the paddy purchasing programme (Chap. 2). However, Ricks (2018: 404) reports that 80% of the benefits of the programme went to millers, exporters, political parties, and bureaucrats, with only 20% going to farmers and farm leaders.

With the change to the voting system in the 1997 Constitution, major party politics with regional constituencies came to the fore. In 2001, Thaksin Shinawatra's Thai Rak Thai Party was swept to power on the basis of rural votes in the North and Northeast. A key policy was an enhanced form of the paddy pledging scheme, which was transformed from a price-smoothing device (enabling farmers to avoid selling immediately after harvest at low prices) into a mechanism for substantial price support. Now, most of the benefits went to farmers; millers became dependent on subsidies and income from renting storage to the government, while exporters' margins were squeezed. The political upheaval in the mid-2000s had complex causes, but at one level, it was a conflict between urban middle-class interests, represented by the Yellow Shirts, and rural interests, particularly in the North and Northeast, represented by the Red Shirts.

The military coup in 2006 was followed by further political turmoil and the installation of a government led by the Democrat Party under Abhisit Vejjajiva from 2008 to 2011, with support from the Yellow Shirts. The Abhisit government replaced rice pledging with an economically more rational price insurance scheme, but large farmers, millers, and exporters lost out under this scheme and Red Shirt protests intensified, resulting in a bloody military crackdown, the reaction to which ultimately brought the Abhisit government down.

In 2011, the Pheu Thai Party under Yingluck Shinawatra was swept to power with Red Shirt support on the promise to restore paddy pledging. From 2011 to 2014, the scheme offered paddy prices up to 50% above the market price, buying over half the paddy produced. As described in Chap.

2, this led to the accumulation of up to 18 million t and the eventual collapse of the programme (Welcher 2017). The expectation of a rise in the export price, enabling disposal of the stocks at a profit, was undermined by India's emergence as an exporter. The Pheu Thai government was removed by a military coup in 2014 and replaced by the National Council for Peace and Order (NCPO), which governed until 2019.

In 2016, the military government discontinued the rice-pledging and income insurance programmes and began disposing of stocks at discount prices (Chuasuwan 2018). Instead, short-term measures were introduced to help finance on-farm storage of fragrant and glutinous paddy, make direct payments and offer debt relief to farmers adversely affected by drought, and offset the cost of commercial crop insurance (Welcher 2017)—measures that can be seen as concessions to farmers in the North and Northeast. As described in Chap. 2, the military government has also initiated policies offering direct financial incentives to reduce the area planted to rice and encourage organic and integrated farming systems. It remains to be seen what direction rice price policy will take under the post-NCPO regime.

The history of rice price policy in Thailand since the 1980s thus represents a variant of Anderson and Hayami's (2019) transition from a large, voiceless agricultural sector that is exploited by government policy to a small but influential sector that can be protected at low political cost and low cost to the budget. In the past two decades, rice farmers in the North and Northeast have formed a sizeable and politically crucial bloc that became wedded to a programme of price support offered by an agrarian populist government,[10] the implementation of which proved politically and financially unsustainable—though it took an alliance between the military and the urban middle class to bring the programme to an end.

Rice Price Policy in Vietnam

In Vietnam, "the combination of a strong, centralized bureaucracy and a single-party system has resulted in a relatively stable political environment" (Nguyen and Talbot 2014: 322) compared with that in Thailand. Nevertheless, rice price policy continues to be an important focus for the government. Support for rice producers remains central to the country's socialist programme at the same time as structural transformation is "creating a large, growing, and politically influential group of net food consumers whose real incomes are compromised when food prices rise"

(Nguyen and Talbot 2014: 319). Hence Nguyen et al. (2017: 4) argue that "rice policy formulation, which involves the Party's top leadership, is critical to the Party's political survival, and underpins Vietnam's development story in the past, and likely in the future." They maintain that the "tension between socialist policy legacies and more recently introduced objectives of trade liberalisation" affects the formulation of rice policy (Nguyen et al. 2017: 4).[11] This is despite the fact that, by the mid-2010s, rice contributed only 7% of GDP and 2% of exports.

Control over rice prices is mainly exercised through the Vietnam Food Association (VFA), established in 1989 to guide and administer imports and exports. The VFA comprises food producing, processing, and trading enterprises, mostly state-owned, including Vinafood 2, the large state-owned corporation that dominates the rice trade in the Delta. VFA operates under an inter-ministerial committee comprising the Ministry of Trade (MOT), the Ministry of Agriculture and Rural Development (MARD), the Ministry of Finance (MOF), and other agencies, reporting directly to the Prime Minister (Nguyen et al. 2017: 27). Essentially, when prices are low, the government provides state-owned enterprises such as Vinafood 2 with funds to buy and store rice, putting upward pressure on farm-gate prices. When world prices are high, export quotas (now targets) implemented by VFA effectively reduce domestic prices, harming rice producers (and exporters) while benefiting net rice consumers (Nguyen and Talbot 2014: 330). Thus, rice policies have been "motivated in turn by a desire to cater to the competing demands of distinct domestic constituencies of net-producers and net-consumers" (Nguyen and Talbot 2014: 329).

With respect to the first mechanism—supporting farm-gate prices—the government issued a resolution in 2010 that farmers are to be paid a paddy price sufficient to give them a minimum return of 30% over production costs (Tran and Dinh 2015). MOF determines the floor price or "directed paddy price" based on estimates of production costs provided by provincial People's Committees in consultation with MARD, which has prime responsibility to ensure that the market price is no lower than the directed price. In a 2011 circular, a floor price for rice exports was introduced, with the MOF and VFA primarily responsible. Meanwhile, the government has invested in vastly expanding storage capacity in the Delta to allow for the accumulation of buffer stocks to support these prices. However, it has proved difficult to calculate production costs in order to set the paddy price. The floor price may not account for costs such as family labour, interest, rental, or transport to the point of sale, and in any case will vary from location to location (Tran and Dinh 2015). Moreover, the VFA and its constituent enterprises are not directly involved in buying paddy from farmers, which is almost entirely in the hands of numerous primary collectors who operate with small margins. Hence, as argued in Chap. 18, it may be infeasible to enforce this policy. Nguyen (2013) reports that with the floor price of VND 4000/kg for the summer-autumn crop of 2010, farmers' average return over cost was between 19% and 24%, depending on the quality of seed, well below the targeted 30%. Moreover, as noted above, even though farmers in An Giang averaged a return of 33% over three seasons in 2009–2010, the net income from rice alone left households below the poverty line.

The more effective instrument for influencing domestic paddy and rice prices in Vietnam has been controlling exports; domestic prices directly reflect the manipulation of export quantities and prices (Nguyen and Talbot 2014). Unlike in Thailand, where the export market is relatively unregulated (though directly impacted by the paddy pledging scheme), the government in Vietnam has, since 1990, sought to "ensure strict control of rice exports at the central level" (Nguyen et al. 2017: 24). Initially, only state-owned companies could export rice because they were easier to control. Up to 2000, exports were controlled by quotas issued by VFA, most of which were issued to Vinafood 2 and its subsidiaries. Over time, the major rice-producing provinces in the Delta exerted their power to obtain a larger share to allocate to their own state-owned enterprises. However, in 2000, Vinafood 2 was given a monopoly of government-to-government contracts, which accounted for 50–60% of exports. From 2001, quotas were removed but replaced with annual export targets. VFA approved contracts up to the target, again favouring Vinafood 2. According to Nguyen et al. (2017: 37), "vested interests are evident in the way VFA manages the rice exports market." For example, the chairman of VFA from 2006 to 2014 was also the general director of Vinafood 2 for much of this period.

In the 2007–2008 global food crisis, VFA pushed for an export ban, which turned out to be in the interests of large companies such as Vinafood 2 rather than producers or consumers (Nguyen et al. 2017: 34). The ban resulted in lower farm prices and higher consumer prices due to hoarding by consumers and wholesalers. Though consumer prices were not as high as they would have been without the ban, poor consumers suffered hardship, while small and large rice producers who had a bumper crop were deprived of a windfall profit. However, large exporters were able to benefit by buying rice cheaply and stockpiling until exports recommenced, thus "export companies made bonanza profits while farmers lost out" (Smith 2013: 5).

A decree in 2010 reinforced the control over exports exerted by large enterprises such as Vinafood 2 by requiring exporting firms to meet the minimum requirements of owning a storage capacity of at least 5000 t and a rice mill capable of processing at least 10 t/hr (Tran and Dinh 2015). However, under pressure from smaller private and state-owned enterprises who were thus pushed out of the export market, these requirements were removed in 2018, along with a requirement for certification for exporters of organic rice and other specialised types.[12] This was expected to increase competition and expand exports, particularly of high-quality rice. At the same time, various decrees and circulars have been issued in the 2010s to encourage export firms to invest more in rice-growing provinces and engage directly in supporting farmers (Tran and Dinh 2015). As discussed above, the growth in demand for high-quality and specialised rice is increasing the incentives for downstream actors to reach back in the value chain and integrate with producers, rather than merely aim to buy low-quality rice from traders at the lowest price with no concern about farmers. It remains to be seen how rice price policy will be implemented in this new, more competitive phase.

Rice Price Policy in Cambodia and Laos

Cambodia and Laos do not have the same capacity to move prices that have been demonstrated by Thailand and Vietnam, lacking the resources to enter the market on a scale sufficient to boost farm-gate prices and lacking full control over cross-border trade in paddy and rice. Prices in Cambodia and Laos largely reflect prices in their higher-producing, higher-income neighbours. Nevertheless, the same internal political dynamic is evident, with commercial paddy producers agitating for higher prices, millers, and exporters lobbying for support, and government agencies pursuing sometimes conflicting agendas.

This is seen most clearly in Cambodia. At the time of the food price crisis in 2007–2008, Cambodia temporarily banned exports but the ban was not effective, given the informal cross-border trade in paddy to Vietnam, and in any case it was soon lifted (Dawe and Slayton 2010). Domestic prices rose sharply, affecting many poor households who were net consumers of rice. More recently, the rapid expansion of production and exports has led to the problem of low farm returns, notwithstanding the government's promise of "white gold"—a sharp fall in paddy prices in 2016 prompted farmers in Battambang Province, one of the main producers of export-quality rice, to stage public protests, symbolically pouring rice onto National Road 5 (Kali and Cheng 2016), braving the often-violent state response to protest under the Hun Sen regime (Strangio 2014). The farmers complained that traders were offering lower prices or not buying paddy at all, whereas they had incurred large debts to microfinance institutions for fertilisers, pesticides, and farm machinery (see Chap. 15) and could not afford to store their paddy and wait for higher prices. The traders in turn were responding to reduced demand from millers, who had insufficient working capital to purchase and store paddy and were buying only to fill export orders (see Chap. 16).

Earlier in 2016, the Cambodia Rice Federation (CRF), the peak body of rice millers and exporters headed by Hun Sen's wealthy and influential son-in-law, Sok Puthyvuth, had lobbied the government for concessional loans to enable the purchase of paddy from farmers, as well as action to prevent what it described as illegal imports of rice from Vietnam that were undermining domestic prices (Kang 2016). The Ministry of Commerce agreed to USD 27 million in "emergency loans" to millers. The government-owned Rural Development Bank (RDB), charged with issuing the loans, stated that millers would be offered loans at 8% on condition that they purchased paddy from farmers for no less than USD 218/t, said to be a price that "ensures farmers make a profit on their crop" (Hor 2016). The government's decision to disburse loans directly through the RDB pointedly sidestepped the CRF. A Bank official criticised millers, claiming many sought funding to increase their machinery and storage facilities rather than to purchase paddy, adding that the Bank would take measures to ensure that the loans were not abused by millers seeking to purchase "motorbikes, cars or land" (Kang 2016). Nevertheless, the CRF was ultimately involved in administering the emergency loans to its members.

Again in 2018, falling prices prompted the Ministry of Commerce to consult with local authorities, the RDB, rice millers, and rice exporters to "put an end to the price fall and ensure farmers' livelihoods" (Sum 2018). The problem was again attributed to "a lack of capital from rice millers and exporters to buy rice, and a decline in the price of rice in the international market." The Ministry was considering "disbursing loans to rice millers and exporters so that they can purchase white rice by taking rice as collateral" (Sum 2018), though the CRF complained that the problem was the lack of stocks in the first place.

The CRF has also proposed a "consortium" of large millers and exporters to manage the export price by regulating supply, arguing that "if exporters get a good price, rice millers and farmers will benefit too as it will allow them to sell at higher prices.... Currently, our members compete against each other when foreign buyers come to buy our rice, which forces prices to fall" (Kang 2016). This proposal added weight to the concerns of some policy observers that "instead of developing new products and markets, Cambodia's so-called rice barons will use their power and influence to limit competition. This would mean a return to the large, undifferentiating paddies of Angkor, missing a valuable opportunity to capitalise on Cambodia's unique strengths" (EIU 2014). However, the Ministry of Commerce was understandably sceptical about the feasibility of such a cartel, given Cambodia's relatively small share of rice exports and the increasingly competitive nature of the international market. This competition will be accentuated if Cambodia loses its preferential access to EU and US markets, as now seems likely (EIU 2019).[13] The recurrent problem of low farm-gate prices also raises doubts about whether higher export prices, if they could be achieved, would in fact be passed back to farmers. The self-interested behaviour of CRF members was highlighted in a recent internal report (Boyle and Sopheakpanha 2018).

In Laos, as in Vietnam, "the party legitimates its economic reforms using a wider socialist ideology, [while] socialist ideology defines the framework of reforms... [Hence] the party is still trying to bring the economy under state control and to maintain its political control" (Yamada 2018: 17). This is seen in the long-standing practice of setting targets for area, yield, and production throughout the country that farmers are urged to fulfil as though they were still in a collective economy (though, as noted above, the attention has now been concentrated on more productive "focal areas"). With regard to price, the National Rice Policy issued in 2016 declares that the "rice price will be stabilized and managed against fluctuation in order to provide 30% of profit margin. The Government will intervene pricing in accordance with global market price mechanism of rice."[14] The Policy also promises a 5–10% increase in market price for farmers certified for "clean agricultural practices" for three years.

As noted in Chap. 9, the State Food Enterprise (SFE) has been a player in the domestic market, buying rice at a controlled price and holding rice stocks. However, the state's capacity to control domestic rice prices is limited. Eliste and Santos (2012: 50–54) found that the price of rice in Laos closely follows the price of glutinous rice in Thailand, which produces twice as much glutinous rice as Laos. The informal cross-border trade ensures that glutinous rice prices in the two countries are brought into line (though not with Thai white and jasmine rice, the prices of which are determined by export demand). Thus, price policy in Thailand may have more influence on rice prices in Laos than any intervention by the government of Laos. This probably means that Lao farmers benefited in the early 2010s from the high prices induced by the paddy pledging scheme in Thailand and have suffered since the mid-2010s from the overhang of the large stockpile that resulted.

The imposition of quotas and ad hoc bans on trade in paddy and rice at the provincial and national levels, designed to keep consumer prices low, has in the past harmed farmers and millers and damaged prospects for a profitable export industry (Chap. 9). Indeed, Eliste and Santos (2012) calculated that the effect of trade policy in the 2000s was a net transfer from producers to consumers, which more than offset the implicit subsidy of government programmes to support rice farmers. That is, there was a net tax on rice farmers, placing Laos at an earlier stage than Thailand in Anderson and Hayami's (2019) transition from taxation to protection of agriculture. There is no mention of the use of trade measures to stabilise domestic prices in the 2016 policy statement (nor of any other specific price measures for that matter), but with the increasing size of the exportable surplus and government support to increase milling capacity and sign export contracts, especially with China, discretionary bans seem less likely. As Eliste and Santos (2012) argue, given that 75% of urban households spend less than 30% of their budget on rice, it is more efficient to maintain a small food reserve as a safety net than to continually intervene in an attempt to stabilise the retail price.

Concluding Remarks

The Lower Mekong Basin has long provided a range of suitable environments for subsistence rice farming, from the narrow inland valleys of the Northern Highlands, to the rainfed lowland plains of the Khorat Plateau, to the extensive floodplain of the Tonle Sap Basin and the Delta. Indeed, rice was "the only food staple that could be grown intensively in [this] monsoon-driven agro-climatic environment" (Timmer 2013: 83). Hence, rice has long dominated production and consumption, more so than with other staples in other regions of the world. Moreover, the Basin has a long history of centralised political control based on intensive rice cultivation—from Scott's (2010) "paddy states," of which Angkor was the epitome, to the modern nation-states that now share the region. The intensification of smallholder rice farming and the stabilisation of rice supplies and prices has been critical to the political legitimacy and survival of these states.

Given the special place of rice in both the agro-ecology and the political economy of the Basin, there has been a deep-seated desire to control access to resources, the activities of farmers, and the appropriation of value along the supply chain. Thus, despite the declining share of rice in both the incomes of rural households and the expenditure of urban households, as well as its declining importance to government budgets and national economies, "rice growing has been kept profitable through subsidies, virtually free irrigation water, price support and stabilization programs, and well-developed rural infrastructure that ensured low marketing margins for rice" (Timmer 2008: 4). This may not be the best set of policies to ensure rural prosperity and food security for the Mekong countries in the coming decades. Timmer (2008: 4) argues that the way forward "is to make rice less 'different' to consumers, farmers, and the world market by making it more of an economic commodity and less of a political commodity."

There has been significant progress in this direction, as evidenced by a gradual relaxation of targets and controls, more diversified and profitable rural economies, greater integration with modern supply chains, a better educated and more mobile rural population, and moves towards coordinated international efforts to stabilise the world rice market. However, the "white gold" of the Mekong is likely to be the subject of policy contention for some time yet.

NOTES

1. Smith (2013) reports that in Dong Thap Province, the area of paddy land covered by the "large field" programme increased from 1467 ha in 2010 to 21,218 ha in the main (winter–spring) season of 2013, equivalent to 10% of the area planted. The programme involved 11,205 households. Of the total area in the programme, 16,148 ha were covered by written contracts between companies and farmer groups, of which contracts with the An Giang Plant Protection Company covered 5070 ha.

2. See a report from the Medium-Term Cooperation Programme with Farmers' Organisations in Asia and the Pacific Phase Two (MTCP2) on

"Promoting partnership of smallholder rice producer group and a rice company." Available at http://www.asiapacificfarmersforum.net/laos-promoting-partnership-of-small-holder-rice-producer-group-and-a-rice-company/ (viewed 23 July 2019).

3. Estimate by Alan Potkin in message to Laofab Group, 24 July 2019.

4. The impact on fisheries is more complex but likely to be severe (Hecht et al. 2019). Given that many rural households combine rice farming with fishing, the overall impact of dams on rice-based livelihoods is a matter of serious concern.

5. Brian Eyler, Southeast Asia Programme Director for the Stimson Center, commented in an interview: "There's very little that Vietnam domestically can do. The [Mekong] delta, being a very important agricultural production zone for Vietnam, could be managed in a way where more water is stored from the monsoon season into the dry season, and this would be a way to mitigate what's happening upstream." While supportive of Resolution 120 of 2017, which aims to restore the Delta's natural ecosystem properties, Eyler remarks that "Resolution 120 won't work if all these upstream impacts are still coming down to the Mekong delta." (China needs to put its money where its mouth is and actually release some water to relieve the drought, Interview with Brian Eyler on Radio Free Asia, 25 July 2019. Available at https://www.rfa.org/english/news/vietnam/brian-eyler-mekong-drought-07252019170305.html?utm, viewed 5 August 2019.)

6. See Government of Lao PDR, National Rice Policy for Food Security, Section 2.2 (issued in June 2016).

7. Shu Fukai, personal communication, 1 May 2019.

8. The official zoning of land into "suitable" and "unsuitable" for rice farming can be viewed online at the following site—http://agri-map-online.moac.go.th/ (viewed 3 July 2019).

9. This includes encouragement to reduce the area of the second rice crop in favour of cash crops, green manure, or leaving the land idle.

10. This despite the fact that the benefits were heavily skewed in favour of large farms (Poapongsakorn 2010) and that rural households in the Northeast derived most of their income from non-farm sources (Rambo 2017).

11. According to Nguyen et al. (2017: 4), this accounts for the "stickiness of policy-making institutions."

12. See Vietnam News, New decree removes barriers for rice exporters, 1 September 2018. Available at https://vietnamnews.vn/economy/464972/new-decree-removes-barriers-for-rice-exporters.html#1cqDtu WtjH1KeZu3.97 (viewed 1 August 2019).

13. One of the issues with Cambodia's preferential access to the EU market is the alleged rebadging of rice imported from Vietnam as Cambodian rice.

14. Government of Lao PDR, National Rice Policy for Food Security, Section 23.6, issued in June 2016.

REFERENCES

Anderson, K., and Hayami, Y., 2019. The political economy of agricultural protection: East Asia in international perspective. In K. Anderson, ed., *Asia-Pacific Trade Policies Vol. 1: Political Economy of Agricultural Protection in East Asia*, chap. 2. New Jersey: World Scientific.

Biggs, D., Miller, F., Chu, T. H., and Molle, F., 2009. The delta machine: water management in the Vietnamese Mekong Delta in historical and contemporary perspectives. In François Molle, Tira Foran, Mira Kakonen, eds., *Contested Waterscapes in the Mekong Region: Hydropower, Livelihoods and Governance*, pp. 203–225. London: Earthscan.

Blake, D. J. H., 2019. Damming Isan's last free-flowing river for cash. *Bangkok Post*, 25 July 2019. Available at https://www.bangkokpost.com/opinion/opinion/1718439/damming-isans-last-free-flowing-river-for-cash (viewed 31 July 2019).

Blake, D. J. H., and Barney, K., 2018. Structural injustice, slow violence? The political ecology of a 'best practice' hydropower dam in Lao PDR. *Journal of Contemporary Asia* 48: 808–834.

Boyle, D., and Sopheakpanha Nem, 2018. Cambodian rice body faces own shortcomings. *VOA News*, 19 February 2018. Available at https://www.voanews.com/east-asia-pacific/cambodian-rice-body-faces-own-shortcomings (viewed 5 August 2019).

Bui, C. B., and Nguyen, T. L., 2017. New rice varieties adapted to climate change in the Mekong River Delta of Vietnam. *Vietnam Journal of Science, Technology and Engineering* 60: 30–33.

Cazzuffi, C., McKay, A., and Perge, E., 2018. *The Impact of Commercialization of Rice on Household Welfare in Rural Viet Nam*. WIDER Working Paper 2018/130. Helsinki: United Nations University World Institute for Development Economics Research.

Chapman, A., and Van, P. D. T., 2018. Climate change is triggering a migrant crisis in Vietnam. *The Conversation*, 9 January 2018. Available at https://the-conversation.com/climate-change-is-triggering-a-migrant-crisis-in-vietnam-88791 (viewed 31 July 2019).

Chuasuwan, C., 2018. *Rice Industry Outlook, 2018–2020*. Bangkok: Krungsiri Research.

Clément, C., Vinckevleugel, J., Pavelic, P., Xiong, K., Valee, L., Sotoukee, T., Shivakoti, B. R., and Vongsathien, K., 2018. *Community-Managed Groundwater Irrigation on the Vientiane Plain of Lao PDR: Planning, Implementation and Findings from a Pilot Trial*. IWMI Working Paper 183. Colombo: International Water Management Institute (IWMI).

Cosslett, T. L., and Cosslett, P. D., 2018. *Sustainable Development of Rice and Water Resources in Mainland Southeast Asia and Mekong River Basin*. Singapore: Springer Nature.

Cramb, R. A., and Newby, J. C., 2015. Trajectories of rice-farming households in Mainland Southeast Asia. In R.A. Cramb, ed., *Trajectories of Rice-Based Farming Systems in Mainland Southeast Asia*, pp. 35–72. Canberra: Australian Centre for International Agricultural Research.

Dawe, D., and Slayton, T., 2010. The world rice market crisis of 2007–2008. In David Dawe, ed. *The Rice Crisis: Markets, Policies and Food Security*, pp. 15–28. London and Washington, DC: Food and Agriculture Organization of the United Nations and Earthscan.

Demont, M., and Rutsaert, P., 2017. Restructuring the Vietnamese rice sector: towards increasing sustainability. *Sustainability* 9: 325.

EIU, 2014. Selling "white gold": Cambodia's rice exports. *Economist Intelligence Unit: Cambodia*, 11 July 2014. Available at http://country.eiu.com/article.aspx?articleid=642005848 (viewed 5 August 2019).

EIU, 2019. Burning bridges. *Economist Intelligence Unit: Cambodia*, 25 June 2019. Available at http://country.eiu.com/article.aspx?articleid=1128157096&Country=Cambodia&topic=Economy (viewed 5 August 2019).

Eliste, P., and Santos, N., 2012. *Lao People's Democratic Republic Rice Policy Study 2012*. Rome: Food and Agriculture Organization (FAO).

FAO, 2018. *Thailand: Country Fact Sheet on Food and Agriculture Policy Trends*. Rome: Food and Agriculture Policy Decision Analysis Team, Food and Agriculture Organization of the United Nations.

Friend, R. M., et al., 2019. Agricultural and food systems in the Mekong region: drivers of transformation and pathways of change. *Emerald Open Research* 1: 12.

Grandstaff, T. B., Grandstaff, S., Limpinuntana, V., and Suphanchaimat, N., 2008. Rainfed revolution in Northeast Thailand. *Southeast Asian Studies* 46: 289–376.

Hecht, J. S., Lacombe, G., Arias, M. E., Thanh, D. D., and Piman, T., 2019. Hydropower dams on the Mekong River basin: a review of their hydrological impacts. *Journal of Hydrology* 568: 285–300.

Hoanh, Chu Thai, Facon, Thierry, Try Thuon, Bastakoti, R. C., Molle, F., and Phengphaengsy, G., 2009. Irrigation in the Lower Mekong Basin countries: the beginning of a new era? In François Molle, Tira Foran, Mira Kakonen, eds., *Contested Waterscapes in the Mekong Region: Hydropower, Livelihoods and Governance*, pp. 143–171. London: Earthscan.

Hor Kimsay, 2016. Terms announced on rice lending package. *Phnom Penh Post*, 20 September 2016. Available at https://www.phnompenhpost.com/business/terms-announced-rice-lending-package (viewed 5 August 2019).

IFC, 2015. *Cambodia Rice: Export Potential and Strategies*. Cambodia Agribusiness Series No. 4. Phnom Penh: International Finance Corporation.

IFC, 2017. *Nam Ou River Basin Profile Summary Document: Environmental and Social Characteristics of a Key River Basin in Lao PDR*. Washington, DC: International Finance Corporation.

Johnston, R., Roberts, M., Try, T., and de Silva, S., 2013. *Groundwater for Irrigation in Cambodia*. Issue Brief. Colombo: International Water Management Institute.

Kali Kotoski, and Cheng Sokhorng, 2016. No sign of relief for rice industry. *Phnom Penh Post*, 15 September 2016. Available at https://www.phnompenh-post.com/business/no-sign-relief-rice-industry (viewed 5 August 2019).

Kang Sothear, 2016. Federation mulls supporting rice export price. *The Cambodia Daily*, 18 March 2016. Available at https://www.cambodiadaily.com/business/federation-mulls-supporting-rice-export-price-110081/ (viewed 5 August 2019).

Mellor, J. W., 2017. *Agricultural Development and Economic Transformation: Promoting Growth with Poverty Reduction*. Cham, Switzerland: Palgrave Macmillan.

Molle, F., Shah, T., and Barker, R., 2003. *The Groundswell of Pumps: Multilevel Impacts of a Silent Revolution*. Paper prepared for ICID-Asia Meeting, Taiwan, November 2003.

Molle, F., Floch, P., Promphakping, B., and Blake, D. J. H., 2009. The 'greening of Isaan': politics, ideology and irrigation development in the northeast of Thailand. In F. Molle, T. Foran, and M. Kakonen, eds., *Contested Waterscapes in the Mekong Region: Hydropower, Livelihoods and Governance*, pp. 253–282. London: Earthscan.

MRC, 2017. The effects of Chinese dams on water flows in the Lower Mekong Basin. *Mekong River Commission News*, 6 June 2017. Available at http://www.mrcmekong.org/news-and-events/news/the-effects-of-chinese-dams-on-water-flows-in-the-lower-mekong-basin/ (viewed 29 July 2019).

Nguyen, H. T. M., Do, H., Kay, A., Kompas, T., Nguyen, C. N., and Tran, C. T., 2017. *The Political Economy of Rice Exceptionalism during Economic Transition: The Case of Rice Policy in Vietnam*. Crawford School Working Paper No. 1713. Canberra: Crawford School of Public Policy, Australian National University.

Nguyen Manh Hai, and Talbot, T., 2014. The political economy of food price policy in Vietnam. In P. Pinstrup-Andersen, ed., *Food Price Policy in an Era of Market Instability: A Political Economy Analysis*, pp. 319–338. Oxford: Oxford University Press.

Nguyen Ngoc Mai, and Dao The Anh, 2018. *A Review of Vietnam's Recent Agricultural Policies*. FFTC Agricultural Policy Platform. Taipei: Food and Fertilizer Technology Centre. Available at http://ap.fftc.agnet.org/ap_db.php?id=908 (viewed 30 July 2019).

Nguyen Trung Kien, 2013. *Food Security in Vietnam: Situation and Policy Options*. Paper presented at Regional Dialogue on Meeting Food Security Goals with Good Policy, 26–27 June 2013, Medan, Indonesia.

Noble, A., and Hoanh, C. T., n.d. Irrigation expansion or other opportunities for sustainable agriculture development—lessons learned from MRB. Vientiane: International Water Management Institute.

Paarlberg, D., and Paarlberg, P., 2000. *The Agricultural Revolution of the 20th Century*. Ames: Iowa State University Press.

Poapongsakorn, N., 2010. The political economy of Thai rice price and export policies in 2007–2008. In David Dawe, ed. *The Rice Crisis: Markets, Policies and Food Security*, pp. 191–217. London and Washington, DC: Food and Agriculture Organization of the United Nations and Earthscan.

Pongsrihadulchai, A., 2018. *Thailand's Rice Industry and Current Policies towards High Value Rice Products.* Paper submitted for International Seminar on Promoting Rice Farmers' Market Through Value-Adding Activities, 6–7 June 2018, Kasetsart University, Bangkok.

Promkhambut, A., and Rambo, A. T., 2017. Multiple cropping after the rice harvest in rainfed rice cropping systems in Khon Kaen Province, Northeast Thailand. *Southeast Asian Studies* 6: 325–338.

Rambo, A. T., 2017. The agrarian transformation in Northeastern Thailand: a review of recent research. *Southeast Asian Studies* 6: 211–245.

Ricks, J., 2018. Politics and the price of rice in Thailand: public choice, institutional change and rural subsidies. *Journal of Contemporary Asia* 48: 395–418.

Rigg, J., Salamanca, A., and Thompson, E. C., 2016. The puzzle of East and Southeast Asia's persistent smallholder. *Journal of Rural Studies* 43: 118–133.

SCB Economic Intelligence Centre, 2017. Thai organic foods have healthy growth potential. *Bangkok Post*, 6 February 2017.

Scott, J. C., 2010. *The Art of Not Being Governed: An Anarchist History of Upland Southeast Asia.* Singapore: NUS Press.

Smith, W., 2013. *Agriculture in the Central Mekong Delta: Opportunities for Donor Business Engagement.* London: Overseas Development Institute.

Strangio, S., 2014. *Hun Sun's Cambodia.* New Haven: Yale University Press.

Sum Manet, 2018. Commerce Ministry intervenes to stop fall in rice price. *Khmer Times*, 19 July 2018. Available at https://www.khmertimeskh.com/513263/commerce-ministry-intervenes-to-stop-fall-in-rice-price/ (viewed 6 August 2019).

Sunsuk, D., 2016. Isaan farmers and local officials slam failed rice policy. *The Isaan Record*, 25 October 2016.

Timmer, C. P., 2008. *Poverty in Asia and the Transition to High-Priced Food Staples.* Reducing Poverty and Hunger in Asia Brief 2. Washington, DC: International Food Policy Research Institute.

Timmer, C. P., 2013. Food security in Asia and the Pacific: the rapidly changing role of rice. *Asia and the Pacific Policy Studies* 1: 73–90.

Timmer, C. P., 2015. The dynamics of agricultural development and food security in Southeast Asia: historical continuity and rapid change. In Ian Coxhead, ed., *Routledge Handbook of Southeast Asian Economics*, pp. 89–113. London and New York: Routledge.

Tobias, A., Molina, I., Valera, H. G., Abdul Mottaleb, K., and Mohanty, S., 2012. *Handbook on Rice Policy for Asia.* Los Banos: International Rice Research Institute.

Tran Cong Thang and Dinh Thi Bao Linh, 2015. *Rice Policy Review in Vietnam.* FFTC Agricultural Policy Platform. Taipei: Food and Fertilizer Technology Centre. Available at http://ap.fftc.agnet.org/ap_db.php?id=406 (viewed 30 July 2019).

Vote, C., Eberbach, P., Inthavong, T., Lampayane, R. M., Vongthilard, S., and Wade, L. J., 2019. Quantification of an overlooked water resource in the tropical rainfed lowlands using RapidEye satellite data: a case of farm ponds and the potential gross value for smallholder production in southern Laos. *Agricultural Water Management* 212: 111–118.

Warr, P., 2015. Agricultural growth and rural poverty reduction in Mainland Southeast Asia. In Rob Cramb, ed., *Trajectories of Rice-Based Farming Systems in Mainland Southeast Asia*, pp. 17–34. Canberra: Australian Centre for International Agricultural Research.

Watanabe, K., 2017. Improvement in rainfed rice production during an era of rapid national economic growth: a case study of a village in Northeast Thailand. *Southeast Asian Studies* 6: 293–306.

Welcher, P., 2017. *Thailand: Rice Market and Policy Changes over the Past Decade.* GAIN Report No. TH7011. Global Agricultural Information Network, United States Department of Agriculture Foreign Agricultural Service.

World Bank, 2015. *Cambodian Agriculture in Transition: Opportunities and Risks.* Economic and Sector Work, Report No. 96308-KH. Washington, DC: World Bank.

Yamada, N., 2018. Legitimation of the Lao People's Revolutionary Party: Socialism, *Chintanakan Mai* (New Thinking) and reform. *Journal of Contemporary Asia* 48: 717–738.